The night before, one of the South Korean patrols, continually on duty in the hills ringing Seoul, had clashed with six North Korean infiltrators. They had been intercepted while crossing the DMZ.

After a short, fierce firefight, not unusual along the DMZ, five of the infiltrators had been shot, one dying shortly after. South Korea's CIOC (Counter-Infiltration Operations Command) was reasonably sure the six were from the NKA's 124th guerrilla unit. It was this unit from which thirty-one North Korean commandos had penetrated the southern side of the DMZ on a bitterly cold January night during the infamous mission of '68, armed with AK-47 submachine guns and grenades, with express orders to assassinate President Park of South Korea. On the second day of the mission, four woodcutters saw them, notified local authorities, and the hunt was on. One of the guerrillas, surname Kim, given name Shin Jo, was caught. Trading his life against the certain fate of being shot as an infiltrator, Kim passed over to *Chungang Chongbo-bu* (South Korea's Central Intelligence Agency) all the details of North Korean President Kim Il Sung's plan to assassinate President Park, including the fact that at each of his unit's eight guerrilla bases in the North, there were three hundred volunteers.

This meant that since 1968, over two thousand North Korean agents were being prepared at any one time for further infiltration, agitation, and sabotage against the Americans and the South, that every day at least one agent was crossing the DMZ.

Also by Ian Slater:

FIRESPILL
SEA GOLD
AIR GLOW RED
ORWELL: THE ROAD TO AIRSTRIP ONE
STORM
DEEP CHILL
FORBIDDEN ZONE*
WW III: RAGE OF BATTLE*
WW III: ARCTIC FRONT*
WW III: WARSHOT*

*Published by Fawcett Books

WW III

Ian Slater

FAWCETT GOLD MEDAL • NEW YORK

For Marian, Serena, and Blair

A Fawcett Gold Medal Book
Published by Ballantine Books
Copyright © 1990 by Bunyip Enterprises, Inc.

Library of Congress Catalog Card Number: 90-93134

ISBN 0-449-14562-X

Manufactured in the United States of America

First Edition: October 1990
Ninth Printing: November 1992

ACKNOWLEDGMENTS

I would like to thank Professors Peter Petro, Yunshik Chang, and Charles Slonecker who are colleagues and friends of mine at the University of British Columbia. Most of all I am indebted to my wife, Marian, whose patience, typing and editorial skills continue to give me invaluable support in my work.

That [Russian] withdrawal began last week, when thirty-one Soviet tanks were loaded onto flatbed cars in Hungary. Among those watching the pullout was Ilona Staller, a member of the Italian Parliament and a porno-movie star. Staller kept her clothes on when she posed with Soviet officers, and released a white dove of peace. Ominously, perhaps, the bird was crushed in the treads of a Soviet tank.

Time, May 1989

PROLOGUE

THE MAN WITH the eye patch was the president's pilot. Once, while making love to a beautiful young woman, he had left the patch on, kidding around, pretending to be a pirate of old. But after, in the hush following the storm, she had asked him to take it off. The patch frightened her, she told him, an augury of the perpetual hush that would follow a nuclear explosion, the bomb's airburst, "brighter than a million suns," blinding all aboard the "Doomsday" plane except the man with the protected eye— killing all below. Leaving the president in charge of what? Seeing her distress, the pilot had quickly removed it. Trembling with fear, she asked him to hold her and he did. It would be more than a year before he would see her again.

Before he had met her, he had never heard of her two brothers—or anyone else in the Brentwood family for that matter— but then, people all over the world had never heard of the family, and there was no reason they should have, that is until, like the Brentwoods in America and Major Tae in South Korea, everybody suddenly found themselves swept into the maelstrom when, whether anyone liked it or not, ordinary human beings would be called upon to do extraordinary things.

CHAPTER ONE

Korea—August 14

HIGH ABOVE SEOUL'S Yonsei campus, the moon was white—the color of mourning. Mi-ja Tae felt her heart race from the fear of parting, the moon fleeing a cloven sky, one moment its light turning the ginkgo leaves silver, the next swallowing them in darkness. As it was the evening before the annual Independence Day celebrations, fireworks could be seen now and then bursting above the old '88 Olympic Stadium south across the river. And tonight, the television news had told them, there was an added reason for celebration. In Europe the Americans and Russians had announced further arms and troop reductions. The prospects for peace, commentators proclaimed, had never been better.

Turning away from her lover, though still in his embrace, Mi-ja told him, "We cannot meet again."

He was stunned. "What are you saying? Why—"

"If my father knew what you are doing," she said, "you know he would forbid me seeing you."

"He doesn't know."

"It's his job to know these things. Sooner or later he will find out."

"How?" asked Jung-hyun. "He's at Panmunjom. We're here."

"Each time he comes home on leave, it is more difficult."

"What is more difficult?"

"To deceive him," said Mi-ja sadly. "I love him very much. If he knew—"

"That's unreasonable!" said Jung-hyun.

"Not to him," answered Mi-ja. "He fears the North. For him, you would be a traitor."

"Then you're not coming on the march?" It was more accusation than question.

"I didn't say that. But can't you see how he—"

"But *you*," Jung-hyun pressed, pushing her away, looking down at her beauty, the nape of her neck revealed in fleeting moonlight. "*You* see, don't you, that North and South should be united? That we should be together?"

"He wants that, too," she said.

"Ah—" Jung-hyun said, turning from her, "he is a *chinmipa*—pro-American."

"He doesn't *hate* the Americans," she said, looking up at him. "If that's what you mean. He says if it wasn't for them, we would all be slaves."

"And you believe that?" Jung-hyun said dismissively.

"I'm—" She shook her head and came closer to him, her arms around his waist. He could smell her perfume, feel her softly weeping against him. "I don't know," she said, her voice trembling. "I don't know what to think. Father says the North is looking to make war before the South becomes too strong. He says that is why it's so dangerous now."

"Rubbish!" snapped Jung-hyun. "The North will never attack us. They only want peace." He pushed her roughly away now, his hands in fists of frustration. "You think I'd be in the movement for reunification if I thought the North wanted war?"

"No," she said.

"Well," he said, "there you are."

The moon was lost in cloud. Slowly he drew her back to him. He could feel her heart beating. Stroking the sensuous curve of her neck, he pulled her still closer. She could feel his arousal. "I love you," he said softly. "You must not worry so. Your father is wrong. There'll be no war."

CHAPTER TWO

"WHAT'S THAT?" ASKED the elderly woman in one of Northwest's Boeing 767's starboard window seats. The flight attendant, on her first trip from Seattle to Shanghai, lowered her head to look out into the "wild blue yonder," as she banally called it. The man from Texas sitting next to the elderly woman didn't care what the attendant called it so long as she kept bending over him for a better view. He was married, but his wife said she didn't mind where he got his appetite so long as he did his eating at home.

"A seabird probably," said the attendant, fresh out of training school in Atlanta.

"At thirty thousand feet?" said the elderly woman. "I don't think so, dear. It would need an oxygen mask."

"Oh," the attendant replied, "then it's probably another plane."

"Same altitude as us?" the elderly woman rejoined.

The attendant peered out again, the man from Texas loving it. If only the plane were empty and they were alone.

"I think you'd better find out," said the woman, a trifle schoolmarmish in her tone. "Would you mind?"

"I'll ask the senior flight attendant. He'll probably—"

"You ask the *pilot*, dear." As the attendant forced an accommodating smile and stood up, the Texan studiously watched her walk away. She had to squeeze past the drink trolley. The elderly woman was anxiously looking out the window. The Texan, James Delcorte, smiled at her. "I wouldn't worry about it, ma'am. Probably just a jet fighter."

She turned, looking sharply at him. "That's what I'm afraid of."

"Well, there's no need," he said reassuringly. "We're in the Corridor. If it's a jet, it's a South Korean. Or American. We're on *our* side of the lane. Besides, there's a radio beacon plum in the middle to guide civil aircraft."

"It could go out."

"Hardly think that's—"

"Well, it did. During the eighty-eight Olympics. 'Course," she continued, "that'd be before your time."

"I'm not *that* old," he replied good-naturedly. He was looking out the window now. "You see? It's gone."

"Behind a cloud," she said. "Remember double oh seven?"

The man was nonplussed. "James Bond?"

"James nothing," the woman said irritably. "Korean Airlines double oh seven. Shot down by the Russians in eighty-three. You can't trust any of 'em. Especially Pyongyang."

The Texan moved uncomfortably in his seat. Pyon—it sounded familiar to him. Some Communist leader.

"North Korea," she explained.

"Oh."

"You a businessman?"

He was glad of the change of subject. "Yes. And you—on vacation?"

"Of sorts. Daughter-in-law lives in Shanghai. Works for La Roche Chemicals. Husband owns it."

"La Roche?" The Texan sat up in his seat. If it was *the* La Roche the old lady could be as irritable as she wanted. La Roche was one of the world's biggest chemical/cosmetic conglomerates—*Fortune*'s top ten. "*J. T.* La Roche?" he asked.

"Yes," said the woman. "He's a fool."

"Oh—?"

"All think they'll make a fortune in China. A billion people. A billion customers, that's the way Jay looks at it." She shook her head. "Won't get anywhere in China. I told him—they'll have to get their distribution system organized first. Lord—you ever fly China Air?"

"No," answered the Texan.

"Well, don't," said Mrs. La Roche. "Love 'em, but Lord, are they disorganized. That's their problem, y'see."

"Sounds like you know a lot about them."

She turned toward him. "Henry—my late husband—and I lived in Hong Kong before the Communists took it over. Moved back to the States when the British left. That's where Jay met the Brentwood girl. She'd been doing some courses in college—to be a nurse. Gave that up and went back to China with Jay. She has a brother out here in the navy—another one in the Atlantic. Don't think it'll work."

The Texan wasn't sure what wouldn't work: the navy for the Brentwood brothers or her son's marriage to the Brentwood girl.

"Lovely girl," continued Mrs. La Roche, "but oh my. Can't sit still. Neurotic as all get out. Low self-esteem." She was glancing out at the clouds again. "Course, Jay loves that. Ego." She turned to the Texan. "You remember that Donald Trump?"

"Sure."

"Compared to Jay's ego, Trump's Mother Teresa. Good boy, Jay, but too big for his britches. Don't know where he got it from. Too much money. Wants to own the world, Jay does."

"Well," smiled the Texan, holding his hand up for another double, "he's well on the way."

"You got family?"

"Yes, ma'm. My son, Walter. With the air force."

"Uh-huh. Where's he stationed?"

"Germany. You been there?"

Mrs. La Roche didn't answer, still peering suspiciously at the cumulus towering thousands of feet above them, its ice white turning creamy in the fading light. "No use fretting, I suppose," she said. "It's a dangerous world wherever you go." She paused and sat back. "I'd worry if I had young 'uns though." She turned to the Texan. "Especially now. Everyone's getting jittery. Lana's folks—that's my daughter-in-law—want her and Jay to go back to the States. Nice people. Navy man, too."

"Uh-huh," said the Texan uninterestedly. Why was it that people told you things on planes they'd never dare bore you with anywhere else? A captive audience, or maybe they thought they'd never see you again. Which was true. He was getting impatient for the double Scotch.

"I told them," she kept on. "Use your brains. It was Gorbachev this, Gorbachev that. Lord—worst thing could've happened."

"Why's that?" queried the Texan, the trolley edging closer.

"Raises expectations," said Mrs. La Roche. "Biggest bully

in Europe for sixty years suddenly smiles and we go ape. And everyone in the Eastern bloc starts agitating for independence. You just knew there was going to be trouble. Think about it, I told Henry—think about it. You really want the Poles and Hungarians to start trouble? Drag us into it? Yugoslavs are just as bad. Coming apart at the seams, that country is. Gorbachev encouraged them, too—everybody'll have more freedom. Pretty soon someone's going to try taking a bit more than they're allowed. Ukranians, Georgians, Armenians, Tuvans, Buryats. You name it. Least with those bullies, Andropov, Brezhnev, we knew where we stood. East was East and West was West. Stay off the grass.'' She glanced up at the trolley attendant. ''I'll have a brandy, dear.'' As she took the drink neat and sat back, the Texan saw a glint of silver coming out from the boiling mass of cumulonimbus.

As they came in over the East China Sea, the serpentine curve of the Huangpu was a river of burning gold.

The Texan held back to let the rush of eager tourists go before him. As he passed the young attendant whom Mrs. La Roche had first spoken to, he thanked her for the flight and asked whether she'd found out anything about the other plane.

''Yes,'' she said. ''The captain saw it.''

''Whose was it?''

''South Korean,'' the chief steward put in hurriedly.

''Hmm. They fly that close.'' It was said more as a comment than a question, but the steward got right onto it. ''Actually, they're always much further away than you think. Air distances are very deceptive.''

The Texan saw Mrs. La Roche walking down the concourse past the glass display cases of China dolls, foreign cigarettes, and battery-operated panda bears that moved if you clapped.

In the waiting crowd beyond customs, the Texan could also see a beautiful brunette in a black and white silk dress, a sloppy-looking chauffeur in gray beside her. She was looking eagerly around, as a bored member of the People's Liberation Army stared at her from the customs exit. When she saw Mrs. La Roche she waved excitedly and pointed her out to the chauffeur, and the Texan knew it must be Lana La Roche.

Damn, he thought, wishing he'd gone out with the old lady after all and wrangled an introduction. You never knew where these things could lead. He tried to hurry—nothing to declare at customs—but by the time he passed through, the swarming mass of people engulfed him, bodies so close together that all he could think of was escape. Surrounded, wall to wall, by Chinese, he found the push of bodies frightening, the noise deafening, and he began to panic. He was so far from home, the crowd so huge, so oppressive, unstoppable—like a world going mad—and for a terrifying moment he feared he might never get out, might never see his son or America again.

At 0400, the moon behind them, coming from the east toward the Rhineland's Eifel Mountains and picked up on one of the NADGE—Nato Air Defense Ground Environment—radars, the four East German fighters, Russian-made Sukhoi-15 Flagons, came straight for the American patrol of four F-16B Falcons out of Hahn, each Falcon's Avgas receptacle open, ready for the refueling exercise with a KC-135 tanker. USAF Col. Walter Delcorte, leader of the patrol from the Tenth U.S. Tactical Fighter Squadron at Hahn, ordered the wing to close refueling vents and drop to five thousand, breaking off west, well away from the "trace," the ten-meter-wide border strip that, despite what had happened to the Wall, still ran for five hundred kilometers between the two Germanys along NATO's Central Front. The Falcons broke as ordered, the moon-bathed quilt of German farms sliding away beneath them at over nine hundred miles an hour as they sought to avoid any confrontation from the Warsaw Pact fighters. Four minutes later the Sukhoi-15s came in again. Nose to nose.

"Break west again," Delcorte instructed the other three Falcons. They did so.

"Bogeys have jinked again," reported Delcorte's wing man, seeing that the four Sukhoi-15s, armed with an under-fuselage cannon pack of twenty-three-millimeter weapons and air-to-air "Anab" missiles, were coming in to try it on again.

"Okay, that's enough," said Delcorte, reporting to Hahn base. "Flagons have jinked again for second time. Coming for us now seventeen miles. Closing Mach 1.1." Even though the twenty-one-meter-long Sukhoi-15s were capable of sixteen hundred miles an hour at over ten thousand feet, their present subsonic speed was still very fast for such a low altitude, and all the pilots knew that with the seventeen-ton F-16s able to match them, life or death could be a matter of milliseconds.

"Get under them!" ordered Delcorte. "Four thousand."

With this the F-16s slid into radar advantage, their electronic look-up of the Sukhoi-15s uncluttered by ground fuzz which would, however, interfere with the Sukhois' radar look down images of the American Falcons.

Ten seconds later all four American RIOs (radar intercept officers) armed their air-to-air Sidewinder and Sparrow missiles. Hahn ground control, NADGE, and a NATO rotodome early warning Hawkeye aircraft out of Geilenkirchen all had tracks of the Sukhoi-15s as well as intercepts of East Germany's Dresden control ordering the Sukhois to confront the F-16s again. CO-MAAFCE—Commander Allied Air Forces Central Europe—at Börfink, in NATO's command bunker HQ, ordered two more evasive actions.

"Bogeys jinking again," came Delcorte's voice. "Twenty-three miles."

"Bogey jinked on me," reported another plane. "Noses on at twenty-two miles. Angels four." This meant the American F-16s were still at four thousand feet. Again Börfink ordered evasive action and again the F-16 Falcons reported the Sukhois had jinked again. At 0411 Börfink gave Delcorte weapons-free—independent decision—authority.

"Master arm on!" Delcorte ordered, his voice, but not the faint growl of the activated missile, on tape at Börfink.

"Master arm on," came his RIO's confirmation. "Am centering the T. Bogeys jinked fourteen miles. Centering dot. Fox One. Fox One." The twelve-foot-long Sparrow missile was off, streaking at over two thousand miles per hour, seen on the early warning Hawkeye's radar.

"Ten miles," came in another American pilot. "Fox One. Fox One."

Another pilot's voice cut in, "Tally Three. Tally Three. Af-terburners," indicating he could now actually see the crimson eyes of three Sukhoi SU-15s' afterburners. "Six miles. Select Fox Two. Five miles . . . four miles . . . lock 'im up . . . lock 'im up . . . shoot Fox Two. Fox Two." Now the *shoosh* of the shorter, lighter air-to-air Sidewinder missile could be heard on the tape together with the burst of one of the F-16's twenty-millimeter port side cannons, adding to the general confusion of sound, overriding voices. "Good kill! Good kill!"

At Börfink the tape-connected oscilloscope hiccuped twice. There was a jolting sound, one of the Falcons hit, a scratch of static on the tape, pilot and RIO ejecting. And then in low stratus, above the invisible spread of Germany's southern Palatinate, the thunder of another two explosions, two of the Sukhois disintegrating at fifteen hundred feet.

East Germany said nothing. Further arms reduction talks were under way in Geneva. In the West, newspapers picked up the story released from NATO's headquarters in Brussels: two more F-16s had gone down during low-level practice runs over southern Germany. The F-16, one story reported, was fast taking over the Starfighter in Germany as the "widow maker." There were more complaints from both the Green Party and even some conservative Christian Democrats that the noise level of the low-flying jets was intolerable. From the villages and towns along the border, a petition was presented to NATO headquarters in Brussels and to the Tenth U.S. Tactical Fighter Squadron, complaining that the noise of the jets was in excess of 123 decibels, "like a train roaring through your bedroom," one farmer said angrily, and never mind the effects on the children and the farm animals. The latter were so jittery that farmers had long claimed it was adversely affecting the output of milch cows and increasingly devaluating already depressed real estate values along the border. Both claims were true.

NATO HQ said it would seriously consider the protests and reevaluate the low-level attack program, but did neither. Since 1980 NATO had lost 97 fighters, 61 of them two-crew F-16s. In all, 103 dead. Few pilots and radar intercept officers managed to bail out at such low altitudes. But NATO kept up the dangerous low-level flights. The reason was brutally simple. Along the five-hundred-kilometer border of NATO's central front there were 2,781 NATO aircraft against over 3,050 Warsaw Pact aircraft and a further 4,000 Russian fighters and fighter/bombers between Moscow and the German border. In the event of war, NATO's only hope against a Soviet "surge" would be to have better planes, but even then they could be overwhelmed, the problem of lost pilots, as in the Battle of Britain, becoming the crucial factor. NATO's hope resided in developing better pilot and ground crews than the other side's. In the event of war the NATO/Soviet–Warsaw Pact sortie rate had to be two to one against the Soviet–Warsaw Pact air forces if the West was to stand a chance. In low-level combat, in the battle of razor-sharp wits and split-second radar blips, in dogfights with G-forces so great that they literally drained you of four to seven pounds of fluid a sortie, the concentration needed was enormous.

That afternoon Colonel Delcorte was instructing a new batch of pilots from the States, cautioning them that at such low altitudes the slightest mistake on the touch-sensitive joystick, the slightest reverie, with speeds approaching Mach 1 under two thousand feet, would bring the ground slamming up to kill them. "Any questions?" Delcorte had asked.

A hand shot up.

"Yes?" said Delcorte.

"You said five hundred feet, sir."

"Yes."

"Ah, when's the next flight home?"

It was a joke, of course. Despite such "probing" incidents as the one this morning, deep down no one in the air force, nor anyone else for that matter, least of all Lana La Roche, expected a war. Since Gorbachev, everything had been looking good.

CHAPTER THREE

AMONG HER FRIENDS Lana Brentwood had always been described as a perfectionist, with energy to burn, a good student who always went the extra distance for the straight A's and to please. And a good listener, often seeing things in others that they thought were hidden and yet having the ability to convey to whomever she was speaking to a sense that for that moment she wanted to hear only from that person. It helped to fend off, or hold at bay, her secret and enormous dread of failure. It made no difference that she had done well at school, won a scholarship to Harvard's medical school. She wanted to do everything at once, and everything wouldn't have it, and one day it all tumbled in on her in one exam, an avalanche triggered by one question in her prerequisite arts course: "Man was born free, yet everywhere he is in chains. Comment." It wasn't that she felt she couldn't answer it, but here were so many reasons Rousseau was right, so many that he was wrong. Like a shopper who, feeling suddenly overwhelmed by the sheer number of choices, in the end walks out and buys nothing, Lana broke out in the perspiration of quiet terror, made some panicky notes with which to start the answer, erased them, wrote them down again. Finally she began to tremble and asked if she could go to the washroom. Worse than her fear that everybody was looking at her was her realization that no one was—all too busy, heads bent, answering the question.

She never went back to the room, and following a six-month stint as a student nurse, from which she again dropped out, the rush of failure after came like a series of storm waves battering down an exhausted swimmer as she tried desperately for the safety of the shore, never making it. Her father, John Brentwood, a sixty-five-year-old ex–naval captain, once retired with high honors from the navy and now on the brink of retirement from his post-naval New York Port Authority job, was a man who had led what he called a "no-nonsense" life, one of duty, discipline, and duty. He was for Lana, however, extraordinarily kind when he sat down with her, putting his arm about her, comforting her, telling her that there was more to life than exams and that perhaps she should think about something else where the pressure wasn't so great.

"Thank you, Daddy, but that isn't what you told Ray or Robert when they had troubles. You said when the going gets tough—"

"Yes, yes," he answered. "But with men, Lanny, it's different. They're expected—"

"Oh, *Daddy!*" she said angrily, storming from the room. He made to follow after her but couldn't get up from her bed as quickly as he used to. "Now, I didn't mean you're not equal, honey," he called out.

"Yes you do," she said from the kitchen, snatching a tissue, hating herself because for her it *was* different than for the boys. They hadn't broken down. Besides, these days women were supposed to be as tough.

"All I'm saying, Lanny," continued her father, "is that you don't have to beat your brains out. You can . . . well, do something else for a while."

Lana, now that the floodgates were open, that her uptight, high school "girl most likely to succeed" image had been shattered, steered the discussion self-destructively back to her exam failure, indulging in a whining catechism of other petty and/or imagined failures, about how she really hadn't done anything with her life, about how she was finished before she'd even started. She'd never even had time for boyfriends, no "serious," lasting relationship. John Brentwood looked over at his wife, Catherine. In an age when condoms were as easy to get as gum and an aging Geraldo Rivera was wrestling on TV with near naked women in mudpits, they were thankful there had been no serious relationships. But they understood well enough what it was she was telling them—that she wasn't even experienced as a woman. Still, John Brentwood turned it over to his wife to deal with, to tell Lana there would come a time when she was more settled, when she knew better what she wanted, that there'd be plenty of time for men and "that kind of thing."

"Take a year off," her father said in his wife's silence. It was announced with the surprise of a captain with a wounded frigate suddenly recognizing the virtue of a retreat. "Return to port," he said half-jokingly. "Get your surface vessel out of range of the sub, eh, Lanny?"

"Oh, John," said Catherine Brentwood. "We're not in the Persian Gulf. Last thing she needs is to 'return to port,' as you put it."

"Catherine, I was only trying to—"

"I know, sweetie. But what Lana needs is something to do, something concrete. Take herself out of herself. That's what you used to tell the boys. You still tell David the same thing. Not that he takes any notice."

"Well," John Brentwood began, but stopped himself. Lana was flushed, in that difficult mood, for self-pity and vengeance and for asking where was God anyway?

"There you go again," Lana told her mother, her eyes liquid-bright. "The boys again. The boys are different. The boys are *special*!"

"We don't mean that," said the ex-rear admiral lamely. Sons were so much easier to deal with. "I never meant that."

"You don't think I can handle it alone," snapped Lana. "Well—" She hesitated, lips quivering. "I can. You'll see."

What they saw was Lana dropping out of college altogether, going back to her old love of horseback riding, spending hours at the stables.

"I hope," said John Brentwood, despairing of his daughter's future while leafing through the evening newspaper—reading of more trouble in Yugoslavia, Serbs against Croats—"that Lanny doesn't turn into one of those 'horsey women.' "

"And what," asked Catherine, her hand steady with crochet needle, "is a 'horsey woman'?"

"You know," replied John, turning on the remote control TV. "Look at that. Montreux Convention distinctly guarantees us right of passage through the Dardanelles and Bosporus." There was an inset map behind the TV announcer showing the narrow straits at Istanbul that lead into the Black Sea and the coast of Bulgaria and the USSR. A Bulgarian destroyer had "bumped," the announcer said, a U.S. destroyer off Odessa on the boundary line of the twelve-mile territorial zone.

"Those Bulgarian bastards," said John Brentwood, peering over his bifocals. "Ivan snaps his fingers and they play the monkey. Ah—" He waved his hand disgustedly at the TV. "We won't do anything. Diplomatic notes. Now, if that had happened in Reagan's day—"

"It might just have been an accident," his wife commented.

"Catherine," began John Brentwood exasperatedly. "A naval vessel doesn't accidentally 'bump' another naval ship, for Chrissake!"

"You did once."

"Damn it, woman!" He tore the newspaper away from his lap. "I did *not*! How many times have I told you that that son-of-a-bitch sub was snooping on us—trying to get a good noise signature for their goddamned mines—and I surprised him. Cut engines and the bastard couldn't turn quick enough. *He* bumped *me*, goddamn it!"

"Don't swear. Well, whatever. All they said was it was a Bulgarian ship. We're getting on fine with the Russians now."

"*Now*, yes. I wouldn't trust those sons of—"

"What were you saying about Lana?"

"What—oh, yes. Well, I just don't want her turning into one of those horsey women. That's all."

"I know lots of men who like horses. And they—"

"Don't bait me, Catherine. You know perfectly well what I mean. One of those women who won't go anywhere near anything unless it farts and eats hay."

"Jay La Roche doesn't do either. Far as I know."

"Jay who? That perfume guy?"

"Yes. You remember. You met him at the equestrian ball. The night you were so grumpy. You and that admiral busy jawing about the president's cuts in defense. I think he's rather glamorous. And he's well connected."

"She's been seeing him? I mean—seeing him a lot?"

"Quite a lot."

Brentwood grunted. He *was* glamorous in a way. Do his little girl good to get out and around away from those damned horses. On the other hand, there was something about La Roche he didn't like. Haughty—that was it. Millionaires' club haughtiness. Or was it because the cosmetics magnate didn't like career service officers? He certainly gave that impression despite the ultrapoliteness. Lot of those people around. Thought being in the service meant you wanted to kill everybody. Young David at college in Washington State had noted the same thing. Some liberal artsy-fartsy types baiting him about "mucho macho." And with the new palsy-walsy public relations between Moscow and Washington and the defense cuts, it made it even more fashionable to put down those in uniform. Christ, they were still on about Vietnam, and that was an age ago.

"Lanny say anything to you about him?" he asked Catherine.

"No. But I can tell. She spends a lot of time getting dressed."

"Good God, that's no criterion. Half the women I know take most of the day getting dressed."

"Oh, and how many have you known?" she teased. "Quite a few, I expect."

He ignored it, watching the TV, shaking his head disgustedly. "You see, that's the sort of thing I mean. This Supreme Court business—women in future being ready for combat. Take 'em half an hour to get their war paint on.''

Catherine was hooking the crochet needle in an end stitch. "You're a dinosaur,'' she said simply. "But I love you all the same.''

"What happened to that pilot she was going out with from Andrews? Nice young fella.''

"Shirer. I thought you warned her off servicemen—if I remember correctly. Too many billets. Strain on the marriage?''

"Maybe. But I prefer a pilot to a perfume maker.''

"Now, John,'' said Catherine, putting down her crocheting. "That's unfair and you know it. What if I told you La Roche Industries makes a lot of munitions for defense?''

"They do?''

"I've no idea.''

"By God, you can be irritating. Wouldn't make a skerrick of difference anyway. Even if he was making rockets. I don't like his snooty manner.''

"He's well-bred.''

"Officers from Annapolis are well-bred, too, but they don't act like that.''

"You leave him alone. Lana needs all the self-confidence and affection she can get right now. Said so yourself.''

"I won't interfere, I won't interfere,'' he said, holding up his hands. "All I want is for my little girl to be happy.''

"Good. But she isn't so little, and sometimes you just have to let go. Let them make up their own minds. Remember what you always said about children—they're on loan to us. They're not ours. Best we can do is point them in the right direction and pray.''

"All right, but what happened to that pilot?''

"A fling, I think. It wasn't long after her breakdown.''

"Wasn't a breakdown, it was just—''

"Whatever you want to call it. He was someone she liked at the time, that's all.''

John Brentwood lowered the sound on the TV. "You don't think she—'' There was a long pause. "You know. Do you? With him?''

"I don't know and I don't intend to pry.''

On the TV the announcer was reporting more trouble in El Salvador, or was it Nicaragua, more killing during elections—

shots of blood-splattered hostages murdered in Beirut in the crazy war that Brentwood had never understood and was plain weary of hearing about. There was also growing support for the president, predictions of more defense cuts about to be announced—even more popular than pollsters had previously thought—and a lot of combat training money being diverted to less expensive electronic "closed-helmet" simulators. And reporters were predicting a major reduction in the number of stealth bombers, from two hundred to one hundred, possibly fewer.

Brentwood was shaking his head—a snappy young female Annapolis graduate was telling a reporter how she thought that the possibility that someday the Supreme Court would go all the way and allow women in front line combat role was just "terrific."

"Sure," said Brentwood. "And what happens when you get pregnant? 'Excuse me, sir, could I please leave the war for six months?' " He turned to Catherine. "Honestly, I don't know what the world's coming to."

There was a brief mention that there was a shooting somewhere on the Korean DMZ. "Should've let MacArthur over the Yalu," said Brentwood, getting up from his recliner. Snapping the TV off, he saw under the streetlights a long car—a limousine—pulling up outside his house. He stared at it in disbelief through the blinds; then he saw the chauffeur opening the rear door and Lana getting out, her dark, shiny hair in sharp contrast with the long, white gown and white wrap, imitation fur—she wouldn't have real fur. "Look at this," he said, but when he turned around he saw Catherine had left, and heard her getting ready for bed. When he entered their bedroom, newspaper clutched in one hand, the other whipping off his bifocals, Catherine was changing into her nightdress.

"It's pink!" he announced in horror.

"What is?" said Catherine, taking the comb out of her hair. "Oh, you mean the limousine? Cute, isn't it?"

"*Cute!* Like a New Orleans whorehouse!"

"Oh, don't be such an old stodge. It's cute. Just a gimmick for his cosmetic company. Shocking pink."

Brentwood stood there looking at her. "Sometimes, Catherine . . . I just don't understand you." She wouldn't be baited and kept brushing her hair.

"Now there's this thing in Korea."

"What thing?"

"Another row."

"That's not my fault."

"Ray's out there somewhere with the Seventh Fleet, remember."

"There's trouble everywhere, John," she said. "I worry about it, too, at times. But we can't do anything about it, so that's that." She slid off her slippers and drew up the covers, patting the bed invitingly beside her. "Come on now, stop fretting. I swear you worry more out of the navy than in it. Besides—" she reached for the lamp's dimmer switch "—the experts say that despite all the little wars that always seem to be going on, there's now a greater chance of peace between the superpowers than ever before."

"Experts," said Brentwood dismissively. "Experts told us nuclear arms would put a stop to war forever. Everyone would be too scared to let one off. They're right—so far. Only trouble is, now everybody's so scared to push the button, we need more conventional arms than ever. So what does the president do? Talk about cuts. I don't know. It's crazy."

They heard Lana come in and then go out again.

"What's she up to?" he asked.

"Go to sleep. Honestly, you're like an old woman."

"Ah—you see? Discrimination. I'll take you to the Supreme Court."

An hour later Lana woke them up to show them the engagement ring—Catherine said the diamond was the biggest she'd ever seen. John held his temper, just wanted her to know, he said—and he broke for a moment before going on—just wanted her to know that he'd loved her from the first moment he'd held her as a baby and always would, no matter what happened. As long as she was happy. She threw her arms about them both.

"Thank you, Daddy," she said, and they were all in tears.

In her bed, Lana dreamed her dreams of the exciting life that lay ahead, while in her parents' bedroom, John Brentwood struggled to control his temper as Catherine hushed him, ordering him, imploring him, to keep his voice down.

"But goddamn it!" he said in a hoarse whisper. "He never even asked me. Goddamn it, I don't even know the man."

"I know, I know," said Catherine, ever philosophical, trying to calm him. "I'm sure he intends to. But I agree—it wasn't very thoughtful. But what's done is done. The main thing is, she's happy. Besides, that's the whole idea of an engagement. A trial period."

"For *what*?" John Brentwood asked darkly.

"To give her time to think." There was long silence between them. "It's a beautiful ring," said Catherine.

"Goddamned thing's big as a missile."

"Well, I'm sure it won't kill anyone."

That had been eighteen months before, and since then Lana had jet-setted about the world, on the society pages from *The New York Times* to England's *Country Life*—it seemed that Jay T. La Roche had an interest in horses after all, at least in buying and selling them, and had acquired some of the best stables in Europe. His ownership of stables, however, was not confined to horses—it also extended to a mistress in Paris and others "flown in" upon request, a shattering discovery that Lana had made only when, being mistaken for his umpteenth secretary, she had been given a telephone message to give to Herr La Roche that Fräulein Bader was *völlig gesund*—"perfectly clean." Perfectly clean, Lana discovered through a private detective whose assurances that she was doing the "correct thing" only made her feel dirty herself, meant that Jay's one-night stands were carefully screened by one of a bevy of doctors, retained solely by La Roche to insure that whoever he was bedding aboard his Lear jet at twenty thousand feet was free of AIDS and/or associated viruses.

Believing he still loved her, Lana had tried to tough it out, hoping he would settle down. She even performed for him in ways disgusting to her but which he insisted upon. He kept upping the ante during the foreplay, an extended ploy which, though she didn't realize it at the time, was a vicious psychological game he couldn't lose. If she refused to debase herself further, he told her he could claim sexual incompatibility, a label he made it clear he was personally unconcerned about but one that he could easily use with his army of lawyers to smear her in every scandal sheet and tabloid he owned, and even in some of those he didn't. It was a label, a potential smear campaign that, for her family's sake, Lana dared not risk. The very idea of having such things revealed in public was unthinkable to her, so that suing for divorce was simply not an option.

Not long after Mrs. La Roche had left Shanghai, Lana told him she was leaving. While Mrs. La Roche had been there, her son had reined in his more outrageous sexual habits, such as having everything from young schoolchildren to "perfectly clean" whores flown in from Tokyo and Hong Kong.

For Lana, things deteriorated when Mrs. La Roche returned to the States. Jay seemed bent on a catch-up orgy of sexual indulgence the likes of which she had never even imagined and which, she now understood, was one of the reasons why he favored Shanghai and Hong Kong. The Communist police were in his pocket, and the Beijing government needed his hard foreign currency as much as any of his other customers. So confident was he that no challenge she could make would stand up in court unless she was prepared to debase herself in the witness box as well, that he became increasingly contemptuous of her protestations. Her fear only drove him to further cruelties. He had all the money, all the power. He told her he would sue for libel if she said a word, and ruin her family as well in the process.

The end of the marriage came late one night when he had returned home high on a mixture of cocaine and booze, taking her into the bathroom and insisting she do what she could not do, and the beating began. In the morning, still drunk but knowing he'd gone too far, that if she was seen in public like this, his business might suffer, probing questions asked, he had his best cosmetician brought in—the one who had worked on a lot of film stars in his virtually tax-free Hong Kong studios, where his tax loophole umbrella of production companies churned out films for the Chinese market, all about upstanding Communist heroes of the revolution overcoming corruption. The cosmetician spent six hours on Lana, so that by the time the dark-windowed Mercedes took her to Shanghai Airport, her transit through customs and past a sleepy PLA guard merely a formality guaranteed by Jay's power, Mrs. Lana La Roche, with sunglasses and high scarf to hide the rope burn, was looking none the worse for wear.

He had said he'd send her enough that she could live well on condition that she keep her mouth shut. She said she didn't want his money.

"All right," he'd told her. "Then you won't get it. But—" and then he'd lifted her chin with his lily-white, perfumed hand "—remember this, Lana. You say a word, one word, I'll have Mommy and Daddy Admiral in the *National Enquirer*. I'll make it look like you left me but I wouldn't eat *your* shit." He dropped her head. "Bon voyage. I love you."

It was even more disgusting than what he'd made her do. Yet, hateful as he was, she somehow knew that the moment he had said it, he had actually meant it.

She stayed by herself in the Manhattan apartment to give herself enough time for the bruises to disappear but knowing she was going to need a lot more time before she made any contact with her parents, with anyone she knew. Unable to sleep, she went to see a doctor, who prescribed sleeping pills. It was the darkest period of her life, and more than once she had the vial of Seconal in hand, staring at the person in the mirror she no longer recognized. Even so, a faint voice, from where she did not know, always stayed her hand.

CHAPTER FOUR

Panmunjom

ON MAJOR TAE'S calendar hanging on his Quonset hut office wall, beneath the old black-and-white of him as a young ROK lieutenant in the honor guard for JFK, Major Tae had marked the next day, August 15, in red: Liberation or Independence Day for the South. Apart from *Chusok* (Thanksgiving), Independence Day was the most important holiday of the year, full of parades, color, and pride as dancers in traditional costumes, exotic dragons, and military march-pasts celebrated not only the republic's birth in 1945 after thirty-five years of Japanese occupation, but also South Korea's astonishing progress in the league of industrial nations.

Turning his gaze from his sector of the two-and-a-half-mile-wide demilitarized zone that snaked for 152 miles west to east, the small, slim, immaculately uniformed ROK intelligence officer for Panmunjom pushed the "play" button on his VCR to watch a rerun of the day's meeting, the 917th between the North Koreans and the UN delegation since the cease-fire way back in 1953. Flipping open his notepad, he prepared to take direct quotes from the North Korean delegation, part of his daily report to ROK-U.S. HQ in Seoul.

The meeting had started badly, the tension from the long-forgotten war in which fifty-four thousand Americans and three million Koreans had died as palpable as the muggy summer heat that hung low and sullenly, smelling of dung, in the valley around Panmunjom. As the first frames flickered on the TV screen, the North Korean delegation was already in the process of accusing the UN commission of the "ninety-eight thousand one hundred and fifty-third 'violation' " of the DMZ. Apparently a child's kite, its string broken, had been found by the North Koreans on their side of the DMZ twenty-five miles east of Panmunjom, near what on the Americans' map was called Pork Chop Hill. The aides to the head of the North Korean delegation, Major General Kim, were charging that a microcamera had been found mounted under the kite "to take pictures of 'militarily sensitive areas' in the Democratic People's Republic of Korea."

"That is ridiculous," replied the UN head of the commission, U.S. Army Gen. George C. Cahill, who sat in the middle on the south side of the long, emerald-baize-covered truce table, his arms folded, right forefinger hooked about the stem of a briar pipe. Unlike the red and blue circle of the South's flag, its halves joining, a symbol of harmonious duality between opposites— the *um* and the *yang*, male and female, night and day, fire and water—everything in the negotiating room stood in hostile opposition to one another. The uniforms alone were strikingly different: The NKA's—North Korean Army's—General Kim and his four-man contingent, sitting rigidly on the northern side of the table, were in stiff, high-necked white summer uniforms with red and yellow collar tabs. Kim was unblinking, his only discernible movement a short, studied ejection of a Sobrainie cigarette butt from a white bone holder before lighting another, acrid bluish smoke rising and curling back off the ceiling. Opposite him, on the southern side, the UN uniforms were of a mix. The American general, Cahill, a tall, thick-set man, appeared more relaxed, smoking his pipe, sitting back in open-necked and short-sleeved summer khaki. His South Korean compatriot on his right, more formal-looking, wore the ROK's light blue air force uniform with tie, while to Cahill's immediate left a British brigadier sat resignedly in dark green summer drill. The green-baize-covered table between the two sides was itself marked by division, an inch-wide white ribbon running its full length, the line continuing as a meticulously painted strip up each wall.

Even the beverages were different, the North Koreans' stony expressions during the long, strained silence broken only when they sipped steaming glasses of hot green tea, the UN delegation taking ice-cold water from a silver decanter carefully placed the same distance away from the ribbon as the North Koreans' tea. Directly opposite the North's red-starred flag stood the gold-fringed blue of the UN standard, again both equidistant from the ribbon. The strained silence continued, and Major Tae could see a number of visiting U.S. officers outside the viewing windows of the hut growing restless, pacing on the white cement strip, cameras dangling forlornly.

It was then that General Kim, breaking the silence, leaned forward, crisp white uniform creasing against the edge of the green baize, his malevolent smile, Tae noted, a rare departure from his usual carefully practiced stare.

"All Americans are taught to lie!" Coming from outside the hut there was the sound of armed soldiers, North and South, heels clacking on the concrete as the guard changed. The NKA officer to Kim's left was tacking up black-and-white photographs on a map stand, purportedly showing the "violated area" on their side of the DMZ as the North Korean general continued his accusation. "On the southern side," said Kim, "you have allowed all vegetation to grow wild to camouflage your flagrant acts of aggression against the Democratic People's—"

"I unequivocally reject these charges," cut in Cahill, "as do my fellow members of the United Nations Commission." He indicated them with a wave of his pipe.

"We are not charging the *United Nations* with violation of the territory of the Democratic People's Republic of Korea," Kim spat back. "We are charging the *United States of America* for its imperialistic warmongering. . . ."

"The United States," Cahill replied, his tone controlled, unflustered, "has no intention of—"

Now it happened. Kim leaned forward again, smiling. "Be careful," he cautioned Cahill, stabbing the air with his cigarette, his eyes like dark glass, "or all you Americans will end up like the Kennedys—shot down like dogs in the street."

This was a favorite phrase of Kim's. To Tae's relief, Cahill, who had only been on the job for two months, refused to take the bait, calmly asking instead what evidence the NKA had for their accusation about the kite.

"I will show you," replied Kim confidently. With this he rose. Immediately there was a loud scraping of chairs as he was followed by the entire North Korean delegation. A second later the three Chinese PLA—People's Liberation Army—officers who had been sitting in the rear as observers also rose, the yellow shoulder boards of their new "ranked" uniforms catching the light. Wearily Cahill and his colleagues followed suit; it was part of the ritual, both sides heading outside to the enclosure of hard, mustard-colored earth within the joint security area, where they were to examine the "evidence." In the background a voice, one of the visiting U.S. officers, was asking, "Why do we have to take that shit? That rotten insult about President Kennedy and . . ."

For Tae it was the most frequently asked question by Americans who had bothered to come to Panmunjom, and he didn't even mention it in his report. Before the video jerkily followed the two delegations out of the room, Kim's parting shot was, "You Americans do not realize the effect of your defeat in Vietnam. Now everyone in the world knows you can be beaten."

Privately, Tae conceded that Kim had a point. The danger, as Tae saw it, was that the United States had tried so diligently to forget the war, it was apt to forget its lessons as well. What had Santayana said?—those who ignore history are doomed to repeat it. Most of the Vietnam vets were now dead or too old to pass on, to anyone who even bothered to listen in America, the knowhow of fighting a war in Asia.

For the next two minutes Tae advanced the video, as there

was not much to see, the North Korean and UN delegations standing for over half an hour in the broiling sun. The North Koreans apparently didn't mind, or if they did, weren't showing it, taking what General Cahill later described as a "typically petty satisfaction" in keeping the UN team sweating and waiting in the stifling, fly-infested heat. Kim was the only man Cahill had ever seen who could tolerate swarms of flies crawling all over his face, across his lips, in and out of his nostrils and eyes, without once allowing himself to blink.

"Have you got something to show me or not?" said Cahill. "If not, my colleagues and I propose an adjournment until . . ."

Some signal that Tae didn't see must have been given by Kim, because the evidence, or rather a battered-looking, carp-shaped kite, a dirty orange color, about four feet in length, four to five inches across, was carried solemnly into the cordoned-off compound by two NKA soldiers, two lines of North and South Korean guards grimly facing each other.

"Where's this camera?" Cahill demanded sharply.

"It is being analyzed," said Kim. "It is made in Japan."

"May we see it?"

"I said it is being analyzed. The film, which we have now developed, shows it was a blatant act of imperialist aggression upon the sovereign territory of the Democratic People's Republic of Korea by U.S. aggressors and . . ."

Cahill turned to his aide, a Captain Jordan from the joint U.S.-ROK command. Jordan rolled his eyes skyward. Ignoring Kim, Cahill took a handkerchief from his pocket and proceeded to wipe beads of sweat from his forehead, asking Jordan, "Haven't I heard this somewhere before?"

"Word for word, General. On Pyongyang radio. All this week." When Kim had finished his diatribe, Cahill poked the carp's ugly, gaping face with the toe of his boot. "This . . ."

Kim stepped forward menacingly, as did his entire delegation. "Do not touch the evidence!" he shouted.

Suddenly the entire compound was electric, both lines of border guards stiffening at the ready. Hands on holsters. Each man fixing his opposite number. Cahill smiled, having forced Kim to react.

"Evidence?" snorted Cahill. "It's a child's kite."

"Yes!" shouted Kim's aide, shifting his gaze from Cahill to

the ROK officer, General Lee. "And made in the U.S. lackey
state of South Korea."

"A child's kite," repeated Cahill contemptuously. "Blown
over to your side by the wind. Why, it's not big enough to hold
a camera, and besides . . ."

"We have the spy machine!" Kim shouted, pointing his finger. "You cannot deny . . ."

"Spy machine? You've got nothing," said Cahill, turning on
his heel, leading the UN delegation out of the compound.

"We have the evidence!" Kim shouted after him. "We have
the film showing that . . ."

"Showing," said Cahill, still walking away, "the photos
you're busy taking today so you can fabricate a case."

"You be careful!" Kim shouted after him. "Be careful, you
Americans. You will end up like the Kennedys. You—"

"*You* be careful," said Cahill, but in a voice he knew neither
Kim nor the North Korean delegation could hear. "Go back to
Pyongyang, you running dog turd!" Cahill turned to his aide.
"Christ! I'm getting too old for this nonsense, Jordan. I'd like
to shoot that son of a bitch."

"Me, too," put in the South Korean general.

The video over, Tae dragged the day's SIGINT—Signal Intelligence—and IR—Interrogation Reports—toward him. There
were several Blackbird American infrared SR-71 reconnaissance photos taken from eighty-five thousand feet. No change
in the NKA's unit dispositions except for another surface-to-air
missile site being built in the Taebaek Mountains that ran like a
spine down Korea's east coast.

Tae made himself a cup of coffee and for a moment sat ad-
miring the desk photo of his children. Mi-ja, eighteen, was re-
splendent in a watermelon-pink *chima*, the traditional flowing,
high-waisted skirt, with matching *chogari*, or short jacket.
Dyoung, eight, stood proudly in the loose, white, pajamalike
uniform of Taekwondo, the ancient Korean hand and foot combat sport of self-defense. It was their future he was most worried
about.

He opened the interrogation folder. Another NKA infiltrator
caught south of Munsan near the west coast. As usual, he'd
refused to talk. The report stated the prisoner had "fallen down
stairs." Even so, he had still not revealed the specific target of

his mission—they never did—and so he was put on the train for the prison camp on Koje Island in the Southeast. Tae hoped that these days Koje was more secure than during the Korean War, when the American guards had been so accommodating in assuring prisoners' rights under the Geneva Charter that the NKA had actually managed to kidnap an American general and stage a massive breakout. Even so, Tae wasn't naive enough to think the present ROK guards hadn't been infiltrated by "sleepers," NKA agents already "in place" in the event of hostilities. For every NKA agent they caught, Tae suspected there were at least four or five who managed to slip across the DMZ undetected.

Only one thing of interest had emerged from Tae's quieter, more democratic interrogation of infiltrators captured in his zone that week, and it wasn't until after three sessions involving the usual cursory examination of the prisoners' possessions that he'd even noticed it and decided to add a note to his daily report to Seoul. He had observed that the chopsticks the NKA infiltrators had on them when captured—no Korean would travel without them—were seven inches long instead of the standard nine.

Tae typed his SITREP, situation report, quickly, for he knew that as soon as the general had finished inspecting "nighttime readiness" of the local U.S.-ROK fortifications, he and his aide, Jordan, would leave for the more civilized environs of Seoul HQ.

CHAPTER FIVE

SHORTLY AFTER 10:00 P.M., heading back to Seoul, Jordan decided to take a chance. He'd been thinking about it ever since the farce in the joint security enclosure. "Ah, General—" he began hesitantly.

"Yes?"

"Ah—I've got a buddy—a pilot—stationed down south at Osan. They say he's a real genius with aerial reconnaissance, camera-to-speed ratios—all that sort of hi-tech stuff."

"So?"

"I was just wondering, sir—say we got a really small Nikon and one of those big kites, like the ones they have in some of the temples . . ."

"For Chrissake . . ." said Cahill.

"Well," began Jordan defensively, "it was just a thought. We can't see what they're up to at night. They could be tunneling—again. Our infrared overflights didn't detect any of those we found last year."

"We found them, didn't we?" challenged Cahill.

"Yes, sir, but more by good luck. An infiltrator spilled the beans, but normally they never reveal . . ."

"We've got seismic probes," replied the general. "Ground sensors."

"Yes, sir, but the problem is the moment the NKA go on maneuvers, we can't distinguish a tunnel being dug from any number of other noises—heavy trucks, road-working equipment. All we get is tremor graphs."

"You're worrying too much," said Cahill. "Winter—maybe—when the ground is hard enough for their armor. But not now in the monsoons. Rice paddies all flooded. Everything'd bog down."

They were now completing the U-turn near the "Bridge of No Return." On their left, a U.S. army truck sat fully gassed, motor running twenty-four hours a day, ready to back up and block the bridge. To their right stood the Y-shaped tree that had been trimmed for a clearer view across the DMZ and where thirty-one NKA regulars had come across in '76 to club and hack two Americans to death.

"Anyway," continued Cahill, "we're constantly patrolling the DMZ. We've got minefields right up to the wire. Fighter Command's on a moment's notice. Attack choppers are ready. Isn't like it was in fifty, Dick. Moment Kim or any other gook puts his cotton-pickin' finger over that ribbon, I'll chop the goddamned thing off. What's our G-2 say?"

"No unusual movement."

"There you are."

"But," Jordan pressed, "there is this report from Major Tae. The ROK intelligence officer up here . . ."

"Yes, yes," said Cahill impatiently. "I glanced at it before we left. Goddamned chopsticks aren't as long as they used to be. Timber in Korea's in short supply—always has been—so he figures all the wood saved goes into building more *chiges*— A-frame backpacks. Infantry buildup."

"Yes, sir. We know their infantry, like the Vietcong used to, carries everything on their backs."

"So do ours, Dick," said Cahill. "That's why we call 'em 'grunts.' You'd grunt, too, with eighty pounds weighing you down." But the general knew what Jordan meant. "Yes, yes, I know," he said irritably. " 'Red Army's two legs better than Americans' four wheels.' Right?" The car had passed the Munsan checkpoint. Suddenly its brakes squealed, the car skidding sideways in an earsplitting screech, headlight beams roiling with talc-fine dust. As custom decreed, the driver was allowing an old man in traditional white "pajamas" to cross the road, the man's tall, wide-brimmed stovepipe hat a symbol of his status of country gentleman.

"I hate those stupid hats," said Cahill, venting his fright at the car almost hitting the old man. "Ever see anything so goddamned ridiculous in all your life? What's it good for? Damned horsehair won't keep out water—won't keep off the sun." Jordan made vague noises of assent but was more interested in getting the general to think about Tae's hunch than elders' hats. Cahill anticipated him.

"Lookit, if an attack comes—to have any hope, any hope at all—it's got to be hard and fast. Right?"

"Yes, sir."

"Yes, and all those backpacks of yours are going to be worth squat—all unless armor clears the way. Right?"

"Yes, sir."

"Any sign of armor—visual or ground sensor?"

"No, sir."

"*No*—because it's too damn wet. Moment a tank goes off a road, it's in rice paddy. Sinks in the mud." He paused. "How about the canaries—they okay?"

"Far as I know, sir." Cahill was referring to the tunnels the

NKA had dug beneath the DMZ during the seventies, which were later found and cemented shut except for a small, wedge-shaped peephole in each cement bung. A few yards away from each bung there was an ROK machine-gun post manned around the clock, and right by each peephole, a canary in a cage. As in the mines, the bird's death would be an early warning in the event of gas attack.

"Tell you what," said Cahill. "I'll issue a bulletin on KBS TV and radio announcing that despite tomorrow's Independence Day celebrations, South Korean defense forces remain vigilant to any incursion of the North."

"General!" said Jordan, happily feigning shock. "That's using the airwaves for willful propaganda."

"Damn right," Cahill smiled. "It's true, too. We'll run a few tanks through town tonight. Keep you and Tae happy, *and* it's a darned sight cheaper than a general alert." Cahill smiled. "General Accounting Office'll probably give me a medal—in addition to my KBM." KBM was Seoul HQ's acronym for "Kim Bullshit Medal," awarded for patience and restraint "above and beyond the call of duty."

Jordan laughed, but the worrier in him remained—as persistent as an allergy one can do nothing about. "So you think we should forget about the chopsticks?"

"Forget the chopsticks. There's nothing to it," said the general. "Chopsticks aren't going to start a war."

The general was right, chopsticks had nothing to do with it. The fuse that led to World War III would be lit by a misunderstanding over red ants.

CHAPTER SIX

TRADITIONALLY A KOREAN delicacy when slightly sautéed, the ants were sold downtown in glass jars by vendors scattered throughout the myriad alleys and side streets of Seoul's brightly lit and bustling Myongdong district. Here, amid the spicy odors of *kimchi* (pickled cabbage) and the cooking of marinated meat, throngs of workers from late night shifts were hurrying home before the midnight curfew, past the variegated plastic canopies of the pushcart stalls, their owners hawking everything from *soju* (octopus) and *pin daeduk* (pork-and-vegetable-garnished mung bean pancakes), to the favored tangerines and oranges from the southern island of Cheju. The crowd's shadows flitted through islands of fiercely burning carbide lamps that illuminated the vendors' faces as if they were polished china, their voices rising, bartering becoming frantic in the race against the clock. Above the alleys, in the polluted and unusually cool summer air, neons flashed with accompanying urgency in the collective frenzy before the blackout drill, which tonight would precede the usual midnight-to-4:00 A.M. curfew, the blackout's wailing of air raid sirens yet another reminder, as if any of the twelve million inhabitants of the city needed reminding, that they were only twenty miles from the border between North and South, and two and a half minutes from North Korea's bombers, and within range of the NKA's long-range artillery.

Only the night before, one of the South Korean patrols, continually on duty in the hills ringing Seoul, had clashed with six

North Korean infiltrators. They had been intercepted while crossing the DMZ.

After a short, fierce firefight, not unusual along the DMZ, five of the infiltrators had been shot, one dying shortly after. South Korea's CIOC (Counter-Infiltration Operations Command) was reasonably sure the six were from the NKA's 124th guerrilla unit. It was this unit from which thirty-one North Korean commandos had penetrated the southern side of the DMZ on a bitterly cold January night during the infamous mission of '68, armed with AK-47 submachine guns and grenades, with express orders to assassinate President Park of South Korea. On the second day of the mission, four woodcutters saw them, notified local authorities, and the hunt was on. Even so, the guerrilla unit reached Pugak Mountain on Seoul's northern outskirts, and charged the Blue House—official home of the president. Twenty-eight of the thirty-one North Koreans were killed in the ferocious gun battle that followed, rifle and machine-gun fire and bursting grenades echoing like strings of firecrackers in the hilly amphitheater around the city, causing several small brush fires. Two of the guerrillas managed to escape, but one, surname Kim, given names Shin Jo, no relation to the present General Kim, was caught. Trading his life against the certain fate of being shot as an infiltrator, Kim passed over to *Chungang Chongbo-bu* (South Korea's Central Intelligence Agency) all the details of North Korean President Kim Il Sung's plan to assassinate President Park, including the fact that at each of 124th Unit's eight guerrilla bases in the North, there were three hundred volunteers.

This meant that since 1968, over two thousand North Korean agents were being prepared at any one time for further infiltration, agitation, and sabotage against the Americans and the South, that every day at least one agent was crossing the DMZ.

It wasn't that the North Korean agent in the Myongdong area this evening before Independence Day had been badly trained. On the contrary, he had received high commendation from his base commander in Kaesong. Not only did he know the military dispositions, weapons, and insignia of all South Korean and American units, especially those along the eighteen and a half miles of the 155-mile-long DMZ guarded by elements of the U.S. Second Infantry Division, but in addition, before being sent south, he had been carefully instructed in those local habits and customs that can so often trip up an agent. He was made well aware, for example, of the new words and phrases creeping into the language. He was told to remember that even though South Korean men, like their North Korean counterparts, expected total obedience from their womenfolk, in the South one should no longer use the word *sikmo*, calling a maid a maid, but *kajongbu*—"homemaker" or "home manager"; in much the same way, his NKA instructor told him, as garbage collectors in America insisted on being called "sanitation engineers." And the agent knew about the red ants. Savored by most Koreans, especially by those from the South, the insects were collected from their favorite habitat in the hills around Pusan and trucked to Seoul on the 270-mile highway that ran almost the entire length of South Korea. Once in Seoul, the ants, like so much other produce, were auctioned off to the highest bidders among the street vendors, who in turn sold them to shoppers off the fashionable Myongdong.

What better way for an agent to indulge his weakness for the delicacy that he could ill afford in the North and at the same time reinforce his cover as a genuine South Korean? Approaching the pushcart, he realized he had only forty-five minutes to get back to his safe, cheap *yogwan*, or "inn," two miles from the city center, before the start of the midnight curfew and air raid drill. Still, he would have ample time if he used the subway. The misunderstanding that was about to occur was largely due to the fact that the agent, having only just slipped across the Han River near Kumchon fifteen miles northwest of Seoul the previous night, had been so busy avoiding ROK patrols and settling into the *yogwan* that he hadn't yet had a chance to sit down and read a newspaper. This meant he'd missed the two-paragraph story in most of the dailies, except the *Korea Times*, which didn't publish on a Monday, about the brush fires in the hills around Pusan, fires that had killed off large numbers of ant colonies. Reduced supply meant higher prices. Unaware of the sharp increase in price, the agent gave the vendor a ten-thousand-won bill, about ten dollars, for an eight-ounce jar. The vendor waited politely, the glass of red ants the customer asked for now costing twice the usual amount. His customer waited, expecting the jar of ants and at least two thousand won in change. Then he realized he hadn't given the vendor enough. *"Olmayo?"*—"How much?"

"Ee-man"—"Twenty thousand." The customer dug deep into his jeans pockets, joking weakly that it would wipe out his subway fare. The vendor, though annoyed, did not show it and got a good look at the man, remembering the posters, as common as theater billboards throughout the city: "If you see a stranger who does not know the exact price of things, or spends a lot of money and hasn't got a job, or calls you *tongmu*, which means 'comrade' or 'friend'—grab him! He is a spy."

Well, he mightn't be, thought the vendor. Then again . . . The vendor had lost both parents when the NKA had invaded in 1950.

"Sorry," apologized the customer. "I haven't got enough."

"That's all right," replied the vendor, already starting to pack up his stall with the speed and deftness of long experience. "I'll still have some of these left tomorrow—or another consignment will come in."

"Thanks," said the man, taking his leave.

The vendor turned off the carbide lamp, quickly asked a colleague to watch his cart, and followed the would-be customer out onto Sejongro's sixteen-lane-wide avenue, where he saw the man walking north, drawing level with the huge statue of Admiral Yi. The admiral, in ancient armor, left arm akimbo, right hand gripping his enormous battle sword, had also been vigilant in his time, the vendor recalled, alert to foreign invaders, defeating the great Japanese invasion fleet of 1597.

The vendor, however, while full of the spirit of Admiral Yi, couldn't see the yellow light of a police station, let alone a policeman. Where were they when you needed one?—always sniffing around when they wanted free samples from the cart. The only official in sight, her smart blue U.S. Navy–style cap barely visible amid the Hyundais and the noisy red and white buses roaring past, was an immaculately dressed and beautiful woman traffic director, her white gloves moving with the suppleness of doves in flight. But the vendor knew by the time he weaved his way through the river of oncoming vehicles and reached her, the man might disappear. Seeing a taxi sign above the crowd, the vendor dashed out to flag it down. But it was brown, only for military personnel, so he had to wait a second until he saw a green cab approaching. Barging ahead of others in line, he jumped into the backseat, glimpsing the yellow-uniformed woman driver as a blur, quickly instructing her to have her dispatcher alert the nearest police cruiser to meet up with them. To the vendor's alarm, he heard the flag drop and the meter ticking.

"What are you doing that for?" he asked. "This is a public duty."

"So?" She shrugged. "Someone has to pay." He was astonished, but she was very young, and he knew that the horrors of the Korean War, so vivid in his childhood memory, must be nothing more than dead history to her generation.

For an anxious moment he thought they'd lost the stranger as they turned right into Yulgog Street, heading east near Changdok Palace and the zoo, but the young woman told the vendor to relax. She still had the man in view and was going to pass him, just in case he suspected he was being followed.

"How could he know?" asked the vendor. "With so many people about?"

"If he's an infiltrator, he'll have been trained in such things."

After a few more minutes, amid the usual honking and insults to various ancestors, a beaten-up, off-white Sinji sedan drew alongside the green cab—two KCIA agents, the one driving telling the cabbie they'd take over, the other asking the vendor to point out the stranger in the crowd. Suddenly the man disappeared into an alley off the Sejong a hundred yards behind them. Without hesitation the vendor told the cabbie to stop, got out, and headed back toward the alley, the agents swearing, pulling sharply into a no-parking zone and following suit. It was now 11:45—fifteen minutes to blackout.

Five minutes later, at the end of the alley as the air raid sirens began their wailing, the two KCIA agents caught up with the vendor, who was now gasping, out of breath. "We're in luck," said the younger of the two.

"How d'you mean?" asked Chin Sung, his older colleague, a shorter man in his midfifties.

"He never got on the subway after all. He's gone through Donhwamun Gate, so it's either Changdok Palace or the Secret Garden."

A strong wind hit them full force in the alley, kicking up dust and litter, forcing the shorter, older agent to lower his head, the grit bothering his contacts. "In luck," Chin growled sardonically. "The garden alone covers seventy-eight acres." Candy wrappers and fallen ginkgo leaves, their small, polished green fans turning black under a dim pole light, swirled scratchily about the men's feet. For a moment the older agent felt nostalgic for the Olympics of '88—*then* the city fathers had made sure there was no garbage to be seen anywhere on the city streets, like the cleaner cities of Germany, where Chin had once been stationed, attached to the ROK embassy in Bonn and trade legation in West Berlin.

"Well, it's just about curfew," said the younger agent optimistically. "He isn't going anywhere. Has to stay in there or risk being picked up the moment he leaves. And if he tries scaling the walls, we've got him!" Chin grunted, looking through the gate across at the pavilions of Changdok Palace, the home of the surviving royal family, and toward the lighted, wing-tipped roof of the pavilion by the *Pandoji*, the Korea-shaped pond, pathways radiating from it through maples, the wind moving through the trees like rushing water.

Chin took a small walkie-talkie from the inside of his coat. "All units—we're going to lose him over the wall, whoever he is, if we don't surround the whole area immediately."

A voice crackled from somewhere on the other side of the gardens. "We're cordoning it off now. You want us to send in the dogs?"

Chin shook his head in disgust. "No—I want to keep him in there. Trap him, not panic him." Retracting the walkie-talkie aerial, Chin turned to the vendor. "You *sure* he acted suspiciously? Could have been a young buck hurrying to the gardens to meet his girlfriend—keep her warm during the curfew?"

"Yes," said the vendor, "I'm sure. I'm telling you, it was *nunchi*." He meant "eye measure"—beyond mere sight, a sixth sense. "And he didn't know the price of the ants," continued the vendor. "Everyone's been reading about the fires down—"

"All right," cut in Chin gruffly. "Where do you live?"

"In Chamshil." It was the area of dozens of huge, look-alike cement high-rises clustered several miles south across the Han River near the Olympic village.

"We'll get a car for you," Chin told him, pulling out the walkie-talkie's aerial again.

"Will you let me know what happens?" asked the vendor.

"Yes. Certainly," said Chin, giving the vendor his card and signing a "pass through" chit for the blackout drill and curfew.

"Thanks for the tip," said the younger agent. Soon another unmarked car quietly appeared at the far end of the alley.

"There's your ride," said Chin.

As the vendor walked away, five minutes before the onset of curfew and blackout, the younger agent tried to find out what his older colleague's plan was without wanting to appear stupid. "He must know he's being followed."

"Not necessarily," answered Chin. "Unfamiliar with the subway maybe, mistimed it. Rather than get caught in the curfew—probably decided to hole up for a while, stay out of sight till the morning. Garden's as good a place as any, and there's a lot of pavilions—in case it rains. Which," he said, looking skyward, "I think it will."

"I think we should go in and get him."

"And if he is an infiltrator, what will you find?" asked Chin.

The young agent thought for a few seconds. "Maybe he won't swallow it. While there's life, there's hope. Right? Look at Kim Shin Jo—came down to shoot Park, gets caught, and ends up with a nice suit and eating out. *Peking duck.* Some of the boys tell me that when he wanted, he even got to go to the Angel Cloud House, the *kisaeng* girls pampering and singing to him. Nice work if you can get it."

"Yes, well, this guy isn't Kim Shin Jo," said the older agent, pausing, unscrewing the top from a Dristan bottle and tilting his head back, his voice more strained. "If we rush, our boy might pop the pill. Then where are we?" He paused, snuffling back the nasal spray while screwing the lid back on. "Damn summer cold." He turned to the younger agent. "But let's say there's a slight chance he doesn't know he's being followed, that he simply ran out of time. Maybe he remembered the curfew but not the blackout—after all, blackout only happens once a month. Might be doing what a lot of others do—just sitting it out. So while he's still got light to see by, he heads north on Sejong to Yulgog—then straight to the garden before blackout begins."

"So?" asked the younger agent.

"In the morning we follow him home. That's what we want. Why grab him now if we can get the whole cell?"

"You believe he doesn't know we're on to him?"

"No," answered Chin. "But it's a possibility. Old Kim Shin Jo didn't know those woodcutters were on to him either. Did he?"

"You have a point."

The air raid sirens were reaching full volume and the lights were going out all over the city, huge skyscrapers that dwarfed the Secret Garden's gnarled pines and tile fluted walls suddenly appearing twice as big and brutish in the moonlight. "You want some gum?" offered the younger agent.

"Gives me gas," said Chin, turning toward his younger colleague but unable to see him, the moon now enveloped by cloud. Even so, the younger man sensed the other tensing.

"What is it?"

"Unless," began Chin, his voice dropping, "the *garden* is the meeting place and he has a set planted."

The younger agent heard his colleague take out the walkie-talkie, his voice in whispered tones requesting the RDF—radio direction finding—truck to move in to pick up any signals coming from the garden.

"Ah," said the young agent. "He'd be a fool to transmit from here. In the *heart* of Seoul!"

"You remember Sorge?" asked Chin. "Germans' top Communist agent in Japan—the best of them. Told Moscow Japan wouldn't be attacking through Siberia, so the Russians were able to move a million fresh troops from Siberia to Stalingrad. Changed the war. You know where Sorge transmitted from, my young friend?"

"You're going to tell me. Right?"

"Used to give parties for all the big shots in the Japanese military aboard his yacht in Tokyo Bay. He'd slip away from the party—transmit from the cabin right below them."

"He had balls then."

"You think we have a Communist with balls here?" asked the older man, snuffling the spray again. "Not what you'd expect, is it—transmitting right under our noses?"

"No," the young man conceded. "It isn't." After a few seconds he spoke again. "Shouldn't you wait longer with that stuff?"

"What?" asked the older agent.

"Nose spray. It can screw up your sinuses if you take it too often."

"Whose nose is it?"

Soon, in the darkness, they could hear the RDF truck rolling softly down the alley toward the entrance to the Secret Garden, crushing the ginkgo leaves blown down by the summer wind.

CHAPTER SEVEN

IN WASHINGTON IT was morning, and the president, James R. Mayne, was about to go jogging along the Camp David trails for the TV crews to get their clips for the evening news. But there was a problem: the forest-green jogging suit, which would blend in well with the woods and was being insisted upon by the Secret Service, was objected to by the president's press aide, Paul Trainor. Trainor was advising the president to wear the white jogging suit, which would stand out more in contrast with the woods. It was silly, and normally Jean, the first lady, would have settled it on the spot, but she was away campaigning for the president in the Northwest. In her absence the chief executive left media tactics to Trainor. The election was only eight weeks off, all polls showing the race would be a cliff-hanger. Mayne was still getting high marks for what looked like another arms reduction treaty with the Soviets, but his challenger, Sen. J. D. Leyland from Texas, was batting well, too, with his promise to trim the "federal fat" from "overused, overabused" social programs so that he could in fact "reduce taxes" without weakening national defense.

Mayne's election platform was based on his cuts in defense spending, directly related, as his campaigners pointed out, to his much-lauded success in having kept the United States from becoming embroiled in "other people's wars," particularly in Central America. He had also been successful in keeping down the costs of maintaining U.S. bases throughout the world, such as the forty-two-thousand-man force in South Korea.

Senator Leyland, on the other hand, was running on a platform charging that America was becoming "gravely weakened" by her cutbacks in defense and that the president's nonintervention in the "wildcat" fires of Central America represented not so much a saving in America's defense budget as a "bankruptcy" of national policy, which "ignores the demands of U.S. national security and global obligations."

"Mr. President!"

It was Trainor, handing him *The New York Times* and *Washington Post*. "Latest polls, sir." They confirmed it was still "neck and neck," but increasingly the president's "age factor," sixty-one, against that of his challenger, fifty-one, was commanding more attention from the press.

"Okay—I'll wear the white suit," the president told Trainor. He didn't like playing the media game, but he knew it would be a heck of a lot easier getting things done for the country if he was reelected.

After the photo opportunity in front of Aspen Lodge's kidney-shaped pool, the president and Trainor headed out by limousine to Andrews for the long hop to California. On the way they saw a banner: "Reelect Mayne—the peace president."

"That," said Trainor with conviction, "is what's gonna beat the ass off Leyland! Seems a terrible thing to say, Mr. President, but in the long run, Vietnam may have turned out to be a blessing in disguise for this country."

"Well, Paul," said the president, on the lookout for more groups of supporters, "you're going to have to explain that one to me."

"I mean, Mr. President, that this country is going to think twice before they let the drum thumpers send our boys out to get slaughtered for a piece of real estate that means squat-all when you come right down to it."

A group of Leyland supporters flashed by, holding an enormously long banner reading, "Vote Leyland. And make America great again!"

"That's one hell of a drawn-out message," said the president, glancing back. "Take you half an hour to read it."

"Yeah," agreed Trainor. "Look great on TV, though. They'll have to pan wide to get it all in. More exposure."

"By making America great again," reflected the president,

"I suppose they mean it's time we bombed something. Flex American muscle?"

"Something like that," responded Trainor.

"Well, if that's what they want from me, they're going to be sadly disappointed."

CHAPTER EIGHT

BEFORE DAWN, INDEPENDENCE Day, the pungent odor of breakfast *kimchi* and gasoline fumes filling their interiors, the riot buses, twenty of them, wound their way down through the early morning traffic and pollution to central Seoul, disgorging a squad here, a squad there, at various strategic positions throughout the city. Subway entrances in particular were favored by those retreating students hoping to grab a train out of the fray when the "riot ritual," as the police call it, got too rough for them. And it would get that way soon enough in the exhausting and exhilarating business of baiting the police amid temperatures that AFKN, the U.S. Army radio network, predicted were going to climb well into the nineties with matching humidity, creating a fifty-fifty chance of late thunderstorms. But now it was still cool as the police quietly took up positions throughout an H-shaped grid running north-south through the city's western sector down Sejong and Taepweong, and in the eastern sector running down Chang-Gyeong. Joining the two arms of the H were platoons stationed along Chcong Gye, the largest concentrations in the left half of the H around City Hall. Of the squads allocated to the protection of U.S. buildings, most were stationed outside the U.S. Chancery, as it, unlike the heavily fenced-in embassy, was immediately accessible from the street.

"Maybe it'll be too hot for them," joked a platoon leader in one of the rear buses.

"It's too hot now," replied one of those standing in the aisles. "This gear's killing me."

There was a chorus of mock sympathy and a punching of shields.

"You'll have rest soon enough, Chun," said the platoon leader. "Some student'll put you on your ass."

"I'll break his head first," said Chun, lifting his truncheon.

"You monster!" cried out another. "What if it's a woman?"

"Then I'll stick it somewhere else," Chun retorted.

"Someone'll have to show you where it is."

"I know where it is." Chun was twenty-four, son of a janitor, and a policeman who hated university students with a passion. It wasn't only the fact that he had come from a poor family and had missed out on the opportunity to go further than high school that made him feel so hostile toward the students—rather it was what he saw as their blatant hypocrisy. The same hooligans who would be throwing rocks and insults at him today would, in four years time, be executive trainees for Hyundai, Samsung, or some other giant *chaebol* while Chun would still be a policeman fighting a new generation of students shouting their obligatory anti-U.S. slogans. The latest were "Drive away the American bastards!" and "Down with the government!" banners scrawled in their own blood. Oh, some of them, the leaders, were genuine Communists, and Chun hated them the most for inciting the more gullible in the giant "reunification rallies," thousands of students deafeningly applauded the Democratic Reunification Party's belly-crawling overtures to Pyongyang. But Chun believed most of them were simply out for excitement under the pretense of it being serious political protest. It was a *lark*, a time to vent all the teenagers' pent-up rage against parents forever pressuring them to *Kongbu haera!*—"Study! Study!" A chance to lash out at police, teachers—against all the Confucian-bred respect for authority.

Chun filed out with the rest of his platoon near City Hall, but designated as a "flying wedge," his platoon would not remain at any particular junction. Instead it would be on standby—ready to move quickly to reinforce weak spots in the H. Chun took great pride in knowing he was a member of the most experienced riot police in the world. Never mind all the "Cherry

Berets''—the old ''Olympic Police''—sliding headfirst down ropes like monkeys for the evening news crews, or the blue-denimed National Police; when the big battalions of protesters came out, when it went from bricks to Molotov cocktails, it was Chun and the other ''Darth Vaders,'' the black-helmeted riot police, who settled it. A squad of neatly turned out National Police passed by, their white helmets wonderful targets for any projectile. One of them waved. Chun nodded with stiff formality; the riot police remained aloof. Someone in Chun's platoon said the Catholics and Protestants were coming out in support of the students.

''So?'' a rookie asked.

''Protestants!'' replied a corporal. ''That's how they got their name, right? *Protest-tants.* Means *golchikkori*—troublemakers!'' As far as Chun was concerned, the Catholics were no better. And if it was true the Christians were going to get involved, it *would* be a long, hard day. Students might then win middle-class support. The worst possible combination.

He heard a crackle of radio static; another three platoons, a hundred men in all, were being requested by the officer in charge of policing the square around Myongdong Cathedral. Catholic nuns were forming a human chain, swaying and singing hymns. Then there was a call from police HQ diverting two platoons to Yonsei University in the west. Less than a minute later an urgent plea came in from Korea University campus in the city's northeast. This was unusual, the students normally favoring inner-city streets for their protests, where they could best be concentrated to gain maximum TV coverage and where if you ran out of paving stones, there were always construction sites—plenty of loose brick. Besides, the Molotov cocktails, made mostly with empty OB and Crown beer bottles, were much more effective against closely packed police in city streets than on open campuses. What was behind the new tactic? Chun wondered.

A ''most urgent'' call came in for a ''wedge'' at the corner of Yulgog and Donhua about four blocks northeast of the U.S. Embassy.

The first shower of projectiles thudded against the bus's thick window mesh as it passed Changdokkung Palace on the left. Chun could see the students, about two thousand, he guessed, overwhelming a hundred or so National Police, white helmets dotting the huge crowd as it swarmed about the entrance to the Secret Garden—a phalanx of placards demanding reunification. Soon the crowd of students, half already inside the gate, was expanding, contracting, and expanding again, at once controlled and uncontrollable, pushing and pulling, its waves surging through the gates, spilling into the gardens.

"Beautiful!" said Chun. "Those bastards are bottled up inside by the wall. Perch in a pond." He pulled out his club. "Boom! Boom!"

"Chun!"

"Sir?"

"You're on tear gas."

"Yes, sir," said Chun, cursing under his breath. He badly wanted to use the stick. He broke open the short, wide-barreled gun and plopped in a canister of "pepper" gas, the most acrid, then snapped the gun shut. Next he tightened his gas mask and flipped down the steel mesh face guard. The platoon, now in its "Darth Vaders," was ready.

"Don't be disappointed, Chun," said a muffled voice. "Fire 'em off quickly. Then you can go in with the butt."

"Perepi"—"TV!" somebody shouted. *"Migook"*—"American!" Chun pulled the stock into his shoulder, aiming high as if readying for a lob shot, but the moment the cameras were gone or their view blocked by the wall, he intended to fire straight into the crowd. And if it zapped one of the protesters full force on the head—"Tough tit!" Accident.

"Wedge forward!" came the command, and Chun fired.

Four of the half-dozen leaders of the riot, all members of the DRP, Democratic Reunification Party, were at the Secret Garden at the time, but this wasn't known until film of the incident was replayed several days after by police. By then it was too late. The two KCIA men at the garden entrance never stood a chance, swept along in the irresistible tide of students and riot police—much of the crowd lost from view in the gardens, where giant billows of tear gas enveloped the ginkgo trees and evergreens like morning mist, many of the students, eyes burning, stumbling blindly into the lily ponds, the riot police now in "free run," clubbing as they went, leaving the dazed and fallen to be arrested or clubbed again and dragged away to the buses by the more lightly equipped National Police regrouping in the rear.

The younger of the two KCIA men, his voice drowned in the cacophony of screaming students fleeing from the gas, was knocked down, still holding his coat, his other hand reaching into his shirt pocket, extracting his KCIA card. The next instant he was trampled underfoot, warm blood streaming down his face, unable to see. He lost consciousness.

By the time he was picked up, his KCIA card lost in the stampede, and put in an ambulance, the only available emergency ward was at Severance International Hospital; all the others were already overflowing from the citywide riots that now included Songan University as well. Badly concussed, the agent also had six ribs broken, minor cuts and abrasions, as well as a split on his left forehead requiring ten stitches. In plain clothes, being mistaken for a protester, he was put in a room under police guard, after X rays were taken.

A UPI—United Press International—stringer, celebrating Independence Day with relatives from Kwangju in the southwest Cholla province, had taken his visitors up on the cable car to the top of Namsan, or Nam Hill, to escape the muggy, teargassed atmosphere of the city, relishing the cooler temperatures atop the observation tower that rose another four hundred feet above the nine-hundred-foot hill. With the vast bowl of the city stretching all around them, he pointed southward to the site of the old 1988 Olympics on the far side and to the nineteen bridges that spanned the Han River.

Looking northward through the coin telescope, they could just make out the tearoom and restaurant atop Pugak Skyway, and beyond it, blue, smoky ridges that obscured the DMZ. The wind shifted, clearing parts of the city previously hidden by tear gas, and off to the southwest, they could discern the suburban sprawl running either side of the Han spreading westward to the harbor and industrial clutter of Inchon twenty miles away. Beyond Inchon there was a metallic glint, the Yellow Sea, separating Korea from China, broken here and there by the gray slivers of American warships.

The stringer excused himself from his relatives and called the four emergency wards closer to the city center before phoning his contact at Severance International around 9:40 A.M. Yes, said the contact, a nurse's aide, there was something she could tell him: A young man—name on the security card Lee Sok Jo—brought in about an hour ago had just died. Brian hemorrhage, they thought. After repeating the name to make sure he'd got the spelling right, the stringer, exiting the booth and adopting the fourth level, or tone, due the most elderly of his relatives, excused himself to make one more call—to UPI's downtown office. He could feel his pulse racing. With a civilian dead, he knew it was no longer just another riot. It was now, as his American colleagues would say, "a whole new ball game."

Within minutes the name "Lee Sok Jo" was on the wire services all around the world and being simultaneously broadcast on Seoul's four major radio stations and the U.S. armed forces network in Korea.

By midafternoon, students at all eighty-seven colleges and universities throughout the ROK had declared "war" on the government for its "massive brutality." Within two hours the riots were nationwide. By 3:15 P.M., despite official denials, Lee Sok Jo had become the latest martyr of the struggle against the "oppressive imperialistic regime of the South Korean government—puppets of Washington." When fumbling bureaucrats finally discovered that Lee Sok Jo was not a student but had in fact been a KCIA agent, this information was deliberately withheld by the government, for fear it would be seized upon by the students as further evidence that security agents were being used as agents provocateurs and spies against them.

It was a gift to both the leftists and the Democratic Reunification Party. The police could not contain the riots, beaten back by hails of pavement stones and Molotov cocktails. Scores of police and students were injured, some seriously, the worst fighting occurring around Myongdong Cathedral, eight blocks southwest of the Secret Garden, the gardens themselves now all but deserted.

Agent Chin called the vendor in from Chamshil and replayed as much news and police video as he could get his hands on. Not surprisingly, no matter how many times Chin "froze" the film, zooming in for a close-up, the vendor couldn't pick out the stranger he'd seen in the Myongdong from the crowd, particularly as many of them were wearing either blue or white gauze masks as protection against the pepper gas. Whether the

North Korean agent had reached a phone in the gardens during the night and enlisted the leftists' help in getting him out of the gardens, or whether he'd simply lucked out, Chin would never know.

By 4:00 P.M. the situation, especially around Myongdong Cathedral, was rapidly getting out of control. Exhausted riot squads, Chun among them, charged repeatedly through choking, riot-strewn streets, only to find themselves reeling under new onslaughts of stone and fire, many driven back so far, they ended up crashing into the long "congo" lines of arrested students who, heads bowed, holding one another's waists as ordered, were snaking through the rubble, herded by National Police to waiting paddy wagons.

At 4:15, a momentary hush, not unlike those experienced in the midst of a village shaman's incantations, suddenly descended upon the feuding students and police. Even the endless swirl of humanity about the pagoda-shaped Namdaemun, or Great South Gate, slowed to a crawl, as a Buddhist monk, in his early twenties, assumed the lotus position, poured kerosene over himself, and struck a match. For a second his saffron robes were as one with the flames, his charred torso curling into the fetal position like burnt paper.

Immolations by several other monks in Kwangju and in the always politically discontented "Cinderella" province of South Cholla injected more tension and violence into the increasingly chaotic scenes of Seoul. Finally at 9:00 P.M. that evening, South Korean President Rah felt he had no option but to call in the army. For the first time in ten years, rioters were confronted with live ammunition.

The troops were ordered to fire overhead and did so, but several, firing just as a new hail of projectiles rained down upon them, instinctively lowered their weapons, their volley tearing into the panicked crowd. Two students were killed, seven badly wounded. Several Democratic Unification Party leaders were arrested, along with most of the known leftist leaders, a few of whom the riot police had nabbed in their sweep through the Secret Garden. The government announced it would put them on trial for insurrection.

KBS, the Korean Broadcasting System, Yonap, and Kodo,

the Japanese wire service, together with the three American networks and the BBC foreign service, had it all live.

In Washington, it was now 8:00 A.M.—too late for the visuals and sound bytes to run on "Good Morning America" and its ilk, but in plenty of time for the networks' much more influential evening news broadcasts.

The arrests triggered further riots. Leftists and Reunification Party members under house arrest now called for general strikes, but Rah's government was determined not to give in.

At 10:05 P.M. Radio Pyongyang reported that it was deeply distressed by the situation in the South, and described in glowing terms the earlier arrival at Panmunjom of the more than ten thousand "peace" marchers from the South. Pyongyang television then showed pictures of the North and South Korean students joyously greeting one another, and then as dusk had fallen, bidding each other farewell in a moving, nostalgic rendition of *"Uriuisowon'un-tongil"*—"Our Wish Is Reunification."

Major Tae and his guards, having watched the rally from the southern side of the DMZ, were convinced that more students were now heading south than had arrived in the DMZ that morning. A perfect opportunity, Tae thought, for the NKA to slip infiltrators across. Accordingly he ordered the ROK's DMZ unit at Panmunjom to halt everyone on Unification Highway after they had cleared the DMZ and to carry out a thorough identity check.

The students, objecting vehemently, as he knew they would, were incensed enough to fight, but the ROK troops stationed along the DMZ were heavily armed and less tolerant than riot police. Besides, now that it was dark, student leaders knew television coverage would be minimal and so advised their fellow protestors not to resist the U.S.-ROK search but rather to show dignified solidarity in the face of the "imperialist lackeys." Despite the downpour of a thundershower, the single file stretched out for over two miles, inching forward, each student being searched for arms and false papers.

The thing that most struck American commentators at the time as well as the South Korean reporters was the fact that despite its obligatory use of Communist rhetoric, Pyongyang radio had for once shown some political sophistication and even, perhaps, goodwill, in publicly counseling the students during the Liberation Day meet at Panmunjom not to provoke a confrontation, clearly intimating that the North did not wish to do anything that might undo forthcoming negotiations concerning the possibility of peaceful reunification between the two Koreas.

At midnight, as usual, "Pyongyang Polly" came on the radio announcing the evening's reading: verse by "the venerable and much honored grandfather of our great and respected leader, Kim Jong Il," the poem "Pine Trees on Namsan," ending with, "I will be unyielding while restoring the country, though I am torn to pieces."

"Major Tae!"

There was a long silence, Tae busy with paperwork as the last hundred or so students were being processed. "Yes," he asked, pulling yet another file toward him. He was tired but relieved that, after all his apprehension, another Liberation Day had come and gone without any military incursion from the North to shatter the fragile peace.

When he looked up he saw a guard, drenched by the rain, reluctant to enter, water still dripping from helmet and boots. But as Tae rose, a smell, or perhaps it was the way the soldier moved, told him something was wrong.

"Well—what is it?" demanded Tae. The guard turned, motioning to someone outside.

A figure appeared. Mi-ja. She pushed back a wet strand of hair. It was a small gesture, but Tae could not tell, her eyes in deep shadow, whether she was looking directly at him or not. But in his fury, his humiliation, he interpreted the gesture as one of defiance. He made as if to speak, stopped, then turned away. "Search her." He paused. "She's no different from the rest of them."

"Yes, sir."

When the guard had taken her away, Tae sat staring straight ahead at the small map of old *Chosun*—Korea, "land of the morning calm"—but he could not see through tears.

CHAPTER NINE

AS LIBERATION DAY had ended, a spectacular sunset of huge, towering, cream-white cumulonimbus edged with gold, the men of the American Second Infantry Division Platoon manning OP (Observation Post) Fort Dyer were at "stand to," normal procedure at sundown all along the DMZ and a drill that was as old as the Roman legions, as soldiers stood armed, silent, straining to hear or spot any movement in the rich green valley of rice paddies below that was turning soft black in the dying light.

Sgt. Elmer Franks, standing on the trench's wooden duckboards, was looking through the periscope binoculars. He could see nothing beyond the wire. Soon he would move over to the big infrared scope; no color in the picture, but contrasting black and white shapes were enough. The problem with the infrared scope, however, was that it was not passive, so that rather than simply picking up infrared emitted by a target, it needed to project an infrared beam, which in turn could be picked up by the other side—if they had the right equipment. To cover their bets, the Americans at Fort Dyer also had a "TI," or thermal imager, which *was* passive and through which radiant heat from beyond the wire would show up in opaque sections, like white blobs on an X-ray film.

Overhead storm clouds began crashing into each other, lightning spitting in the distance. Franks looked through the TI.

CHAPTER TEN

TWO HUNDRED SEVENTY miles south, night mist shrouding her, the fast guided-missile frigate USS *Blaine*, part of the Seventh Fleet's carrier screen, sliced an oily calm between Korea's southernmost point and Japan's Tsushima Island off her starboard bow. As the ship left the warmer waters of the East China Sea, heading north into the Sea of Japan, the sweet smell of land wafted by, momentarily subduing the rubbery smell of the bridge. It was a routine patrol, and Ray Brentwood, the tall, thirty-seven-year-old Annapolis-trained captain of the *Blaine*, was on the bridge halfway through the eight-to-midnight watch.

For a moment Brentwood found himself thinking of his wife, Beth, their two young children, and of Lana, his sister in New York, whose last letter to him was full of unhappiness about her marriage. His eye caught sight of the sign—REMEMBER THE STARK!—taped to the bulkhead, and he immediately put all thoughts of family out of mind. He'd drawn the sign up himself and had copies posted throughout the ship. The *Stark*, a U.S. guided-missile frigate of the Perry class, like the *Blaine*, had been attacked in the Straits of Hormuz in '87 by an Iraqi jet firing two Exocets. Only one of the missiles had gone off. One was enough. Thirty-seven U.S. sailors killed, ass-kicking all down the line, the captain court-martialed, and behind all the official inquiries, families and friends devastated by the loss.

The sign on the *Blaine* was the young captain's exhortation to his crew to keep sharp, not to let the boredom or fatigue of a Far Eastern patrol dull efficiency, to remember that for all the wondrous "gizmology" aboard the USS *Blaine*, and wondrous it was, the first warning that a missile was going to hit the *Stark* had not come from the ship's state-of-the-art SLQ-32 radar but from a *man*—the *Stark*'s forward lookout—who saw the Exocet's blue exhaust ten seconds before it hit. Above all, the sign was a reaffirmation of man over machine in the most mechanistic age in history.

At the end of his watch, before going down to the ward room for a snack, Brentwood made his way to his cabin, drew the green drape shut, tossed his cap onto the bulkhead peg, and sat down at the bare, gray metal desk to perform his weekly duty of writing home. He smiled at the snap of the four of them, taken a few months before during the spring, when they had visited Beth's folks in Seattle, across from their navy home in Bremerton. They were all in gaudy-colored shorts, young John, four, bribed to grin with the promise of a Big Mac, Jeannie's seven-year-old smile trying to be sophisticated, despite the missing teeth. Beth, petite, brunette, didn't like the photo. "Unfair," she'd proclaimed good-naturedly. "Ray's eyes are so nice and blue. Healthy-looking. Can't see mine for the bags. Aghh—look at my hair!" Said she looked worn out, "too pale . . . a hundred and four," instead of thirty-four, which her mother said was par for the course, seeing as how the navy had moved them four times in five years, and with two young ones. "Divorce," her mother had added ominously, was highest in the navy—forced separations its major cause.

Never mind forced separations, he'd told Beth jokingly; it was tough enough when you did get together. With the kids at this age, trying to make love was like planning D-Day. Impossible before ten o'clock at night, by which time Beth could only flop exhausted in the living room, needing an hour of Ann Landers and anything that moved on TV to unwind. They'd tried getting Ray's mom to come to Bremerton from New York to "see the kids"—run interference—but she said she "couldn't stand the rain." What she meant was she couldn't stand the strain—didn't like baby-sitting and having to read the same bedtime story fifteen times, with interminable cries of "toilet" and "thirsty." Ray didn't blame her, didn't blame the kids, missed them terribly—said so in his letter each week. But next leave, damn it, he and Beth were going to take off. A couple of days up on the Olympic Peninsula, a little place like La Push, cottage by thundering green-white surf, the smell of dead intertidal life giving off that fresh "ozoney" tang, and craggy, pine-covered mountains sweeping down to the sea. He and Beth in bed—all day. Then chilled cans of "Oly"—none of the diet stuff. Make love till they couldn't do it anymore.

He felt the ship alter course slightly.

Strangest thing was, he couldn't picture her face clearly after a month out, her facial expressions blurred in memory, precisely when he thought they ought to be clearest.

The voice tube's hollow whistle sounded.

"Captain here—what is it?"

"CIC, sir—immediate."

Brentwood grabbed his cap from the peg and headed for the Combat Information Center one deck below the bridge.

"What've we got?" he asked, entering the blood-red cave of winking consoles. "Bogey?"

"Not sure, sir," reported the first officer. "A blip from five thousand feet. Hundred miles from us. Very fast, then nothing but scatter at sea level."

Brentwood looked at the situation board. The carrier *Salt Lake City* was a hundred miles behind them. "Downed aircraft?" asked Brentwood.

"None reported, sir."

Brentwood turned to the radar operator. "What do you think, sailor?"

"Don't think it's a black box down, sir," answered the operator, meaning an airliner. "No SOS."

"Is it in the Alley?" Brentwood asked, referring to the civilian corridor between Japan and Korea.

"Yes, sir. Off Cape Changgi."

"One of the radio beacons acting up?" suggested Brentwood.

"Negative, sir. First thing I checked."

"No steady beeps since first sighting?"

"No, sir. Sporadic."

Brentwood studied the chart in the CIC's subdued light. Cape Changgi was the eastern tip of Pohang Harbor, two-thirds of the way down South Korea's eastern coast. "How far are we from the area?"

"Around a hundred miles, sir."

"Exactly?"

"Ah—ninety-three point seven miles, sir."

"Very well. Flank speed."

"Flank speed," came the confirmation.

"Steer two seven five."

"Two seven five."

"Announce Condition Three," ordered Brentwood.

"Condition Three, sir," replied the officer of the watch, informing the crew over the PA system. The ship was now in the middle of five stages of alert, a third of its complement ready for immediate action.

Brentwood rang the galley for a mug of coffee and glanced at his watch. He estimated they should be off Pohang shortly after dawn.

CHAPTER ELEVEN

AT FORT DYER, Sergeant Franks checked the thermal imager again. No movement, but the usual noise of North Korean loudspeakers from which Pyongyang Polly's seductively soft, warm voice wafted through the rainy darkness, as usual exhorting the Americans to desert. The reward for turning over a radio, Franks noted, had held steady at two thousand won, while the amount paid for defecting with a helicopter had increased slightly to sixty thousand. There wasn't going to be any war, she told them, unless the "gangsters in Seoul" and their *"guerae migook"*—"American puppets"—invaded the Democratic People's Republic of Korea, in which case the loyal workers and peasants of "our great respected leader and teacher" would repel the aggressors.

"Are you lonely, Private Long?" asked the sexy voice. "Who wouldn't be? You are forced to stand guard for the corrupt lackeys in Seoul and Washington. While you are standing in the rain, your capitalist bosses are sitting in their warm mansions, enjoying their women and their wine. They send young men like *you* away from your families. You think they would come in your place?"

Private Long, a newcomer from Yongsan base, had been warned about this "personal approach bullshit." He grinned at his buddy, but he was shaken. It was an unnerving experience, no matter how often you'd been told to ignore it—particularly when she started in with personal details that you thought were unknown to anyone else, least of all the North Koreans. To make matters worse, the volume of the speakers was now so loud, to compensate for the thunderstorm, that Long knew she could be heard all over. She mentioned his fiancée, Joan. Would she remain faithful to her soldier? "So far, far away, Private Long. Of course, your friends will look after Joan—make sure she won't be lonely. Why don't you come over? There are nice girls in Pyongyang. Sweet dreams."

For Private Long, everything about the Korean night was alien. The monsoon was different, the stench of human feces, used as fertilizer, rising from the rice paddies, revolting. Even the sounds of the insects were different. He ached to be back home.

Sergeant Franks called the platoon corporal to take over the TI scope so he could go and talk to young Long. Halfway down one of the sandbagged trenches radiating from the bunker's hub, Franks heard an explosion.

"Minefield!" someone yelled.

Franks cocked his M-60, swung it onto the sandbagged parapet. "Lights!—Two o'clock!" Three searchlights swept and crisscrossed the DMZ. He saw tracer arcing leisurely toward their position. "Mark and fire!"

Shooting erupted from both sides. A searchlight tinkled and died. The firing kept up.

"Flare!" ordered Franks. This was followed by a loud pop, and night became flickering day. It was an errant musk deer, one of its hind legs missing, the other caught in the fence, frantically thrashing against the barbed wire—the animal's huge, soulful eyes bright beneath the dying flare, its doglike bark growing louder.

"Get a light on it," said Franks, and taking careful aim, shot it dead.

Soon everything was quiet again except for the reassuring hiss of the monsoon's rain. Following normal procedure, Fort Dyer reported the "false alarm" to "Big Blue," the situation board deep in the underground headquarters in Seoul twenty-five miles away.

In Seoul the meteorologists forecast more rain.

CHAPTER TWELVE

IN THE PREDAWN darkness off Korea's east coast a hundred and eighty-five miles south of the DMZ, six bombers were coming in low. Their undersides white, the remainder of the fuselage a navy blue, the Russian-made TU-22M Backfires were traveling at Mach .9, the sea a leaden blur beneath them, their wings fully extended rather than in the swept-back position, sacrificing speed for greater maneuverability. Each bomber, its nine-ton load the equivalent of several World War II squadrons, carried two seven-hundred-pound CBU cluster bombs, each of these containing 150 smaller three-pound bombs, set for a DP, or dispersal pattern, saturation bombing; three one-thousand-pound "iron" or high-explosive bombs; and ten one-thousand-pound concrete-piercing bombs—the remainder of each bomber's load, around eleven thousand pounds, made up of fifty two-hundred-pound FAEs (fuel air explosive), a close relative of napalm. All six planes in tandem V-formation were under strict orders not to activate either their "Bee Hind" gun-control radar in the rear barbette of each aircraft or the "Down Beat" attack-and-navigation radar situated in the nose. Both sets involved using active signals, and any emissions from them could bounce high enough off the water to be picked up by low-horizon scanners, some of them on the coast about Pohang's seven-mile-wide harbor, and one at Cape Libby, near the eastern terminal of the trans-Korea oil pipeline, another near the barracks of the U.S. Marine Corps Advisory Group south of the bay.

An ROK destroyer in Pohang Harbor managed to get off several bursts from its quad radar-controlled 127-millimeter cannon before its radar was jammed. The Backfires made only one pass over the harbor, airfield, and environs, thus denying ROK ground control a second chance to lock on to any of the planes' infrared signatures on a second run. One pass was all the Backfires required, their cluster bombs splitting open, releasing eighteen hundred three-pound high explosive minibombs and incendiaries and six FAE pods. The pilots, now miles beyond the crater-pocked airport, could see the FAEs' thickened-gasoline canisters bursting on impact, filling the air with a fine aerosol, covering Camp Libby and the entire dockside area. Then, in the few additional seconds it took for the silvery gasoline mist to ignite over the big oil tank farm at Camp Libby, creating a raging inferno covering several square miles, the six Backfires, one losing speed because of a hung bomb, banked in thunderous unison, streaking inland toward the big air base at Taegu forty-two miles, three and a half minutes, away. Behind them the entire city of Pohang and the storage dumps were afire, flames leaping madly hundreds of feet into the air, fanned to firestorms by the stiff east breeze that had assisted the bombers on their run in.

Of the five thousand casualties in the first ten minutes of the raid, the lucky ones were the twenty-three-hundred-odd who died outright from suffocation due to the sudden loss of oxygen as the fuel bombs exploded. For the remaining victims, flesh melting like plastic, their agony overwhelming any emergency services that were still operating, the final hours were a horror of bloodied screaming shapes, once human, stumbling grotesquely through the ruins, many of them begging the few soldiers who had survived the holocaust to end their agony with a bullet.

Still flying low, the Backfires, lighter now, came in faster over Taegu, each aircraft releasing its twelve one-thousand-pound concrete piercers, the pilots all the time half expecting to be attacked by Phantoms. But there were no American F-4s in sight—or rather none that could fly after the meticulously co-ordinated early morning sabotage attacks by NKA infiltrator cells all across the South, some of whom had been waiting years to hear the last two lines in Pyongyang Polly's rendition of "Pine Trees on Namsan."

The Backfires' three-man crews, pilot, copilot, and electronics warfare officer, could hear nothing but the panicking radio traffic of the ROK's Taegu ground control and the comforting scream of their bombers' twin Kuznetsov afterburners, speed indicators moving steadily from Mach .9 to 1.8 as the planes gained speed and height over the mangled airstrip that had been Taegu base, the soot-colored spars and ashes of the gutted F-4s receding far below like spilt campfires.

Turning hard right, the Backfires climbed northwest up over the black bulk of the five-thousand-foot Sobaek range, its peaks hidden by the monsoon into which the bombers now sped. Finally the hung bomb on one of the Backfires released, disappearing into the wilderness as the bombers dropped low again over Ulchin, tandem V intact, wings swept back for maximum speed, heading fast and low over the Sea of Japan.

As dawn broke on the DMZ, Sergeant Franks could see the dead deer's leg still trapped in the barbed wire. By the time it occurred to him to ask himself why the animal would have suddenly jumped the series of cow fences between the North Koreans and their carefully plowed minefields, it was too late.

Several miles to the west, driving wildlife frantic before them, ninety-five of General Kim's I Corps dark green tanks, "upgunned" Chinese versions of the Russian T-55s, were racing toward Kumchon seven miles south of the DMZ, preceded by the turretless, ungainly-looking, but highly efficient plow-and-roller T-54 mine clearers that had startled the deer and other wildlife. The deep tank traps in which ROK-U.S. command had placed so much faith were easily forded by T-54 bridge-laying tanks, their unfolding steel lattice spanning sixty-nine feet across the traps—eight feet more than ROK-U.S. intelligence had thought possible.

Behind Kim's armor came support battalions of motorized infantry, their 120-millimeter heavy mortars already putting down deadly fire on the flanks of the roads that the tanks used to breach the DMZ on a five-mile front. But this attack, for all its shock, was only a feint, Kim unleashing the main NKA thrust a quarter hour later farther east down the western prong of the wishbone whose junction was Uijongbu, only twelve miles north of Seoul. From here the NKA's "historic" Fourth Division, in recognition of its blitzkrieg attack of 1950 that had so astonished the world, was assigned the task of quickly smashing its way down the Uijongbu corridor to Seoul.

And so while all U.S.-ROK attention was being focused on the smaller and totally unexpected breakthrough around Kumchon in the west, Kim's Fourth Armored Division of 12,400 men, with the First, Second, and Third Motorized Rifle Divisions in support, half choking from the subterranean dust, streamed out of five undetected fume-laden tunnels six miles south of Fort Dyer and three other U.S.-ROK forward observation posts, overrunning the eastern road of the Uijongbu "wishbone."

Major Tae's sector didn't come under attack until after the first exchange of fire three miles south of the DMZ, around Munsan, at the bottom of the big "U" formed by the Imjin River before it flows into the Han estuary and then to the Yellow Sea. With the exception of a lieutenant who stayed behind to help him destroy classified files, Tae ordered his staff to head for the big shelter near the Swiss/Swede (UN observers) hut three-quarters of a mile away only minutes before a mortar shell hit the Quonset hut, ripping open its roof, leaving an enormous hole at least twenty feet across through which Tae and the lieutenant could see bruise-colored storm clouds passing swiftly overhead. In the distance they could hear the rolling thunder of the NKA artillery, Seoul's position to the south behind them clearly indicated by towering columns of black and gray smoke rising to join the rain-laden stratus that had come streaming out of the north.

"They've even got the weather on the run," said the lieutenant bitterly.

Tae didn't respond; he was still gazing up through the hole that had once been the roof of his briefing room, where only a day or so ago he had told a contingent of visiting American officers of his fear that the NKA might invade, that in an age when nuclear war was so feared by both sides, conventional war would, ironically, be the only alternative. The one kind of war for which the democracies, particularly America, were least ready to fight. The lieutenant kept feeding the fire they'd made for the classified documents, but his attention, like Tae's, kept wandering skyward as they waited anxiously for the reassuring "chuffing" sound of "Gnats," the Cobra choppers on twenty-four-hour alert which should soon appear, squadron after squadron of them filling the sky, cannons spitting, their wing pods winking.

Tae told the lieutenant that coming in at over six hundred miles per hour, the effect of a ninety-five-pound Hellfire antitank missile against the armor of the older Russian- and Chinese-made T-55s and lighter PT-76s would be like watching a tin can hit with an ax. But when Tae saw only a few of the helicopters silhouetted against the rain-curtained sky to the south, he feared the worst—that the networks of infiltrators all over the south had struck in deadly unison. Though he tried not to show it, the shock was such that he did not hear the lieutenant, who was now crouching by the radio, telling him that forward American observation posts were reporting more Chinese-made T-55s crossing the DMZ in force, the U.S.-ROK minefields "unzipped" by the NKA's "creeping" artillery barrages. As the Americans' voices faded, radios transmitting only static, until this, too, ceased, Pyongyang's radio signals grew in strength. Giving horrifying descriptions of the riots in the South, Pyongyang Polly repeatedly referred to the deaths of several monks and of Lee Sok Jo as "undeniable evidence" of the "unprovoked violence against our brothers and sisters in the South, who, no longer able to tolerate the oppression of the American warmongers, have called for us to come to their aid in their hour of need under the leadership of our great and respected leader, Kim Jong. . . ."

Amid deafening, earth-shattering noise, dust, and screams that signaled the NKA's advance on his sector, a strange calm visited Tae. It was as if now that the worst, certainly *his* worst, prediction had come true, a great weight was slipping from him. Now, whatever the danger, the terrible, nerve-eating suspense of all his years on the DMZ was finally over. His persistent requests that U.S. radar stations not be lit up at night against the American assertion that the whole point was to let the North see what they were up against, his part in the long, hard battle, almost lost, against Carter's intention to pull out altogether from Korea—all had been to no avail. It was the end.

"Sir!" The lieutenant was pointing outside.

Tae turned his hopeless gaze from the turbulent sky to the brown, dusty window of his office. Across the white cement strip that marked the old demarcation line and where several U.S. marine guards lay dead, he saw NKA assault troops, green-mustard-splotched figures fanning out a hundred yards off, encircling the hut. He cocked the M-60 and opened the canvas bag of grenades, motioning to the lieutenant to help himself. The lieutenant, a much younger man in his early thirties, was drained of color, a mere ghost in khaki, trying to talk, his throat dry as leather with fear. Tae adjusted the sights, knowing this was it—as ready as he ever would be. All the names of the ROK's informers were in Seoul headquarters, Tae keeping only situation reports from the forward OPs in his sector. The names of the ROK's top five counterinsurgency agents, two each for Pusan in the southeast and Kwangju in the southwest, one for Taegu, were in his head—too important to put on any piece of paper. Not that they'd seemed to do much to thwart NKA sabotage. So intense was his calm that Tae felt as if he would momentarily nod off even as the lieutenant, almost apoplectic with fear, continued racing from window to window as if a different view would present a more comforting reality. The thing that kept Tae from total surrender to the impending catastrophe was the thought of his family.

There was a crash of glass. The lieutenant spun around and fired a long burst at the far window, the hut ringing with the sound of the bullets piercing the corrugated metal. It hadn't been the NKA at all but a piece of loose guttering giving way. The lieutenant, sweating profusely, tried to stuff more grenades into his pockets, fumbling, dropping two, scrabbling frantically for them under the desk as if they were gold. Tae felt sorry for him, but it was the detached feeling of an observer, as one felt for a trapped fish, that is, until the major thought once more how neither of them would see family again, of how none of those trapped would ever see loved ones again—unless they could somehow hold. Hold until the U.S. Air Force out of Japan two hundred miles to the east, out of the Yellow Sea to the west, out of anywhere, could launch air strikes on either side of the DMZ and drop ammunition and supplies to the beleaguered troops until ground relief was possible. It was a fantasy for his sector and he knew it, but any delay he and the lieutenant could inflict on the NKA might save someone further down the line.

Outside there was an increase in the din of the battle all through the joint security area, clouds of dust from grenades exploding—whose, he couldn't tell. He caught a glimpse of a mortar being set up two hundred yards off, opened up with his M-60, and saw the bullets kicking up dirt about the mortar crew. The NKA soldiers disappeared in a small depression. For a second or two all Tae could hear was the steady thumping of artillery. Then came the high-pitched rattle of AK-74s, their rounds going too high to do any damage, only raking the hut's eaves but frightening the lieutenant very badly. The next bursts were lower but still too high to be dangerous. Next Tae could hear the *pomp . . . pomp . . . pomp* of either T-55s' or lighter PT-76s' multiple grenade ejectors spitting out smoke bombs that quickly covered the entire joint security area with a dense white cloud. The NKA platoon nearest the hut was obviously waiting for one of the tanks to save them the bother and simply roll through the Quonset—probably not even bothering to waste a shell.

For some inexplicable reason, Tae found himself noting the time, seven minutes to ten, and it struck him how it was an illusion that such situations as he was now in take a long time to resolve, that, in fact, most of the firefights over the relatively open ground would be over quickly—it was only in the hills and mountains, where the terrain lent itself naturally to defense, that a single engagement could stretch into hours, days, and even months. He heard a flapping noise, then more thunder in the distance—artillery or real, he couldn't tell. It began raining heavily and the flapping noise ceased. It had been the huge 100:2 scale map of Korea in the hut's briefing room, shredded by the wind but now sodden with rain and flattened against the wall, a large strip, where the Yalu River had marked the border between China and North Korea, missing.

"They're gone!" said the colleague in astonishment. "I don't see anything mov—"

"*Soryong?*"

It was coming from where the NKA had set up the mortar, but like the lieutenant, Tae couldn't see anything, and there was something odd about the voice. For a moment Tae's ears, ringing from the sounds of battle, couldn't tell what it was.

"*Tae Soryong!* Major Tae!"

"How . . ." began the astonished lieutenant. "How do they know your name?"

"Everyone knows my name," said Tae.

The lieutenant wasn't sure whether to laugh or to be more terrified than he was, for it suddenly dawned on him that in the battle swirling about them, presumably all along the DMZ, this brief pause about the lone Quonset hut might have been a conscious decision by the NKA. They *wanted* Tae. Tae admitted it was possible. It would explain the poor shooting, the AK-74s going high into the hut—merely to keep heads down until the political officer reached their position. But how about the roof? the lieutenant asked Tae. "The mortar shell?"

"A lucky shot," said Tae. "Or *unlucky*. Depends on your point of view."

The NKA officer's voice was starting up again, and Tae realized what was strange about it—it was coming over a battery-operated megaphone, or some kind of loudspeaker mounted on a vehicle. The voice was explaining in English to any U.S. or ROK soldier within or without the hut that the Army of the Democratic People's Republic had no wish to hurt Major Tae. They simply wanted to talk to him. "Comrade to comrade." And if his friends cared about him, and themselves, they would stop "all resistance." It was all quite hopeless anyway, the voice told them—the entire sector was surrounded.

"Give us the major, Comrades! There will be a big reward. The major will be safe. Otherwise . . ." the voice warned, it would be "very bad" for everyone.

Inside the Quonset the lieutenant laughed nervously. "I could earn a few won," he said, adding just as quickly that, of course, he didn't mean it—it was only a joke. Tae nodded understandingly. He knew it was a joke.

Tae heard something clattering above them on the roof and lifted the M-60, ready to fire.

"Loose guttering!" the lieutenant said hurriedly.

Tae was watching the jagged circle of steel-gray sky. Why would they want him? Surely the names of the ROK's top counterinsurgency agents in Pusan, Taegu—wherever—wouldn't be any use to them now. *Unless*, like a long line of conquerors before them, they wanted to teach a quick lesson to the occupied population—to show unequivocally that whoever opposes the Party in thought, word, or deed, whoever dared oppose the beloved leader, would be publicly denounced and executed. To demonstrate conclusively that underground resistance was futile. Tae vividly recalled old villagers telling him how the British and American soldiers had feared the North Korean guards more than the Japanese, the cruelty of the North Koreans so infamous that the mere threat of handing prisoners over to the NKA had been enough for the NKA's allies at the time, the Chinese, to quiet their most intransigent American and British POWs.

"They're waiting," said the lieutenant, sweat trickling down his neck.

"I know."

"*Tae Soryong!* Major Tae!" came the tinny voice. "You have five minutes to surrender—or your comrades will die."

CHAPTER THIRTEEN

IN A REVETEMENT area six miles south of Uijongbu, tank troup commander Lt. George Clemens, field glasses scanning the luxuriant green of paddies and blue mountains beyond, felt his skin itching, an infallible sign, as if he needed to be reminded, that he was in an acute stage of excitement. After all, this moment was what he had been trained for, dreamed of, and wanted all his adult life. Since he was a boy, the behemoths of the battlefield, from the huge, cumbersome monster pillboxes on tracks that were the first tanks in World War I, lumbering across the fields of France, to the blitzkrieg Panzers of Rommel's North Africa Corps churning up the sand in the Western Desert, had awed him. They were for him like a ship, self-contained, an island of war—above all, free to move. And everyone knew the tanks would decide the ground battles, despite what all the air boys said about the deadly saturation fire they could unleash from choppers and ground support fighters. He was sure that once the battle was joined, the confusion of smoke and dust cloud would mean, particularly at night, that for all the fancy arsenal of the air-to-ground missiles, the fight would end up like that of a dogfight in the air, tank against tank, the very kind of dogfights the experts said would never happen again after World War II because the planes were so fast, so modern. Until it happened all over again in the skies over Vietnam. One on one.

His M-1 tank, the Abrams, was the best main battle tank in the world. The Brits had tried to better it with the Challenger, the West Germans with the Leopard, but in the last five years it had been the M-1 that had consistently won NATO's Canadian Cup, the top tank gunnery competition. It was enough for Clemens that the tank's design, from its long 105-millimeter gun to its low profile and sloping armor plate hull which mitigated all but a direct hit, put it ahead of the others in its class. Other features that made it exceptional were the CO_2 laser range finder and air-conditioning to insure longer crew endurance times. Clemens had grown up with the M-1, from the early days of congressional heat because of cost overruns, the months of the temperamental test engines, till now. He had kept the faith. Above all—and this, rather than any technical explanations, is what had won most congressmen over—there was the experience of sitting in an Abrams, its fifty-four tons accelerating from zero to twenty miles per hour in six seconds, moving over rough terrain at forty miles an hour, dust flying, the engine roaring but not screaming, and that turret, steady as a billiard ball on green baize, its low body, hence low silhouette, riding on a cushion of independent suspension, the likes of which had never been seen before. It was a thrill not easily forgotten. Apart from the sheer ascetic beauty of it, it meant that the M-1 at top speed, fifty miles per hour, could fire as if it were standing still when in fact racing at a speed that had once merely been a dream in what some experts had thought was a mad designer's eye.

As Clemens waited, then saw the green-brown camouflage of the first P-76, the "tin can" of the NKA, appear on the Uijongbu road, he almost felt sorry for them, until he remembered how murderously and mercilessly they were shelling Seoul, killing American and Korean women and children indiscriminately.

* * *

Tae had always been prepared for invasion, and as he had a plan for his family, he had one for himself, his plan predicated on the assumption that in most men's lives the glass would be half-empty rather than half-full. Even so, it was no comfort amid the rage and dust of battle that he had long predicted an all-out attack by the North, for in the end each man who had taken the oath of loyalty to the ROK would have to make his own decision, the line between surrender and cowardice often so nebulous in the split second of combat as to have little meaning. He saw the lieutenant in front of him quaking at the thought of surrendering. Now only one hope remained: not that the ROK or U.S. Army would counterattack—both were in disarray—nor did he suppose the NKA would treat him kindly. The war within a family was the most ferocious, the most unforgiving.

His only hope, he believed, lay in the cyanide issued to all front line intelligence officers. The safest place for this, Tae had discovered, was not the teeth—most people ground their teeth at night—or in the wristwatch, for the Communists would take this, not so much for its material value but because a watch in solitary confinement was a comfort. Often it was the only thing by which one could measure the passing of the hours and seasons—sometimes the years. The very act of measurement, of recording a day at a time, was a way of staying sane. The orifices of the body, too, were unreliable, not only because they were often searched by the enemy, but there would always be danger of a capsule or, in this case, a small chewing-gum-like strip of concentrated potassium cyanide, breaking in the body, releasing its poison accidentally.

Tae had studied the problem as diligently as he had the matter of the different lengths of chopsticks. In this case it was a German research project into tunneling, mine, and avalanche disasters that had caught his eye, wherein German scientists had tried to find out the most reliable place to insert a small microchip "beeper" to send out a signal in the event of a cave-in. It had been discovered that the most reliable place on the body was a worker's boots, for unlike shoes or any other form of clothing, these nearly always remained, no matter how many tons of debris or harsh treatment the body had suffered. It wasn't a foolproof plan, of course; in battle, boots were often claimed as bounty, particularly in peasant armies, and besides, a man without footwear was more a prisoner than if you put him behind barbed wire, unable to run very far in bare feet. And Tae also knew that footwear was sometimes removed for purposes of torture, but generally, except for their laces, which, like a prisoner's belt, were removed as possible instruments of suicide, boots, he knew, were left to prisoners of war for a very down-to-earth reason—namely that in the heat of battle, moving captives quickly required it.

At considerable expense, Tae had a fake heel made, and in it placed a cyanide capsule, the NKA knowing that such was the prerogative of any ROK intelligence officer privy, as Tae was, to sensitive counterinsurgency material. Tae then had a small, flexible, gumlike sliver of potassium cyanide, developed by the Americans, sewn within the double-layered tongue of the boot, where, even if his NKA captors searched the boot, flexing it, feeling for hidden razors and the like, the gumlike strip would bend as one with the leather. It was for this reason that, despite the shock and smell of battle, the clouds of phosphorus tear gas and the hiss of spent shell casings in the monsoon rain, an extraordinary island of calm lay within him, for the major carried with him, if it became necessary, the ultimate choice in any man's life, the choice of the moment he would die. He prayed he might not have to, of course, but if, as American colleagues were fond of saying, things got "too rough" and he felt he couldn't keep the names of the top antiinsurgency agents in the South from being given to his NKA captors, then it would be *his* moment, not theirs.

He looked at his watch. The five minutes they'd given him were almost up. Of course, they might shoot him there and then, but he doubted it. Then again . . . He placed the M-60 on his desk, glanced up at the surly sky for a moment, then walking to the door, asked the lieutenant to open it. Frantic with fear and the sense of urgency, the lieutenant jerked the door handle toward him as Tae raised his arms. The lock on the door, always temperamental, wouldn't give. When finally it gave way and Tae walked out, hands high, he heard the lieutenant coming behind him, murmuring in shame but more in relief, "It's the right decision, sir."

The NKA squad leader, in dust-covered green and khaki battle uniform, his face a smudge of camouflage paint, waved for

Tae and the lieutenant to come over toward them quickly. The NKA were very professional, no friendliness but not the rage that Tae knew the front line troops often exhibited under fire along the DMZ. The squad leader—up close he reminded Tae a lot of his younger brother in Seoul, an accountant—had Tae manacled, hands behind his back, and scribbled a note on the label that one of the others had looped loosely about his neck. The squad leader detailed two men of the squad to take Tae to Fourth Division HQ immediately. Now four other men, two Americans, two ROK who had only reached the slit trench a hundred yards or so behind Tae's hut before they came under fire, took their cue from Tae and surrendered. These four, as well as Tae's lieutenant, were also manacled, and the NKA squad commander bayoneted them, their severed heads stuck on posts around the Quonset hut. It was very important, the squad commander reminded his eight men, that they use the bayonet as much as possible, as a round not fired in action was a round wasted.

He then ordered them to collect all weapons and ammunition in the area that had been abandoned by the "imperialists," mostly ROK soldiers hastily retreating, ridding themselves of anything that might inhibit them as they ran to catch the last trucks and jeeps out of the area. While most of the machine guns had been spiked and were of no immediate use, they would be collected by mop-up teams and sent back to Pyongyang to be melted down to make new weapons for the Democratic People's Republic. The M-16 rifles, many of them intact, along with ammunition clips, would be immediately used to arm comrades overrun in the South as they were given the simple choice: join the NKA or be killed.

News of the decapitations at Panmunjom, the squad leader's superiors knew, would quickly spread through the fleeing ROK regiments and bring the NKA many new recruits. In this the NKA was aided and abetted by General Cahill, who ordered news of the atrocities to be beamed via the three U.S. high-resolution K-band satellites to show the world the kind of people he was up against, believing that it would help galvanize the American public's support for Korea's defense. What in hell was Washington doing anyhow?

As far as his shattered local communications allowed him, the general also beamed news of the atrocities to his retreating

American regiments. The message here was more brutal: Apart from total victory, the only way out of Korea would be in a body bag.

CHAPTER FOURTEEN

SUPPORTED BY 2,063 guns, including over two hundred Soviet-made 203-millimeter Corps-level howitzers which the South Koreans did not even know the North possessed, the invasion front now stretched seventeen miles from Kim's tank and rifle regiments in the west to his four divisions east above Uijongbu.

A *Newsweek* reporter filed a story about how the big Russian-made 203-millimeter howitzers—part of the price Moscow had paid Kim Il Sung for sending a small but politically highly symbolic contingent of NKA "volunteers" in 1979 to aid the Soviet invasion of Afghanistan—were all mobile, with self-contained tractor-trucks and sixteen-man crews, who were laying down huge variable-bag charges and high-explosive shells at a round a minute, hurling each two-hundred-pound projectile over seventeen miles into Seoul. When the story reached *Newsweek's* Tokyo office, en route to the United States, an editor did not consider the part about the circumstances under which the USSR had given the North Koreans the guns to be of "general interest," and that section was cut as being too dry. In fact, the circumstances under which the guns were given by the USSR to the NKA would have profound implications within the next seventy-six hours, not only for the beleaguered U.S.-ROK forces in South Korea but for the rest of the world.

Meanwhile the world looked on, stunned by the rapidity of the NKA advance and the concomitant humiliation of the Americans. In one brilliant move, primarily through the use of the tunnels, the NKA had not only succeeded in launching an attack beneath and beyond the DMZ in several places, but in doing so, had now trapped over ten thousand "forward troops" of ROK-U.S. command, which had included Major Tae and his intelligence unit. In what was already being called the "squeeze box," caught between Kim's troops, who were overrunning the DMZ, and his four crack divisions—over fifty thousand attacking farther south from the tunnels—the zone between the two NKA armies was to be a killing ground unless the surrounded Americans and South Koreans could somehow fight their way out through an escape corridor blasted out by the U.S. Air Force.

It was a dim hope, for at the same time as the Backfires' attack on Pohang was taking place, all the major air bases throughout the South, including those at Ulsan and Pusan, thirty and sixty miles south of Pohang, were attacked by small, mobile heavy mortar units of activated NKA infiltrators trained in the 124th guerrilla units. At Kwangju in the southwest, an infiltrator group was caught in the process of setting up an eighty-one-millimeter mortar for a concentrated triangle of fire—fifty rounds for the mortar neatly stacked in carefully prepared and camouflaged dumps. This early alarm saved the six F-4 Phantoms normally parked at Kwangju, but the other airstrips came under withering twenty-six-rounds-a-minute heavy mortar fire, effectively destroying the core of the U.S.-ROK's air interceptor defense, a defense that in the 1950–1953 war had blunted the NKA's dash southward to Pusan.

The '50–'53 war, however, and Vietnam, particularly the North Vietnamese Communists' siege of Khe Sanh, had taught the NKA and the 350,000 Chinese volunteers in that war that while enemy air power alone could not win a conventional war, its ability to play havoc with your supply lines and to resupply its own troops, giving them valuable breathing space even to the point of enabling them to mount a counterattack, could be formidable. For General Kim the only answer to this was the present *byorak kongkyok*—"lightning war"—in which the primary objectives, the enemy's air bases, had been taken out by the NKA's own air strikes or, as in most cases, had been rendered inoperative so swiftly *from within* that the United States, even with its massive reserves in Japan and its B-52 bases in Okinawa, would be unable to catch its breath before Korea was lost.

CHAPTER FIFTEEN

AT DAWN, HEADING south down Korea's rugged east coast, a supertanker, the MV *New Orleans*, had requested ROK coast guard assistance. Several miles north of Pohang she reported that whatever it was that had struck her, a ship-to-ship or air-to-ship missile, she was badly holed in her stern, had lost control of her rudder, including the auxiliary, and was now drifting. While one of the three ROK destroyers in Pohang Harbor was ordered by ROK's Area Five's southern command to remain behind on picket duty, just beyond Cape Changgi, the other two destroyers were sent to assist the tanker out of fear that a massive spill off South Korea's famed beaches would feed the inferno on Pohang's waterfront, or rather what was left of it after the Backfires' sneak attack.

The tanker was now a mile from the two ROK destroyers, themselves a quarter mile apart, their captains deciding the best way to harness a towline before trying to deal with the problem of the hole in her stern, not yet visible because of the head-on angle of the tanker. As dangerous as the salvage procedure would be in the deteriorating weather, both skippers were greatly relieved at having the opportunity to haul the leviathan away from the coast. It would recoup part of the acute loss of face, which they were suffering after having failed earlier to detect the approach of the low-flying Backfires, though they guessed, correctly, that at that moment there were many others throughout the length and breadth of South Korea who could see courtmartials stretching before them. Indeed the NKA infiltration units had been so successful that the problem facing U.S.-ROK

HQ was that if everyone were to be court-martialed who should be, particularly those who had parked their fighters in rows instead of "staggering" them, the armed forces attorneys general would have enough work for the next ten years. However, morale was so low that Seoul HQ decided that to call so many to account at once would only further erode the army's confidence and add to the already acute embarrassment of Seoul and Washington.

Proceeding slowly through increasing fog, the two ROK destroyers were a quarter mile from the tanker when she dropped the hinged panels on both sides of her flying bridge, launching four Exocets at what was effectively point-blank range.

The first destroyer, hit both on the bridge and in its gun-control radar antennae, managed to get off one of its eleven-hundred-pound Harpoons, the missile's aim thrown off, however, by the cluttering images of its own bridge's flying debris. It missed the tanker by a wide margin. The second ROK destroyer was already a mass of flame, both Exocets having exploded at the waterline, the fuel tanks ruptured, men drowning in oil. Undeterred, the tanker, jamming the picket destroyer's radio calls for help, maintained a steady heading toward Pohang Harbor, using the southern curve of Yongil Bay, not hit by the bombers earlier in the day, as a reference point, its quaint, old-fashioned jumble of hotels rising higgledy-piggledy above neat rows of advertisers' red, white, and green beach umbrellas.

On the golden crescent of Yongil Bay, confused and terrified tourists, including members of the local "Pohang Pelicans" who earlier had been readying for their monthly predawn dip when the Backfires had appeared, were now stunned as the city proper across the big bay and now the destroyers continued to be consumed by the fire. As the tanker appeared, emerging from the early morning haze, it appeared to be disintegrating, as if, one of the waiters from the Sun Day hotel commented, "pieces of it were peeling off."

No information could be gained from Pohang Central, telephone communications and roads having been cut by the bombers' raid. Only two inhabitants in all of Yongil Bay could remember anything like it—the great fires lit by the retreating U.S.-ROK forces over fifty years ago when, in pell-mell retreat from the pursuing NKA, the Americans and South Koreans had frantically gone round torching massive supply dumps, denying them to the Communists. This time, however, Pohang had been attacked not by NKA artillery but by its air force, the Backfires flown by North Korean veterans who had served in the Syrian air force in the Arab-Israeli wars and who, against all prediction, had come in so low on the final run before the "hop" over Pohang and Taegu that they had been no more than forty feet above the waves. It was for this reason that the USS *Blaine*, now approaching the area at 0817, had earlier picked up only one of the bombers as a faint and inconclusive blip on its radar.

At 0823 a fisherman, trying his luck six miles off Cape Changgi in wind-scattered mist, saw that what those on Yongil Beach had thought were pieces of paint, or scales, flaking from the giant tanker were in fact swarms of APCs, amphibious personnel carriers, and other assault boats, carrying two thousand NKA marines and flanked by two Nanuchka guided-missile patrol boats. All had slid effortlessly from the tanker's roll-on, roll-off stern cavity, each patrol boat armed with a twin fifty-seven-millimeter AA gun, a single seventy-six-millimeter AA gun, one thirty-millimeter general-purpose Gatling, and two triple-loaded N-4 (NATO-designation "Gecko") air-to-surface missiles with a range of six nautical miles. It was not realized at the time, but these were the first Nanuchka *Class IIIs* the West had seen.

As soon as she understood what was happening, her radar and radio signals to the U.S. Naval Advisory Group in Pusan still jammed, the lone ROK destroyer off Cape Changgi joined battle, engaging the two patrol boats at a range of eight nautical miles. She destroyed one of the Nanuchkas with a Harpoon missile, but the other patrol boat closed in a fast "weaving" pattern to six nautical miles, putting the destroyer within range of the Nanuchka's SA-N4s. Firing all six missiles, the patrol boat hit the Korean destroyer with three of them. It was not the explosions themselves that did in the destroyer so much as the resulting fires amid the massive structural damage, fires that could not be fought effectively, as most of the water lines had been severed or punctured by white-hot splinters, the Russian-built missiles slamming into the ship at six hundred miles an hour. The fires heating the aluminum superstructure caused massive blistering on and between decks, filling the air with highly toxic fumes from incinerated plastic moldings, the fumes alone responsible for half of the destroyer's eighty-three casualties, thirteen of whom had been killed outright by the missiles' impact.

Disgorged from their mother ship like sharks from the belly of some great whale, the armada of landing craft, 140 in all, wasted no time heading toward Pohang Beach, churning the gray sea white while a flight of twenty bulbous-eyed Kamov-25s rose from the bowels of the supertanker, their distinctive double-decker contrarotating rotors catching the weak morning light as they passed over the wakes of the assault force. Armed with air-to-surface rockets under each stubby wing and 20-millimeter nose cannon, ten of the twenty Kamovs, each carrying twelve SPETSNAZ or special force troops, ferried 120 a mile beyond the beach. The other ten Kamovs took another 120 SPETS twenty-five miles inland, south to the Kyongju junction on the vital east coast rail link between Pusan and Seoul.

To thwart any possible counterattack from what was left of the American base at Camp Libby, or from the USMC Advisory Group garrison two miles south of Yongil Beach, NKA commanders knew the beach must be reached in under twenty minutes. For this to happen, the first wave of commando-trained assault troops plus combat engineers had to be in action the moment their amphibious tracked vehicles hit shore. They would be followed, at three-minute intervals, by a second and third wave, each containing more riflemen and antitank weapons, followed by the fourth and final wave of a battalion of forty PC-76 Plavayushchiy amphibious tanks, each up-gunned to a 105-millimeter cannon instead of its usual 76.2-millimeter gun, and sprouting a standard coaxial 7.62-millimeter machine gun atop its cupola.

But going in with the tide, as planned, the amphibious tanks, capable of ten kilometers per hour in calm water, now reached as much as thirteen kilometers per hour. The increased push of the surf helped conserve precious fuel that would otherwise have been expended in powering the tanks' hydro-steer jets, but the saving in fuel was offset by serious steering problems as spray from the wakes of the amphibious personnel carriers ahead, now too close in front, "salted up" the tanks' extended periscopes. This caused several of the PC-76 tank drivers to steer blind. The result in the heavy surf was over two dozen collisions, five of them fatal, as the twenty-five-foot-long, fourteen-ton tanks tore off each other's trim boards between glacis and nose plates, each of the doomed tanks' 240-horsepower water-cooled diesels driving them under before power could be shut off.

Despite the loss, the NKA commander knew that the advance force of 120 SPETS ferried in by the first ten helicopters held the key to success at Pohang Beach. And already he could see them through the binoculars establishing a small but highly concentrated perimeter of fire fed by hundred-round-per-minute AK-74s, PKMs (7.62-millimeter light machine guns), heavy mortars, and if needed, twenty-five-pound "Sagger" antitank missiles.

In all, over two thousand NKA regulars, a small but superbly trained force, were involved in the Pohang strike on a one-kilometer front. The NKA's big gamble was that if the second 120 SPETS ferried twenty-five miles inland could sever the vital arteries of the Seoul-Pusan expressway and Seoul-Pusan rail link, then Pusan, the east coast's major naval port, would be isolated, infiltration units already having cut the alternate route between Pusan and Taegu farther inland. With the beachhead at Pohang consolidated by additional troops securing it in depth, air strikes could then be launched from Pohang field against the U.S.-ROK naval installations at Pusan. Not only would the headquarters of the ROK navy be in the hands of the NKA, but the vital sea link to Japan and its formidable U.S. garrison would be severed.

As the ragtag U.S.-ROK army unit of less than six hundred men hurriedly assembled from the remains of Camp Libby clambered into armored personnel carriers setting out to counterattack the beach, they were about to become the first victims of what, in the dry technological jargon of ballistics research, was called the development of "improved sleeve design." It was something that the doomed and already demoralized American and South Korean troops could only have known about if U.S. intelligence had penetrated an elite Soviet guard regiment. U.S. intelligence had not done so, and consequently there was no way for the men in the U.S.-ROK counterattack from Fort Libby to know that the new sleeve design for the standard Soviet and Chinese 7.62-millimeter round increased not only the velocity of the depleted uranium bullet but also its penetration capability.

On the beach at Pohang this meant that the very fast eight-

and-a-half-ton American-built M-114 armored personnel carriers, each powered by an eight-cylinder Chevrolet engine and capable of transporting thirteen men, were stopped dead in their tracks. The NKA's Soviet-made 7.62-millimeter rounds fired by Soviet-made PKM light machine guns not only penetrated the M-114s' hulls but traveled so fast that even after penetration of the APCs' hulls, a single ricochet inside the personnel carrier was capable of killing or wounding several men. A full burst often as not put the entire thirteen-man squad out of action.

The bloody scenes described later by a few survivors of the PKM slaughter of the M-114s could not begin to convey the extent of chaos and panic inside the jam-packed carriers, falling bodies and loose weapons often doing as much damage as the ricocheting 7.62-millimeters themselves.

From that time on, even though any objective assessment of the defeat of the American and ROK troops at Pohang Beach showed that this was due as much as anything to the overwhelming force and professionalism of the attacking NKA SPETS, blame shifted quickly to the inadequacies of the M-114s' aluminum hulls. And, though the army would not admit it, after Pohang, U.S. and ROK commanders had great difficulty in persuading men to be transported to the front in the M-114s. The army kept touting the personnel carrier's advantages, including its unmatched speed of fifty-eight kilometers per hour, but U.S. officers in all theaters remained unsuccessful in trying to have the carriers loaded to their full capacity of thirteen, the age-old superstition about the number thirteen reinforced by the disaster at Yongil Bay.

But if the penetrating 7.62-millimeters had surprised the U.S.-ROK forces at Pohang, this was as nothing to the far greater shock that was to come.

CHAPTER SIXTEEN

IN SEOUL, 225 miles northwest of Pohang, deep in the subterranean headquarters of the U.S.-ROK defense force, the extent of the NKA's daring initiatives was only now being fully realized as field reports slowly found their way through the nightmare of a frantic bureaucracy and broken communications. One aspect of the invasion that was becoming clearer with each new report was how successful the coordination of the NKA's attacks had been. One of the most difficult of all the military arts, such superb timing evidenced a professionalism that even dedicated anti-Communists like Cahill begrudgingly admired. For the civilian population, towering columns of black smoke seen all over the South were testimony enough of just how widespread and effective the NKA infiltrators and regulars had been.

However, the most stunning news in Seoul that morning of August 16 was that, following the NKA's breakout southward through the tunnels, both pincers of Kim's army had now reached Uijongbu junction. In all, two hundred thousand NKA troops were massing less than ten miles north of Seoul for the final surge down the Uijongbu corridor.

General Cahill knew that professionally he was finished unless he could buy time to pull a "MacArthur": launch a massive amphibious assault at a weak point somewhere along North Korea's western coastline and push inland, cutting the NKA's supply line, which, due to their present rate of advance, might soon become dangerously overextended.

The initial reluctance of Seoul's state-of-the-art headquarters to believe that a full-scale disaster might befall them originated in a faith, bordering on evangelical-like confidence, in "HiT-R," or high-tech readiness. Cahill, General Lee, and the other commanders in the top echelon had relied too heavily on such indicators as vibration sensors along the DMZ. These, however, as predicted by Cahill's aide, proved as deficient in the spongy terrain of the monsoon as they were efficient in the hard, frozen ground of winter, when any armor-led invasion was supposed to happen. Now Seoul was relying heavily on two last "aces," the first a direct and normally "militarily sound" tactical descendant of World War II: an elaborate system of high-explosive-rigged rail lines, culverts, highways, and bridges throughout the South. Where there were no natural culverts or bridges, enormous concrete slabs had been built either side of the highways. With all roads already assigned "code black," all Cahill had to do was order the RECDET—remote control detonation—units to pull the switch and mountains of debris would come crashing down, forcing the NKA's armored and motorized divisions to a standstill. And if the NKA's armor couldn't move, its infantry couldn't advance in the face of the U.S.-ROK 105-millimeter artillery that was answering the Communists' barrage.

At least this was the theory. The other "ace" Seoul HQ had up its sleeve was the Cobras. Two hundred of them, armed with GAU-8 armor-piercing thirty-millimeter cannons and Walleye antitank bombs, which, once the enemy tanks were identified by the on-ground laser source, would home in, riding down the beam, blowing the tank apart. Fifty were based at Kunsan a hundred miles away on the west coast, another fifty in the center at Taegu, the remaining hundred at Osan thirty miles south of Seoul.

Cahill's disappointment was too much for him to bear when he heard they no longer existed. Many of the captured saboteurs, immediately executed by the ROK, had been "sleepers"—working at the various air bases for civilian contractors—and it wasn't until an hour after the first NKA shells, shuffling through the air in their strangely muted staccato, had started crashing into the South Korean capital that the first reports about the Cobras reached Cahill among the pile of other sabotage reports and assorted debacles.

And it wasn't until the radio message from Fort Alamo, a mile south of the DMZ, came in reporting firing in the distance but no breakthrough in their sector that Cahill and his staff realized the NKA had simply bypassed many American camps on or near the DMZ, leaving the isolated American strongholds for piecemeal destruction later on. It was the delay in such reports reaching him, rather than the contents of the reports as later charged, that made Cahill reluctant to signal Washington with a DEFCON 1 advisory—to put all U.S. forces on a war footing. With his communications in such chaos, verification of reported conditions at the front was at times impossible, and the last thing he had wanted was to be accused of panic. He also knew it was possible the NKA was feeding false information. He had assumed, for example, in accordance with his standing orders, that at least half the Cobras would already be in the air shortly after hostilities began, with "weapons freed" clearance and in action all along the DMZ. What had actually happened, and what Cahill wasn't told until three hours later, was that while the first messages from Osan reported only a few helos afire, in fact, most of the remaining choppers had been destroyed by either mortar splinters or infiltrator sniper fire. What had fooled those reporting the damage was that so few choppers were on fire. But this was not due to any lack of shrapnel or snipers but to the Cobra's plastic inert gas fuel tanks, their honeycombed interior specifically designed not to burst into flame when fired upon. It was only later, after ground crew felt safe enough to go out on the tarmac, that they discovered the extent of the damage, the Cobras riddled by rifle fire that had easily passed through the thin fuselage, slicing and mashing the maze of electronics and hydraulically operated controls.

In Taegu the NKA infiltrators didn't bother to set up mortars because of more frequent patrols operating out of the air base itself and from Camp Carroll ten miles away. Instead they simply hijacked four three-ton trucks, shot the drivers, and crashing the mesh perimeter, drove directly onto the tarmac into the line of twenty-five parked Cobras, demolishing $120 million worth of aircraft and ordnance in less than eight minutes.

At the same time, Osan-ni ammunition dump for the Koonni air range on the west coast south of Seoul blew when two of the NKA's KIS (Kim Il Sung's) "suicide" squads drove a one-and-a-half-ton jeep packed with dynamite through the checkpoint, killing the two MPs on the main gate before slamming into the dump. The shock waves of the Osan-ni explosions were felt several minutes later in Seoul, the pall of coal-black smoke rising ten thousand feet, curdling virgin-white cumulonimbus that were sure to bring more rain.

"Hope the cavalry arrive in time," Cahill said grimly from the control bunker, whose atmosphere had rapidly changed from alarm to near panic as the blue aura normally cast about the OPS room by the big board was increasingly pierced by the flashing of red lights, each one signifying another position overrun by the NKA troops, some of whom were now only nine miles from Seoul.

Cahill ordered the detonation of all demolition charges along the Uijongbu-Seoul highway. It was the first in a series of hard decisions. It meant that eight thousand U.S.-ROK rear guard units fighting their way down the highway were lost. But at least the demolition would force Kim's armored columns off the roads into the flooded soft paddies, and by that time the "cavalry," planes from the Seventh Fleet and B-52s from Okinawa, would be overhead, "pounding the crap," as Cahill put it to General Lee, out of the NKA armored spearheads; the tanks, especially the lighter PT-76s, would be sitting ducks until the NKA could move the rubble of the demolition.

"You think the fleet planes can turn it around?" asked Lee quietly, determined to keep the tremors of impending defeat out of his voice. "How about the NKA's mobile SAM sites?"

"They'll get a few of our boys, no doubt," answered Cahill calmly. "So will their fighters, but the MiG 23s are no match for our F-18s and the 'Smart' bombs. Anyway, General, our fly-boys are the best. And yours, of course. Best planes. Best-trained people in the world."

This was all true, but what Army General Cahill had overlooked, not surprisingly given the myriad complexities of modern conventional warfare, was that the laser-guided "Smart" bombs can only *be* smart when a laser beam can be bounced off enemy targets by "real time" laser designators, or target markers. That is, a mobile laser source, like the man with a mirror signaling the cavalry in the old West, had to be in the area where the aircraft would be attacking. In any event, the NKA, fearful of the U.S. dominance of the air in both the earlier Korean, and Vietnam wars, knew this well, and its artillery, in close support behind its advancing armor, was laying down heavy hexachloroethane shells, producing dense curtains of smoke which played havoc with a laser beam, slicing it up into segments, in effect slicing up the signal.

Outside Uijongbu, Kim's Fourth Armored was encountering stiff resistance from ROK reservists who had been carrying out maneuvers that morning near Camp La Guardia, two miles from Uijongbu. Reports to Kim's headquarters indicated that the unexpected ROK resistance had forced an NKA column off the highway.

ROK rear guard units reported the same thing to Seoul HQ. Cahill was buoyed by the news, by what he called the "first major tactical blunder" Kim had made. Cahill was now convinced that if he could force the other NKA columns off the roads and keep them off until the weather cleared enough for his fighters to zero in, he would halt the entire advance. Stopping Kim's legendary Fourth Division alone would mean blunting the NKA thrust toward Seoul and giving the hard-pressed retreating U.S.-ROK divisions an enormous psychological lift, for so far the NKA's Fourth had penetrated the South Korean defenses with such stunning speed that news reports around the world of the latest NKA advance were virtually outdated the moment they were broadcast.

Cahill was equally aware that should he fail and the capital fall, the impact on the ROK, and U.S. prestige, would be devastating, the prestige of the Communists dramatically increased.

Cahill's hope of help from the air was another dream quickly punctured when a squadron of twelve American F-4 Phantoms managed to scramble aloft out of the chaos and confusion that had once been Seoul's Kimpo Field. Streaking into the rain-thick sky above the DMZ, reaching Mach 2, they suddenly found themselves in combat with fifteen MiG-23s diving on them at Mach 2.2 from the higher, thinner air. U.S. intelligence had known since late 1985 that, as part of building up the NKA's overwhelming advantage in numbers of combat aircraft (750 to the South's 400), Moscow had begun delivering forty-six MiG-23 (Flogger) and SU-7 (Fitter) fighters to North Korea following Kim Il Sung's visit to the Soviet Union the previous spring. And so it wasn't the appearance of the Russian-built fighters that was a shock to the pilots of USAF's Seventh Tactical Fighter Wing, but rather the small blips which turned out to be AS-9s, air-to-surface antiradiation missiles. The AS-9, while something of a "loafer" compared to the speed of other air-to-surface rockets, and no threat to the Phantoms, was nevertheless a potent killer of antiaircraft missile sites, and via their on-board computers, ten AS-9s homed in on the ROK's four Hawkeye and Nike-Hercules batteries. Hurtling in at seven hundred feet a second, the two-thousand-pound missiles failed to take out the ROK's missiles but wiped out the batteries' radars, which provided the Hawkeye and Nike-Hercules with their launch vectors. It meant that in the first day of the invasion, ROK's four surface-to-air missile battalions, as against the North's fifty-four, were rendered useless.

The loss in fighters to the NKA was seven MiGs shot down by the Phantoms. It was a high price for both sides, but in neutralizing the ROK missile batteries, the North's MiGs had opened a window for further AS-9 attacks, leaving South Korea woefully understrength in antiaircraft defense.

On the DMZ, elements of the NKA's Second Armored, held in reserve until Fourth and First Armored had broken out into the South, were now reported crossing the DMZ in force. Despite some determined firefights, Forts Dyer, Cheyenne, and all other forward observations posts in Area 1, from Kumchon in the far west to a point forty-three air miles east beyond Alamo, a third of the entire DMZ, were now overrun. In many of the trenches leading from the U.S.-ROK control bunkers, fighting was hand to hand, and Private Long, so recently wooed by Pyongyang Polly, was one of the first Americans killed in World War III, decapitated by an NKA splinter grenade.

In terms of U.S.-ROK prisoners taken that first day, over seven thousand, the most humiliating of all was the capture of Lieutenant General Hay, commanding officer I Corps, at his Uijongbu HQ while he was in the midst of organizing his eleven ROK divisions and one U.S. division for a counterattack that never materialized.

In Seoul there was utter panic and confusion as the firing of over two thousand massed guns along the DMZ continued, the 225- and 194-pound high-explosive and white phosphorus shells tearing through the rain-gray air, thudding into the cluster of U.S.-ROK targets in and about the South Korean capital.

Millions of panicked civilians, clogging all roads leading out of the capital, prevented ROK armor and infantry from getting through to mount effective counterattacks. The fleeing mobs were soon out of control, terrorized by the thick, acrid smoke which they thought was some kind of poison gas because of its yellowish tinge, the latter in fact a result of Seoul's polluted air and burning briquettes which many households had stored for the coming winter. Trying desperately to escape, many were caught in running battles with squads of riot police who were trying just as desperately and futilely to clear the roads for military traffic, which was now backed up as far as Chamshil Iron Bridge, Olympic Park, and the Sports Complex, the shell of the main Olympic stadium holed in several places and burning, despite the determined efforts and initiative of Chamshil's fire brigade, who, finding water mains severed, coupled hoses and used the Olympic swimming pool as their water supply.

Three of those in the retreating millions heading frantically for the bridges over the Han were Mi-ja, her younger brother, Dyoung, and their mother. For Major Tae's family the possibility of forced evacuation from the capital had always been presented to them as a distinct possibility by their father, and long ago the family had reluctantly promised him that in the event of invasion, they would head south with everyone else. "For civilians," he had told them, "there will be no honor in remaining," warning them to "go—go as fast and best as you can." Where her lover, Jyung-hun, was, Mi-ja had no idea. She tried to phone, but all civilian and most military lines had been cut hours before.

One of the few who were not trying to leave, and whose family was busy at work in the hole-in-the-wall rooms that he grandly called his "factory," was the owner of the Magic Cloud Souvenir Shop off Sejongro, selling North Korean flags faster than his family could sew them.

As the NKA's barrages were a matter of rolling, indirect fire, pinpoint accuracy was not needed, so that changes in the artillery's fourteen "variables," from wind velocity and humidity to gun "jump" due to barrel elevation at the moment of firing, didn't matter as much as they normally would. It was only important for General Kim's gunners to know whether their fire was hitting the city; the only parts General Kim did not want to hit if he could help it were the bridges, as their destruction would delay the NKA's progress over the Han, down the Chengbu expressway to the western plain and Taejon, the southern railhead for the Seoul-Pusan and Seoul-Kwangju lines. For this reason alone, Kim instructed his gunners to be guided by the forward observers, firing no farther south than Yongsan Barracks if possible.

The agent who had escaped from the Secret Garden was one of those whose job it was to act as forward observation officer, and he was doing so from Namsan Hill, where only yesterday in the vibrant sunlight tourists had been enjoying the view of one of the fastest-growing and most westernized cities in Asia. Beyond the city's punch bowl, the flames strangely beautiful against the scudding overcast, the agent could see that a good part of the northern suburbs was also afire, especially, he was glad to note, the area immediately around the Blue House. A few errant shells, like those hitting the Olympic sites, were overshooting, exploding buildings around the mosque, and some fires were starting in Itaewon, whose bars and girls served, or rather *had* served, the Yongsan base. The fact, well understood by the agent, that sooner or later he would die in the barrage, the nine-hundred-foot hill he was on being the central grid reference, or aiming point, for the artillery, was a given. He was only one of many volunteers who, with their powerful shortwave radios, were calmly reporting the dispersal pattern of the guns' fire. Indeed, even as he spoke, he knew there were other agents atop Pugak Mountain on the city's northern perimeter sending in their reports.

By 5:30 P.M. all the bridges over the Han except three, the Panpo and Hannam, leading to the Kyongbu-Pusan expressway, and the Songsan Bridge leading to Kimpo Airport and Inchon, were finally clogged solid with refugees, the air filled with a rancid mixture of pickled cabbage, sweat, and cordite.

The Republic of Korea was teetering on the verge of total collapse, for even if the fighting was to go on, a simple but terrible truth was becoming slowly but inexorably evident to Seoul HQ. It was one that no journalist, and certainly no politician, would utter, let alone a military commander who cared anything for his career. Nevertheless it was a fact that the best troops in the American army, as in all armies, were those who wanted to be where they were rather than those who had merely enlisted with some vague hope of learning a trade or of escaping what academics called socioeconomic ghettos. Those enlisting with the highest educational qualifications got first choice of postings, and Korea, despite all the stories of the easy availability of women in Seoul, came in well after West Germany in the GI's list of preferred postings. And the best of those who *were* in Korea, those who had volunteered for duty on the DMZ, on the front line, from Fort Apache to Camp Pelham, were now trapped, gone from the big blue board in Seoul HQ, where the NKA's overwhelming superiority had now all but turned the board red.

The remainder of U.S.-ROK I Corps's eleven divisions, ten of them ROK, one U.S., now withdrawing from Uijongbu were simply not up to anything like the standard of the "five-year term" soldiers of the North Korean divisions. In the sudden shock of the NKA's highly professional attack, where rapidity of movement had been everything in the early hours of engagement, these U.S.-ROK ground troops had quite simply been outclassed. And in the confusion and contagion of impending defeat, men who had been trained a hundred times in mock battles, often with live ammo being fired overhead, froze in the initial bowel-churning terror of real combat and so did not move fast enough to capitalize on any holes that did appear in the NKA advance. This was especially true in the case of three infantry battalions on maneuvers in the Whiskey Sierra Tango training area ten miles north of La Guardia and in Falling Water on the outskirts of Uijongbu, nine miles north of the capital.

By late afternoon the situation in Seoul had deteriorated so badly, with millions of refugees now choking all nineteen bridges over the Han, that General Cahill was about to make the bravest—to some the stupidest, to others the most militarily sound—decision of his entire career.

Far to the southeast off Cape Changgi, the sea mist took on the aspect of moody ghosts rising one minute, returning the next. Inside the *Blaine*'s combat information center, the radar operator saw two or three more blips appear in the radar sweep. Then more. Soon they were a swarm. "Captain."

"What have you got?" asked Ray Brentwood calmly. "Fishing fleet?"

With news of the NKA invasion flashed to all U.S. ships, the radar operator wasn't sure whether the skipper was fooling. In

any case, he pressed the computer for a readout of the unidentified blips' speed. "Forty knots, sir."

"Patrol boats?"

"Looks like it, sir."

"Satellite confirmed?"

"Satellite confirmed," answered the operator, "but no flag."

Brentwood put down his coffee, looking intently at the small
white squares with the white dots inside them signifying unknown surface ships. "Radio traffic?"

"Negative, sir."

"Range?"

"Fifteen miles and closing."

"Very well—send message. 'Unknown vessels, this is a U.S.
Navy warship on your zero eight five. Request you identify yourself and state your intentions.' "

"Aye, aye, sir."

The message was sent and the *Blaine* waited.

There was no answer.

"Repeat message," ordered Brentwood.

"Aye, aye, sir."

Again there was no answer.

"Call general quarters," ordered Brentwood.

"General quarters. General quarters. All hands man your
battle stations."

"Look for 'skimmers,' " instructed Brentwood.

"Alert for skimmers," repeated the OOD, the lookouts on
the bridge's wings lowering their binoculars, scanning the graying sea, looking for a flash in the distance, anything that would
indicate a missile coming in low under the radar screen.

In Seoul, General Cahill ordered all but three of the nineteen
bridges blown, and with the weather clearing, firmer ground in
the offing, he could finally unleash his heavy fifty-four-ton
M-1s, America's main battle tanks, to buy time for the massive
American reinforcements he was sure would come.

Major Tae had been pushed into one of the long columns of
over seven thousand battle-shocked and bedraggled South Korean and American prisoners of war taken along the DMZ who
were now trudging along in the mud of flooded roads, the forced
march confined to side roads passing lush green paddies and

brownish, shrub-covered hills, seventy miles in all, from Panmunjom east to Chorwon only a few miles south of what, just forty-eight hours before, had been the DMZ. Now they were being herded southwest again, heading back toward Uijongbu. It would have been half the distance to go straight from Panmunjom to Uijongbu, but the speed of the NKA advance was such that the administration of prisoners, always low in the priorities of an attacking army, had gone awry. Tae, like other intelligence officers, with labels about their necks ticketing them for interrogation, found himself pushed from one column to another and witnessed the mounting frustration of the NKA guards. These were fanatical young reservists who didn't seem to know where they were going themselves and took their frustrations out on the prisoners, screaming at POWs too weak to go on as if they alone had been responsible for the guards' confusion instead of the victims of it.

At first there were enough able-bodied men among the seemingly endless columns of prisoners to aid those too weak to go on, putting them on makeshift stretchers of bamboo poles and rain ponchos. But as exhaustion and lack of food weakened the stronger ones as well, the bayoneting began, some of the guards taking obvious relish in killing those South Koreans who had showed any signs of camaraderie with the Americans. Some of the Americans, Tae saw, were obviously being killed for their personal possessions, particularly watches and much-coveted cigarettes. Tae had wanted to help on several occasions, but fear for his own safety made him hesitate.

The warning about the watches swept through the columns but did little to stop the slaughter as prisoners were now being pulled out at random by the AK-47-toting guards and searched. If they found anyone trying to secret something away or not surrendering it immediately, the prisoner's death came slowly and brutally, the guards using rifle butts in a fury that Tae, with his coldly objective eye, recognized as a savage product of Pyongyang's ingrained hatred and envy of Americans in general. It also came from the lingering fear of all guards in all armies that if they aren't tough enough, the entire mud-sloshing column of prisoners might rush and overwhelm them through sheer weight of numbers, the kind of rush the Japanese had traditionally made, preferring death to the ignominy of surrender. The best way to keep control, the guards obviously thought, was to execute any prisoner for the slightest sign of disobedience. The terror of randomly being chosen for death was so palpable in the column that every now and then there was a panicky movement, like columns of ants climbing over one another, as men on the column's edges sought greater safety by pushing farther into it.

To Tae, who had been in the front line of the counterinsurgency war for so many years, the murders of the Americans did not come as a surprise. Of all the Communist countries, the North Korean regime was unquestionably the maddest. What *did* disturb him was the extent of the savagery toward the hated *migooks*, so that by the time the column reached Chorwon, more than forty Americans had been butchered. More than anything else, it told Tae that the Communists feared no retribution—that they were quite sure, like the North Vietnamese before them, that they were going to win, so they feared no reprisal.

An American next to Tae, his left eye bloodied and sodden, the dressing slipping down his face, tripped in a mud-filled pothole. Instinctively Tae's right hand shot out to steady him. The next second Tae heard shouting, the mustard-colored water splashing about him, as prisoners stumbled away from him and the young American. A heavy thud and Tae's head shot forward, a burning sensation in his shoulder blades as he sprawled in the mud. He heard the guard cock the Kalashnikov and, looking up, saw the banana-shaped magazine curving down toward the wounded American soldier. The skin around the American's good eye crinkled in a smile as he fixed his gaze on the South Korean major. "Thanks, buddy—" he began. The Kalashnikov jumped, the sound of the bullet echoing through the lonely, rainy valley either side of the column. The guard swung the semiautomatic toward Tae, about to pull the trigger again, when he saw the label hanging from about Tae's neck and began screaming that Tae shouldn't be here, waving his hand back in the direction of the DMZ, shouting that Tae should have been taken west to divisional headquarters at Kaesong. Tae got up unsteadily from the mud, the white of his eyes so marked in contrast to the mud that he looked like a minstrel clown.

He replied in as nonthreatening a tone as he could that he'd gone where he'd been ordered. During the incident the column had not stopped, only a few heads turning back out of curiosity, the savagery visited upon the American having already become the norm. As Tae knew only too well, people could get used to anything. An NKA sergeant, superior-looking, unusually tall for the NKA, came bustling up, chastening the guard. Hadn't there been an explicit order about conserving ammunition? The guard quietly turned the tables on the sergeant by dutifully pointing out that the collaborationist South Korean major should have been shipped back to divisional headquarters for investigation. The sergeant frowned, Tae realizing the label had even more power than he had realized, glad he hadn't taken it off. At least it might buy him time, perhaps even special treatment, though this, he knew, could end up being followed by much worse than what was being meted out to the column.

Leaning forward, the sergeant wiped the mud from Tae's collar, the patch of newer cloth showing where the major's pips had been before he'd torn them off and thrown them away when the random killing had begun.

"You follow me," the sergeant told Tae.

"Sergeant, may I request a favor?"

"What is it?" the sergeant asked sharply.

"The American's identification tags. Could I have—"

"Dog tags!" said the NKA sergeant in English exuberantly. "I study English at Beijing. Foreign Language Institute."

"Ah," said Tae noncommittally.

The sergeant cut the dog tags' cord with his bayonet. With the tags there was a small gold cross on a slim chain, which the sergeant pocketed. "His God did not help him," he said, grinning, handing the dog tags to Tae.

Tae said nothing and dropped the identification disks into his tunic pocket. The sergeant was now going through the American's wallet, taking out won bills. He saw Tae watching him and suddenly became rigidly officious. "This is for the People's Army," he said.

"I was looking at the photograph," said Tae quietly.

The sergeant handed him the wallet with an air of stiff magnanimity. "You may have it."

"Thank you," said Tae.

The sergeant waved down a motorbike and sidecar, its driver and passenger caked in mud, the engine spitting and coughing as if it were about to give up any moment. Ahead, Tae saw the long column of sodden American and South Korean soldiers, fatigues clinging to them like black wrap as they continued trudging along the narrow, flooded road that disappeared into another misty valley. It was the most hopeless sight he had seen.

The sergeant ordered the driver to take the South Korean major ahead to Uijongbu, explaining that he was an important prisoner, to be debriefed as soon as possible. There was an argument, ending with the passenger in the sidecar getting out grumpily and stalking off with the guard who'd shot the American to the rear of the column. The sergeant ordered Tae into the sidecar and started binding his hands with twist wire. Tae wondered if they'd let him keep the wallet at Uijongbu. He took his only comfort in the cyanide strip hidden in his boot. It was doubtful, he thought, that they'd take his boots away from him.

The bike moved off slowly, sliding at first in the mud, the sidecar ahead of the bike, the sergeant pushing; then it picked up speed and straightened. Tae looked back at the murdered American, a khaki heap in the rain, wondering who he was.

CHAPTER SEVENTEEN

Washington, D.C.

IT WAS BREAKFAST time, the sharks were hungry, and Press Secretary Trainor, having strongly advised the president to spend a day or two at Camp David being presidential, meeting with Arab heads of state, now gladly offered himself for consumption, enjoying the cut and thrust with the pack, some of them former colleagues, a few still friends.

"But shouldn't the president be in the White House?" It was from some young hotshot blonde at the *Post*, her white lace bra clearly etched beneath the tight white silk blouse and black tie that only accentuated her femininity rather than conveying the no-nonsense tough-press-woman image she'd intended. Trainor responded, smiling, assuring them that "the president is receiving regular reports on the situation in Korea." What he didn't say was that the last report had been received at 5:00 A.M., four hours ago—6:00 P.M. in Korea. Trainor saw the young reporter scribbling frantically, about to launch another question, but in all the excitement she'd forgotten to predicate her first question by indicating there'd be a follow-up. Trainor pointed to the *Atlanta Constitution*, who did announce he had a follow-up. "Mr. Trainor, we've had reports from Kodo News Service that North Korean forces are already halfway down the peninsula. Can you confirm this?"

"Well, with the DMZ being already halfway down the peninsula, Mr. Burns, I'd be surprised if they weren't." A chuckle from one of Burns's colleagues. A mistake, thought Trainor—don't alienate them. "I know what you mean, Mr. Burns. No, we don't have any information that would confirm that."

"Or would deny it?"

Get it over quickly, thought Trainor. "No. Ah, Miss Vogel?" She was a tall, older, freckle-faced woman they called "String Bean," who had seen four presidents come and go. Her favorite technique was to let the eager pitchers go first, then she'd step in with the curve ball. "Mr. Trainor—" and no crappola, her expression seemed to say "—when did you *last* hear from Seoul? And I have a follow-up."

"This morning," Trainor answered—you crafty old bitch, and I know your follow-up.

"Precisely when was that?"

Trainor paused. Say you don't know and you'll be assigned gofer status. Tell her it was four hours ago and your information's got whiskers on it. Trainor had never had to deal with a military "incident" this size before, but he knew enough that in modern conventional warfare, let alone nuclear strikes, four hours could turn defeat into victory or vice versa. "Earlier this morning," he answered. "Ms.—" Trainor looked puzzled, trying to read the name on the press pass of a newcomer, giving him time to regroup his defense. She was a honey blonde, good-looking, about five feet four inches, cool blue eyes.

"Miss Roberts," she said. "Does the president plan to remain at Camp David until the New Hampshire primary?"

Trainor struck a thoughtful pose, giving the question its due, both hands on the lectern, remembering that the opinion polls were showing a surge in support of Mayne as the presidential candidate least likely to involve the United States in military actions around the world, especially in another quagmire in Asia. Trainor's strategy of getting Mayne off to Camp David over the weekend before he could make any premature statement about Korea left Senator Leyland plenty of room to lose himself in how he'd deal with the Korean situation. "Stay away from it, Mr. President," Trainor had advised. "Ask anyone in the street where the hell Korea is. They don't have a clue. Over there somewhere. The Olympics one time."

"The president," Trainor answered, "has full confidence in General Cahill, commanding officer of all U.S. forces in Korea, and in the Republic of Korea's ability to repel the incursion."

"That's a new one," a reporter muttered in the first row, " 'incursion.' " Trainor wasn't happy with his answer either, but only because he had mentioned Cahill first and not the ROK. He couldn't mention President Rah of South Korea—even his own people had serious doubts about his commitment to democratic rule. The best you could say was that the "old fox," as they called Rah in Korea, was more liberal than anybody in Pyongyang.

"How big is this 'incursion'?" someone shouted from the back.

"Ah—the information we have, John, is that it's regimental size."

The blonde from the *Post* was on her toes, both hands waving in the forest of other hands, pen mikes, and barely controlled mayhem. "How big is that?" she called out, willing to show her ignorance of military dispositions the moment she sensed Trainor's throwaway nonchalance as a little too cool. "How big is that?" she repeated loudly, stretching to her full height, the lace bra threatening to burst right out of the blouse.

A beefy, middle-aged West German reporter from *Der Spiegel* looked at her as if appraising a good leg of lamb. "Depends," he said to no one in particular but with his eyes fixed on her. "It means two and a half thousand men in the NKA, nine hundred in the U.S."

"Quite a difference," she said.

The German's bottom lip protruded, "*Ja*. I think he wants you think American. Nine hundred."

Trainor was pointing to someone else in the front row as the blonde, sitting down, pressed the German, "How come they're so different in size?" She tried to read his name from his press card, but it was hidden beneath his blue suede jacket.

"Very confusing," he answered. "NKA is based on the Russian regiment, you see. They only have the same firepower, though, as a U.S. regiment."

Now the *Post* was really confused. "So there's no difference really?"

The German held his hand up for her to be quiet, his attention shifting to Trainor. Someone, probably a plant, he thought, was asking for the administration's response to Senator Leyland's accusation that the apparent "debacle" now overtaking one of America's "foremost allies" was an example of the "serious implications of President Mayne's cutting of the defense budget."

Trainor loved it. It wasn't a plant, it was a gift from Heaven. He could take the high road. "If Senator Leyland wishes to disparage our allies and use this incident to inflame sword rattling, then, of course, he's free to do so. This administration, this president, has repeatedly said that the security of the United States has no price. And quite frankly, I'm—er—I'm somewhat taken aback by the senator's apparent attempt—did he use the word 'debacle'?"

"Yes," came a shouted chorus. "No TV at Camp David?" called out another. There was laughter.

Trainor shrugged. "All I can say is that 'debacle' is an odd word to be using only hours after some violations of the DMZ have been reported. But if Senator Leyland is so desperate for votes that he wants to characterize—"

"They're shelling Seoul, aren't they?"

"That's not new," retorted Trainor. They were closing in on him, but he saw a way out. "Gentlemen—and ladies—" He was flushed, but it was anger they were seeing, not embarrassment. "If you people spent as much time in Korea as you did in the—" he almost said, "Tel Aviv Hilton," but stopped himself in time "—in the Middle East, you'd realize that in Korea, as in the Middle East, incursions take place every week along the Korean DMZ and that there have been several false alarms already. As recently as July we had . . ."

The hands shot up again. "Are you saying, then, that this is a false alarm?"

"No—I'm not saying that. There are significant numbers of troops moving, but—ah, as yet we don't know the full extent . . ."

An aide slipped in from behind the podium, keeping his eyes low out of the glare, deposited a note on the lectern, and was gone. Despite the heat, rhetorical and that coming from klieg lights, Trainor felt his gut go cold. General Cahill, the note informed him, had ordered the three remaining bridges leading south out of Seoul to be blown within the hour.

Trainor didn't read the rest of the page or notice the fact that Cahill's decision signified much more than the imminent collapse of Seoul.

Because the subway stations of Hapchong, Yongsan, Ichon, Oksu, and Kangbyon—all on the north side of the river—had been gutted by NKA infiltrators earlier that morning, the three bridges to be blown—Songsan, leading to Kimpo Airport, Tongjak Bridge, seven miles east, leading out of the city from Yongsan and Niblo Barracks, and Chamshil Iron Bridge near the old Olympics site, all packed with high explosives, ready for destruction—were the only escape routes left for millions of civilians still trying to flee the city, and there would be no time to clear the bridges, even if they could be cleared, by the time set for demolition. All Trainor knew was that his and National Security Adviser Schuman's plan to distance the president in order to underplay the situation was collapsing around him. Mayne had to get back to the White House. And fast. The news conference, amid howls of protest, the scratching of chairs and dousing of lights, was called to an end, Trainor excusing himself, smiling, nodding, saying "No comment," trying to avoid the snaking TV cables that, despite his strict instructions to have them coiled and bound with fluorescent tape, seemed as disorderly as ever, waiting to trip him up.

If it was a cold-blooded military decision on Cahill's part to blow the three bridges, it was a hot-blooded affair for the millions of terrified civilians—mostly women and children and elders, their screams heard above the screams of artillery, some stumbling, near death—pressed into three enormous funnel-shaped escape routes converging on the three remaining bridges, many people on the outer edges of the funnels spilling off, others shoved aside down embankments into the now putrid river or trampled to death by those behind in the unstoppable force that was five million trying to flee the razored hail of hot steel.

The NKA's Fourth Division met its first really sustained resistance around Uijongbu. In doing so, the NKA tied up so many of the rear guard elements that the NKA Special Forces Corps on the western flank driving south from Munsan to reach the Han three miles west of Seoul's western outskirts easily forded the Han in an armada of lightweight canvas boats under cover of heavy smoke.

CHAPTER EIGHTEEN

MELISSA LANGE WAS one of Pacific Northwestern's brightest and most beautiful, her swept-back ash-blonde hair and wide, sensitive eyes turning heads wherever she went on campus. She didn't flaunt it, but she knew she had it, which made it doubly hard for her to understand why David Brentwood, down to his candy-cane-striped shorts, was standing in the middle of the room, glued to the TV, its bluish aura intensified by the soft peach lighting of the room and the heavy drapes shutting out the morning sun. "You'll catch cold," she said.

"They can fight, the bastards," he said. "Got to give 'em that." Behind ABC's Sam Donaldson there was a gray high-relief map of South Korea, four wide red arrows curving down from the DMZ converging on Seoul.

"Holy Cow!" said David, shaking his head. "Look at this!" There were pictures, very shaky, bad sound tracks, of the bridges being blown, the air full of black smoke and dirt-cored waterspouts. Then a very wobbly shot, as if the cameraman had stumbled, of the bridges, the Songsan's span now V-shaped, some of the spans of the Chamshil Iron Bridge still standing, others simply not there, the smoke from it strangely yellow, the explosions having set afire a nearby barge of sulfur. Melissa was out of bed, pulling on her panties. Tongjak Bridge was the next to go, the TV screen going fuzzy, interspersed with thousands of flickering dots. "Jesus!" said David, pointing at the set. "They're people!" He moved closer.

At first Melissa didn't answer, her arms reaching behind her and up, clipping her matching black bra against her milk-white skin, but David was now on one knee, fiddling with the controls. "They're people, Mel."

"They can't be," she said, glancing over. Suddenly her anger at his inattention to her paled against what was happening on the screen. A commercial came on for Australian beer, "the golden throat charmer." Only now did David see that Melissa was dressing. "Hey, Mel, what're you doing?"

"What's it look like?"

"Hey, no, honey. Listen, I just wanted to—my brother's out there with the Seventh Fleet."

"Men and war," she said. "You love it."

"No I don't."

"Yes you do. That's why you're in the reserves."

"Come on, Mel. I'm sorry." His hands extended, palms up. "Hey, you're more important to me than any damn war."

"Hmm—"

He hit the "off" button and went down on his knees.

"You idiot," she said.

"At your service, ma'am!"

"Well—"

He clapped his hands together, then opened them wide. "Without you I die."

"Without me you'd watch TV. Lounge lizard!"

He rose and reached over the bed, taking her hand. "Love you, babe."

"I've got geography at one-thirty. Have to prepare for it. Last summer school class."

"What's to prepare?"

"Answers. Spot quiz."

"Ask me. Go on."

"Hindu Kush?"

"What about it?" He pulled the sheets up about them.

"Where is it?"

"Asia."

She punched him softly on the arm. "Big help. *Where* in Asia?"

He slid his hand over her buttocks, reveling in their firmness, and when it happened, he knew they'd be tighter than this, tight as a basketball.

"*Where* in Asia?" She pushed his hand away.

"In the mountains. India." He was nibbling the lobe of her ear.

"I thought you poli sci majors had all the answers," she said.

"We do. I'm giving you one now." He slid his left hand between her thighs, pressing into her. "Means killer," he said. "Hindu killer."

"Are you serious?"

"Dead serious."

"Go on."

"All the way?" he asked.

"*No—*"

"Why not?"

"I promised Daddy."

"I could have a talk with him. Make it official."

"Not till we finish school. Remember—it was your idea. Till then, *frottage*. Don't you like it?"

"Sure." He slid his right hand behind her to unclip the bra. "But I'd like to have it all."

"When we finish schoo—"

"Damn thing's caught," he said.

She arched her back, unclipped it for him, and dropped the bra to the floor.

"Oh God," he said, seeing her breasts, closing his eyes and opening them again—as if it had all been illusion. She laughed and began stroking his hair, lowering her body slowly on his. When he came up for air, gently rolling her over to her side of the bed, his face was flushed with excitement. They kissed hard and longingly and he rolled on his back so she could sit astride him, her breasts firm yet pendulous above him. She leaned forward, gently rocking side to side, her long ash-blond hair falling down like warm rain.

"Mogul emperor—" he said, "in 1672."

She giggled and sat up, blocking his hands with hers, their fingers intertwined, she fending him off. "What about 1672?" She kissed his hand.

"Mogul emperor—sent in forty thousand troops—through the Khyber Pass. Only five came back. That's why they call it Hindu Kush."

"That's awful," she said, sitting up, pushing her hair back.

"Yeah."

"See? I told you. Men love war."

"I'm just trying to help—" His mouth went dry just looking at her. "Any more questions?" he asked raspily. Her nipples were engorged and he felt hard as concrete between her legs. A frown swept over her face and she sat upright, flicking her hair back, a bobby pin in her mouth. "You don't think there'll be a war, do you?"

"What—oh, in Korea. There's already a war."

"No, I mean a world war?"

"No chance. Nuclear weapons'll stop it."

"Nuclear weapons could start it."

"No, before that they would have to—hey, is this a seminar or what?" He pulled her down to him, running his fingernails gently over her back, dragging them lightly back up to her shoulders and down along her outstretched arms, her murmurs of pleasure making him happy. "Love you, babe," he said.

"You, too," she said, and began doing the same for him. He slipped off his watch, hearing it drop softly to the floor. Now she lay down full length on him, moaning softly as he began to move, arching his back, lifting her, all the pressure in her groin. She kissed him wetly, hard, tongue thrusting hard for his. "Don't—don't leave me, Davy," she called softly, gently, lonely as a child in the night.

"I won't," he said.

Now he could feel her buttocks, the hard, rounded silkiness rousing him so he doubted he could hold out much longer. She stopped, perfectly still, sensing his razor-edge excitement as he calmed down. Her smell was overwhelming him, and now for a moment, a man possessed beyond his years, his vision blurred, she came back into focus, and he could feel the blood pulsing through him. She raised herself above him using her elbows and began moving rhythmically again side to side, his mouth like a fish gasping for water. She laughed and it relaxed him, his shoulders slumping back, falling on the pillow. She had to be careful—sometimes the slightest giggle could make him angry, as if he thought she was laughing at him instead of with him—for him.

"You're going—" he stopped to get his breath "—to be late for class."

"Yes," she smiled.

"Sweetie—let's get married now. Today. This afternoon."

She placed her forefinger gently on the tip of his nose. "*No. After we graduate.*"

"That'll be—hell, that'll be the end of next term. Christmas. I can't wait that long. I'll go nuts."

"*No.*"

"You're a hard woman, Melissa Lange."

"I'm old-fashioned."

"This isn't old-fashioned."

"How do you know? Maybe your parents and mine did it."

"My dad?" he said disbelievingly. "You're joking. Mom would never have let him."

"Oh—they didn't have sex in the navy?"

"Shore leave," he said.

"Oh Lord!"

"What?"

"I promised Rick I'd loan him my notes."

"*Stacy?* Let him make his own."

"He was sick last week with the flu."

"He'll be all right."

"What time is it?" she said. She reached over and picked up the watch. "Oh Lord! He'll be waiting for me at the Student Union Building."

"Let him wait."

"I promised, Davy," but she could see he was getting mad. "You know how you are about promises, Davy."

"For Stacy?"

"Oh, come on." She shook his shoulders. "You'd go."

"No I wouldn't."

"You know you would." She hopped off the bed. "I'll make it up to you, sweetie. Promise."

"Why don't you make it up with Stacy?"

"Davy."

He slumped back in the bed, throttled a pillow, relaxed his grip, then threw it across the room.

"Listen," she said, getting dressed as quickly as she could. "Who was the one watching the TV?"

"That was only a minute. You were watching it, too."

"I certainly was not."

"The hell you weren't! You were asking me if they were people."

"*After* I'd been waiting for you to come."

"Don't be dirty," he said. "I don't like it when you talk like that."

"*What?*—oh, for Heaven's sake. You're the limit. You're the one with the dirty—"

"All right, all right. Forget it."

"Okay, I will. See you around. When you grow up."

He flung the bedding aside. "Fucking Stacy. I'd give him notes. Right in the face."

"Well, if you're going to use that language, David . . ." She was tucking her shirt into her jeans.

"Oh Jesus," he said, "Little Miss Muffet."

"You're so stupid," she shot back. "There's nothing between Rick and me."

"God, you're blind. I can't believe it. He wants your *notes*. You really think that's all he wants?"

She grabbed her satchel. "Well, if you keep this up, Bub, he might just get it."

"You—"

"Go on, say it."

"Never mind—"

"Say it."

"Bitch!"

"All right, buster," she said. "That's it! See you around." She stopped at the door and swung about. "And those shorts," she said, glancing contemptuously down at the red and white striped underpants. "You look like a barbershop. Never seen anything so ridiculous." She walked out and slammed the door.

"You hear the news?" asked Rick Stacy, a fourth-year student majoring in commerce and international relations.

"What news?" asked Melissa.

"The fighting in Korea."

"Yes," she said. "Well, now I know how wars start."

"What do you mean?" he said as he gathered up his things from the plush but grubby Student Union sofa.

"It doesn't matter," she said. "Here are my notes."

"Oh, I already got them from Linda. Thanks anyway."

"I could strangle you, Richard."

"What?" he asked, alarmed. "What'd I do?"

"You started a war."

"Uh-oh. Davy Brentwood. Right?"

"Right."

"Hey, don't sweat it. Really. I'll have a talk with him. Set him straight. I *am* in IR."

"What?"

"International relations. Conflict resolution. My specialty."

"Maybe we should send you to Korea."

"Aw, they're just trying it on," said Stacy.

"You see the news this morning? Looked pretty bad to me."

"Sure it does. Right now. Surprise is with the North. Always is with the attacker. You'll see. It'll be over by Christmas. Not like it was back in the fifties. Caught the South napping, that's all. President'll mobilize the reserve maybe—that sends the message to Moscow and China real quick. End of series. They don't want a war."

"Neither did South Korea, but they're getting it."

"Stop worrying. It always takes us a little time to react, but when we do, it's game over. Moscow'll tell them to get their ass out of there. Russia's got enough to worry about. Estonia, Latvia—"

"You think the president will mobilize the reserve?" asked Melissa.

"No question. Doesn't want to seem too weak—not with old Leyland breathing down his neck in the polls. But doesn't want to be seen as a warmonger. But he won't send troops in. Happens all the time, Melissa. You call up the reserve or hold *maneuvers*—that's another good one. Sends the right messages to Beijing, Moscow."

"What if they call our bluff?"

"Hey! Are you serious? China's on overtime just trying to feed itself, and Russia's had one of the worst harvests in years."

"Where have I heard that before?" she said, frowning, unable to pin it down exactly.

"What?" asked Stacy as they walked over past the library to the cedar-hidden geography building. Stacy thought for a minute. "Bastogne?" he proffered. "Thought we had it all wrapped up and bam! Out come the Panzers. But we beat them."

"No," said Melissa, "it was in Korea. MacArthur or some-one said it would be over before Christmas. Then the Chinese came in."

"History," said Stacy.

"And history repeats itself, right?"

"Up to a point. That's an old wives' tale. It's always different really."

"Then it'll be different now," she said. "If the president mobilizes the reserve, maybe it won't work."

"Listen, Melissa—and don't take this the wrong way—I'm no male chauvinist." They kept walking toward the quadrangle, the smell of the cedars strong in the high humidity. "But putting an M-1—that's a tank—"

"I *know* that, Richard."

"Well, what I'm saying is that putting our M-1s up against what the North Koreans have—hell, like a heavyweight boxer against a bantamweight. *No hay compración.* No contest."

"I hope you're right."

"Bet you dinner. The Steakhouse," said Stacy.

"Okay," she answered, knowing the moment she'd accepted, she shouldn't have. It would only antagonize David further, but—darn it, his final words had hurt. The ingrate. Anyway, it would probably be weeks before the Korean business was over. She never did see why so many Americans had to stay there—it was up to the South Koreans to protect themselves. Well, in a few weeks everyone would have cooled off. David would have simmered down—he wanted her as much as she wanted him—they both knew that. And Richard would probably win his bet about the Korean thing being over by Christmas and impress his international relations seminar. She'd gladly pay for the steak dinner and invite David to keep the peace.

From his well-camouflaged revetment area, six miles south of Uijongbu, tank commander Lieutenant Clemens no longer felt the slightest pity for the men he was sure he was about to kill. News of the atrocities found its way into the crackle of radio traffic. One of the four who had been beheaded had been from one of the dual-based mechanized infantry support companies out of California, Clemens's own state. On hearing the news, Clemens steeled himself for vengeance. Now he could see the first of an NKA battalion of PC-76 tanks, the "tin cans," emerging like parts of a long, segmented green snake on the rain-polished highway. His laser range-finder told Clemens that the distance between his six American tanks and the NKA's sixty-four was exactly 1,203.4 meters, well within the M-1's four-thousand-meter range, the range confirmed by the additional thermal sight used in bad weather or at night.

Clemens, elbow resting on the cupola's 12.7-millimeter machine gun, could also see NKA infantry moving up alongside the dark green PT-76s, the ceremonial red stars normally visible on the turrets painted over with slightly darker green camouflage paint. Clemens gave his orders quietly and unhurriedly to the loader and gunner, the gunner's integrated computer display verifying the elevation of the 105-millimeter gun and compensating for crosswind and rain-caused deviations as the M-1's four-man crew waited patiently for the enemy's lead PT-76 to come to a thousand meters so that the whole column would then be within killing range. Clemens's thumb was rubbing the steel guard, ready to press the computerized fire control system that was even now compensating for the effects of wind drift, barrel bend, temperature, and humidity. Clemens had to make the decision whether, at the moment of firing, the tank would be "buttoned up" or he would do what the four men in his tank and the other two tanks of his platoon called an "Israeli," standing up, his head and shoulders out of the turret. Despite the tank's state-of-the-art CO_2 laser range finder, driver's thermal viewer, and the rest of it, an "Israeli" would afford him a better all-round view of the road and surrounding paddies. And so Clemens kept standing, careful not to make any move that would shake the camouflage netting around the fifty-four-ton tank, quietly telling his driver the fallback position once the tank's initial rounds gave its present position away to the enemy column.

The USS *Blaine* was in condition five, its top readiness alert. On its radar the swarm of white dots within the white rectangles that had signified unknown surface ships had now become white squares, hostile ships, identified now as three 180-foot-long Nanuchka and fourteen Shershen-class fast-attack torpedo boats armed with four twenty-one-inch torpedoes and four twin thirty-millimeter machine guns. The Shershens, originally moving at thirty-eight knots, were the faster boats but now held back, knowing the U.S. frigate's more sophisticated electronic defenses could better be penetrated by the Nanuchkas, which were now closing in the rolling fog banks.

"Missile incoming!" shouted the *Blaine*'s OOD, and the Phalanx Mark-15 close-in radar and weapons system with its twenty-millimeter gun opened up together with the seventy-six-millimeter gun aft of the multiple target radar.

"Hard right five degrees," ordered Brentwood.

"Hard right five degrees."

The *Blaine* was now bow on to the oncoming swarm, attempting to deny the NKA boats a wide broadside target in or out of the fog as she "ghosted," projecting fake radar images of herself to decoy the attacking boats even while her six-barrel Gatling gun, with a sound like linoleum, was spitting out a hail of depleted uranium bullets at over three thousand rounds a minute. Any one of the bullets, twice as dense as the normal steel-jacketed kind, was capable of deflecting, or causing detonation of, incoming missiles.

"We've been locked on," shouted the electronics warfare officer, indicating one of the Nanuchkas' "fishbowl" radars had switched to fire control mode. Immediately the Tactical Action Office ordered "chaff" and the torpedo launcher shot out a cloud of fine aluminum chips to hash the incoming missile's radars. Another high tone from the *Blaine*'s SLQ-32 radar indicated another missile had been fired at the *Blaine*. Seconds after Ray Brentwood ordered the four antiship Harpoon missiles fired from the launcher forward of the bridge, he felt a slight tremor from the back-blast and at the same time received confirmation that the *Blaine*'s two LAMPs—light airborne multipurpose helicopters—had taken off within seconds of each other from the stern pad armed with clusters of air-to-ship rockets. The next minute he heard two thumps that reverberated through the frigate as the *Blaine*'s two triple-tube torpedo launchers discharged four MK-48s into the swarm now closing at less than two nautical miles.

There was a bright orange flame forward of the starboard beam about three hundred yards into the fog, an enemy missile hit, and a second later the crash of a destroyed Nanuchka came rolling over the ship. At the same time the TAO reported, "Bogey missiles destroyed." There was a cheer in the combat information center, cut short by the TAO's command to the radar operators to compensate for clutter caused by the *Blaine*'s own chaff and the increasing chop caused by the wash of the remaining sixteen NKA boats. A sharp pulse of light on the radar and seconds later the sound of an explosion told them another Nanuchka had been hit, but Brentwood was worrying about the changing positions of the two remaining Nanuchkas. To maintain flank speed would mean entering the swarm sooner, but to slacken off would give him less maneuverability—the fact that the two Nanuchkas were slowing down didn't abate his fears as they had separated to form the two tips of a bull's-horn formation. Meanwhile the armored shell of the *Blaine*'s combat information center being below the bridge, Brentwood and the others were only dimly aware of the cacophony of firing outside as the Shershen attack boats coming at him broadside opened up with thirty-millimeter fire and began launching their torpedoes from staggered overlap tubes, the bull's-horn-like formation of fast boats now becoming a rough semicircle of a half-mile radius, launching twenty torpedoes at the *Blaine*.

"Fish incoming!" called one of the radar operators. "Bearing—" The operator stopped.

Brentwood swung around, saw the problem—there were so many, a single bearing wouldn't help. "Hard left ninety!" he ordered, reducing the sector of the torpedo attack to a quarter instead of a half circle and hoping to outrun the incoming torpedoes now coming at him from abaft the starboard beam but also putting the *Blaine* broadside to the extreme left half of the semicircle; the lead boat, a Nanuchka, closest to him, he now engaged with another two Harpoon missiles. The blip that was the Nanuchka amid the dancing fuzz of chaff and other clutter suddenly grew very bright on the screen, then disappeared.

"Hard right, ninety degrees!" he shouted, anticipating a cheer or two from the CIC crew, but now everyone was silent, only the hum of the electronics and that tattoolike din outside faintly audible above the radio crackle of the two helicopters, one pilot yelling at the other, "Two o'clock, two o'clock!" and they actually heard the sound of a missile passing one of the helos. In quick succession another three torpedo boats blossomed on the screens and disappeared, taken out by the helos, but now they could hear the scream of the electronics warfare officer aboard one of the choppers as he was hit by machine-gun fire. Seconds later they heard the bang of the helicopter hitting the water, followed by the rattle of machine-gun fire— the attack boats raking the *Blaine*'s starboard side.

"Fish incoming starboard quarter!"

Brentwood knew he could do nothing, the torpedo-tracking radar and digital sonar now malfunctioning, and knew that either hard right or hard left would expose his stern to the torpedoes racing toward him at over fifty miles an hour. Two torpedoes went past, whitish-gray streaks in the fog, the startled starboard lookout informing the bridge a second later.

A machine-gun burst hit the *Blaine*'s stack, puncturing it but doing no more damage as the ship's Gatling gun swung sharply, continuing to fire, causing another missile to explode within a few hundred yards of the ship, but now the *Blaine*'s Phalanx radar was in danger of "fuzzing up" from overload. The sonar operator, ignoring all else, carefully monitored the sea bottom, alert for the telltale ping of mines, while his colleague on the 225-mile-range air-search radar informed the tactical action officer that the radar's dish, between the bridge and the satellite communications dome, was out. Brentwood knew that of all the battle group ships on forward screen for the carrier *Salt Lake City*, the *Blaine*, like her sister ship the USS *Des Moines*, forty miles to the east and closing to assist the guided missile frigate, had never been designed to go it alone in such outnumbered circumstances. The frigate's weapon system had been designed primarily for medium-range escort duty, for what the Pentagon had designated a "low-threat" environment. And the Sea of Japan had been just that—until the NKA had crossed the DMZ.

"Sir!" the surface radar operator began, but then checked his excitement. "Enemy disengaging."

No one in the CIC or anywhere else on the ship eased off, knowing it could be a sucker ploy. But the operator proved right; after losing six patrol boats and only damaging the American warship, though they had downed one of its helicopters, the NKA naval force, it seemed, had had enough. Still, Brentwood kept everyone at their stations, despite their fatigue and the stench of perspiration thick in the air. Even as the enemy was retreating, he ordered another four Harpoons "onto the rails," and in the bowels of the ship the loader pressed the button for the automatic feed, quipping, half in relief, half in celebration, "Four pack to go." Brentwood made an immediate note to enter into the ship's log, together with the tape that, like a civilian aircraft's black box, was set to start recording the moment a U.S. Navy vessel went on "Action Station Alert," how astonishingly ineffective the Nanuchkas' missiles had been. Now it seemed all the peacetime speculation was over. The suspicion among the experts that quantity rather than quality was the central theme of the NKA's, that is, the Russians', strategy seemed confirmed.

"Inbound missile, inbound missile!" All the lookout saw in the fog was a blue sphere of light the size of a basketball, the sixteen-hundred-pound Exocet skimmer coming in amid the radar clutter, hitting the *Blaine* amidships on the starboard side, ripping open the three-quarter-inch armor plating like a fist through glass.

Six miles south of Uijongbu, Lieutenant Clemens saw the line of sixty-four green PT-76s halt. Through his binoculars he could see the enemy tank commander emerge, wearing the Russian-style leather helmet, looking a little like a World War II pilot, big bumps over the earpieces.

"C'mon," said Clemens, "c'mon. God, make him . . ." Then Clemens saw the NKA commander thump the cupola, and the tank, its gravelly roar giving off a thick, bluish exhaust, moved forward again, leading the column.

When the lead PT-76 was well within the one-thousand-meter range, Clemens gave the order to fire. The M-1's gunner pushed the ranging button, activating the split-second computerization, then squeezed the trigger. There was the thud of the recoil, but the M-1 hardly moved because of the superb suspension of the seven-road wheels sprung on each side on torsion bars. The HESH, or high-explosive squash head, round did its job, hitting the PT-76's sloped glacis plate with such force that it blew off scabs of red-hot steel inside the PT-76, creating a lethal shrapnel. Within seconds the PT-76 slewed to a stop, three figures, all on fire, tumbling out of it, the driver trapped in his hull seat beyond the turret. Soon the whole tank was engulfed by fire, the molten scabs of white-hot shrapnel igniting the tank's oil reservoir, creating a thick, billowing cloud of black smoke. To Clemens' dismay the smoke was widening and flowing back over the column in as effective a smoke screen as he'd ever seen laid, many of the PT-76s now adding to it by popping off clusters of smoke grenades left and right of the road. Within minutes the entire tank column was enveloped, hidden from the view of the two American tank platoons, in all six tanks, that were situated in strategic firing positions under camouflage nets either side of the road behind a rise that was now only nine hundred meters in front of the NKA armor. But Clemens and two other M-1s fired at the second and third PT-76s, the HESH rounds stopping them, too, dead in their tracks, littering and blocking the road with their burning wreckage. But though the Americans had stopped them, the PT-76s immediately behind the gutted hulls kept returning fire.

Within fifteen seconds Clemens's M-1 as well as the other five tanks that had also fired into the column and so revealed their positions to the NKA were also under mortar fire from NKA tank support infantry situated in the flooded paddies and from the shoulders of the road. Clemens's M-1 was retreating down a shallow depression to its next selected firing position on a hillock two hundred yards farther back, still on relatively higher, dry ground above the road. The aim of the return fire from the up-gunned PT-76s' 105-millimeter cannons, shooting solid APDS—armor-piercing rounds with discarding sabot—was poor, and even when struck, the tank easily deflected two rounds, the M-1's turret armor so sloped as to deny a "flat-on" impact. Still, the fact that the rounds were coming close told Clemens that the five American M-1s were being sighted through the thick pall of smoke by either laser or thermal imaging, reflecting the NKA's, that is, the Russians', military strategy of equipping tanks, even the older models like the PT-76s, with nighttime ranging capability. It was another indication of Soviet military thinking, their belief that their best chance in any battle with the West was a quick win, which meant keeping your crews fighting twenty-four hours a day before the United States had time to rally the political will to reinforce their troops or others in the Western alliance.

The sound inside the M-1 as it geared down, breaking its speed as it approached the tall bamboo, was like some enormous earth mover. But to the few ragtag platoons of supporting U.S.-ROK infantry outside, the M-1 sounded remarkably quiet, not much noisier than a growling pickup as the fifty-four-ton tank moved at over thirty miles per hour across a corrugated stretch of ground, its gun steady and maintaining its six rounds of accurate fire per minute even as it geared up, heading at full speed, over forty miles per hour, toward the heavy stand of bamboo overlooking a curve on the Uijongbu-Seoul road.

"Incoming missile!" shouted the gunner, Clemens dropping down the hatch, which was seated and secure in a second. It was not a round from any of the PT-76s, which were now breaking left and right of the blocked road as the five American tanks opened up from the bamboo and surrounding hills, but a relatively slow, Russian-made antitank "Spiral," coming in at 560 meters a second, less than half the speed of a tank round. It struck the M-1's turret, exploding against the two-hundred-millimeter-thick sloped armor, failing to penetrate, the shock wave quickly dissipating throughout the composite, or layered, SPC armor. But it caused a ringing that, even through the crew's protective earphones, was so intense that along with all the other internal noise of the fifteen-hundred-horsepower gas turbine, controlling gun elevation and depression, the constant hum of the computers and the steady blowing of the air conditioner, the fume sleeve halfway down the barrel evicting noxious gases from the exploded powder, none of the tank's crew could hear anything on the radio for at least ten seconds. And so they didn't hear the warning transmitted from the other two tanks in Clemens's platoon. A second antitank missile, wire-guided, had been fired at them from ten o'clock, from a fringe of smoke fifteen hundred meters away in the paddy off the left-hand side of the road. The Russian-made Sagger, moving at under five hundred meters a second, came through the black and white smokescreen rolling over the paddies, hitting the right track of Clemens's M-1 as it was nearing the bamboo thicket. There was a muffled explosion, more sound shock, and then a clanking sound, like chains unraveling from a snow tire.

"Damn!" shouted Clemens's gunner as the tank crashed into the bamboo thicket, the top of which rose a good six feet above the hillock immediately to the right of them, blocking them from the NKA tanks' view and placing them in a perfect position for defilade fire, providing the driver could manage to turn the fifty-four tons using only the intact left track and what was left of the right.

"Can you do it, Johnny?" shouted Clemens.

"It'll take the right track off, Lieutenant."

"Do it!"

The M-1 shuddered right, then the driver, in his forward hull semireclining position, used all the skill he had been taught at Fort Hood, coaxing, talking to the twenty-five-foot-long, eight-foot-high tank, as a cavalry man of old might have cajoled his wounded battle horse, to climb that few feet more up the reverse incline of the bamboo hill to the defilade position. In the lowest gear the tank gave a throaty roar, inching forward on the slippery cane of the bamboo almost to the summit of the hill. Here, given the light rain that was now falling and the smoke, the tank was barely visible, and here its gun, the product of the best engineering in the world, could lay down fire with minimum silhouette. The enemy could return effective fire only if the M-1 could not withdraw before the NKA range finders picked it up. With the right track off, Clemens knew he couldn't fire and withdraw with any speed and so asked his wingmen, the other two tanks in his platoon, to "overwatch" him, to feed his tank information about any advancing NKA infantry with antitank weapons or PT-76s looking for a lucky shot from the paddies.

On one hand Clemens felt trapped because it would take several hours at least to fix the unraveled track, if they could do it at all in the field. And yet he was in a perfect position with a wide down angle of fire across the paddies now stretching fifteen hundred to five thousand meters in front of him. He was waiting for the smoke to clear. One of the other two tanks he'd requested help from didn't answer. Whether it was hit hard and out of action or whether its radio had packed it in, Clemens didn't know. In any case, the other tank gave Clemens the happy information that Cahill had released a full batallion of M-1s, thirty-five tanks in all, which were now rolling up Unification Highway. Within twenty minutes they'd be in the area, engaging the remaining sixty lighter-armed and lighter PT-76 tanks now presumably scuttling across the paddies either side of the road under cover of the smoke, trying to find whatever defensive positions they could. In the interim, Clemens decided to save what was left of his original fifty-five rounds until the smoke curtain cleared.

"How many shots we got left, Luke?"

"Forty-two."

"What we got on the menu?"

"Ribs, lobster."

"Never mind the shit—what've we got?"

"Shoes, twenty, HE, ten, and twelve HESH."

It told Clemens he had twenty rounds of solid-shot armor-piercing, ten high-explosive shells, and twelve high-explosives with squash heads.

"Well," said Clemens, trying not to sound too satisfied up against the PT-76 tin cans. "That should hold us awhile."

"I reckon," said his gunner.

As General Kim in his headquarters six miles east outside Uijongbu was informed of the M-1 spearhead rapidly approaching, his face remained impassive, even as his infantry commanders worriedly reminded him of the cardinal rule: that while it was permissible for tanks to move without infantry, this was only wise when you were advancing and using the infantry as your eyes and ears, but in this situation, with a long line of American M-1 tanks, the NKA armored column would quickly be thrown on the defensive, the supporting regiments behind the NKA tanks slaughtered by the cannon and machine-gun fire of the formidable M-1s unless they were withdrawn. Radio intercepts, Kim's officers pointed out, already confirmed that the M-1s, their laser range finders thwarted by neither rain nor smoke, were less than twenty minutes away. And, after the executions at Panmunjom, the *migooks* would surely show no mercy, and would fight eye to eye with the PT-76s. Kim merely nodded. He was already quite aware that he was about to engage in the first massed tank battle since the days of the Israeli-Arab wars.

One of the infantry commanders, a colonel, Russian-trained and in charge of one of the crack NKA Sapper units, had, as Kim ordered, already gone ahead of the beleaguered PT-76 column and blown a huge hundred-yard gap in the road, in effect creating an enormous tank ditch which no tank, including the M-1, could ford. But this, the colonel pointed out to Kim, could only be expected to delay the Americans for at most a quarter hour. And it would not be long, he told Kim, before the American fleet in the East Sea would be near enough to launch aerial antitank attacks. This was dangerous for any tank, the top of the turret being the least-armored part of the vehicle, but for the relatively light-armored PT-76s, it could mean annihilation.

Kim did not respond.

As Kim left the headquarters tent, walking across the squishy ground to his private quarters, two of his chief staff officers tried to fathom Kim's intent. "Perhaps he thinks," proffered a battalion commander, "that once the M-1s engage our tanks, they will be too close. Any aerial bombardment would also destroy any American tanks nearby. I think Kim has something up his sleeve."

"Why?" asked the colonel.

"He did not seem overly concerned about the M-1s. Didn't you notice?" asked the battalion commander.

"I noticed. That is what bothers me. He does not fully comprehend this situation."

"You can't tell with Kim," said the colonel. "He is known for not divulging his tactics till the last minute. Fears a security leak. I've no doubt he's studying the situation carefully. He's up against the American, Cahill. Kim hates him."

"So do I," said the other officer. "But hating is not enough. Hate will not stop an M-1."

"No," agreed the colonel. "But it will help."

"How?"

The colonel shrugged. "In-close armored fighting is not something the Americans—"

"You think Americans are no good at this? Don't you remember Patton?"

"Yes, of course," said the colonel. "But that was a long time ago."

"And what happens when Washington sends reinforcements?" pressed the infantry commander.

The colonel laughed. "They won't."

"If they do?"

"We will have over half the South before they get here. The Americans love to argue. Their democracy," said the colonel comtemptuously, "is all talk. They talk big."

"They have big tanks."

"They have big egos," countered the colonel. "Remember Vietnam, my friend. Once their ego is punctured, they become very depressed, the Americans."

"First," said the infantry commander, "I'd prefer to see the M-1s punctured."

"Be patient," said the colonel.

The infantry commander looked at his watch. "It won't be long. They will be entering the area within five minutes."

"Then," said the colonel, "we don't have long to wait, Comrade. I think we are about to make history."

CHAPTER NINETEEN

PRESIDENT MAYNE DIDN'T favor the world-famous Oval Office in the west wing except for the most official occasions. Most of his work was done in the smaller, more ordinary study next door, and it was here that Peale's portrait of George Washington was moved. In those lonely times that only a president knows, when only he could break the deadlock of advisers' conflicting advice, Mayne would retreat here to mull over the options, the possible consequences, or what Trainor called the "bottom line situation."

For Mayne, the Oval Office had always seemed too big for serious contemplation, no matter how cozy it looked in the narrow-focus television brought to it, hiding by omission the large area between the fireplace and the white leather lounges in front of the desk. The Secret Service thoroughly approved of the smaller room. For the men who protected him, the Oval Office, being on the southern corner of the west wing, was a much more vulnerable target for anyone who might penetrate the elaborate, yet mainly unseen, protective screen of heat and movement sensors that covered every sector of the grounds. The Secret Service had installed a rectangular titanium shell, sandwiched in the paneling of the study, making it even more secure. But above all, the president liked the room because he could darken it completely and keep the secret that only he, his wife, and Trainor shared. This morning he sat down with the Pentagon's updated report of the North Korean invasion.

Early in his presidency Mayne had decreed that situation reports be as brief as possible, no more than two pages, double-spaced, a one-and-a-half-inch margin for his comments. Despite all the words put in his mouth by speech writers and advisers throughout the country, at heart he disliked any kind of verbosity. For Truman it had, as everyone knew, been "The Buck Stops Here" sign that greeted visitors; for Reagan, "It Can Be Done"; for Mayne it was "Get to the Point—Quickly!" He had long accepted the fact, so difficult for others to understand, that decisions from the White House, including those involving life and death, often had to be made without all the facts being in. *All* the facts in any given situation would take a lifetime to uncover, a luxury that only academics and "gunning for you" journalists could afford.

When he'd finished the first page of the Korea report, he pressed the button for his national security adviser, Harry Schuman, to come in, and kept reading with a growing sense of alarm. The most disturbing of the Pentagon's "facts" was that if the American tanks could not hold the line and "substantial U.S. reserves" were not committed "immediately," Korea could be lost within weeks—faster than it had taken Hitler's Panzers to overrun Poland in '39. If this happened, warned the combined chiefs of staff, U.S. treaty obligations and guarantees throughout the world would be considered worthless, of no more use than Chamberlain's piece of paper. And the temptation of the Soviet–Warsaw Pact nations in eastern Europe, particularly East Germany, pushing to reabsorb and effect the "reunification" of East and West Germany might prove irresistible. Mayne simply did not believe the latter; Moscow, no matter its posturing in the post-Gorbachev era, would not endorse such a move in the GDR. The Kremlin, as much as anyone else, wanted to avoid another war—conventional, nuclear, biochemical, whatever.

Harry Schuman, a bushy-eyebrowed southerner whom the White House staff called "Kentucky Fried," entered the office, and wordlessly Mayne handed him the first sheet of the report as he continued pondering the second. The Pentagon in his view was overplaying the concern about NATO versus the Soviet-Warsaw Pact forces in Europe. But they were correct, he believed, about a victory for North Korea weakening confidence abroad, particularly right next door in Central America and in China, where Beijing coveted Taiwan as theirs because of all the mainland Chinese who had gone over with "Cash My Check" in '49 when the Kuomintang had fled the victorious Mao. Most of all, if there was any serious weakening of confidence in America in the Middle East, Iran would be "licking its chops," as Trainor was apt to put it, and Israel, always surrounded, could be attacked yet again. And if Iraq used chemical weapons, as she'd done against Iran in the '79-'88 war, it could well spark a string of firecrackers from the Gulf to the Bering Strait. Mayne picked up the phone to Gen. Ernest Gray, head of the combined chiefs of staff. "General."

"Mr. President?"

"Your people are telling me that if I don't commit more forces to Korea immediately, we're in serious trouble."

"We'll lose Korea, Mr. President."

"The Koreans will lose it, Ernest. We'll be kicked out." Mayne felt uncomfortable with calling the general "Ernest"— didn't sound right—yet "Ernie" invited a familiarity that he didn't like to encourage with the military as their commander in chief. "What I want to know," continued the president, "and this is no reflection on your colleagues, but—are we overreacting?" Mayne had seen the television shots of a few of the bridges going, but TV had a way of making a dormitory riot seem like a whole university was on fire when, as he remembered from his own days as a freshman, most students didn't even know where the dorm was, let alone a riot.

"I support General Cahill's decision to take out the bridges, sir," said Gray. "I know it didn't go down well on the six-o'clock news, but militarily speaking—"

"I have no problem with that, General—awful as it is—but your people worked overtime on the Hill to get more M-1 tanks in Europe so that we could spare several companies for Korea's DMZ. The M-1s would be the 'bulwark,' you said—if I remember correctly—against any possible incursion by the North. Now we have the incursion and I'm not hearing anything about the 'bulwark.' "

"Ah, Mr. President—there's a coded update coming in now. It might be—"

"Fine. Call me right back."

"Yes, Mr. President."

At the Pentagon General Gray and his aides were in a quandary. On the one hand, to admit that the battle of tanks shaping up south of Uijongbu had not been decided was to admit the Pentagon might indeed have been overreacting about losing the peninsula. On the other hand, to paint too gloomy a picture would be to undermine confidence right down the line. What the Pentagon was really doing was hedging their bet—angling for reserves to be in place in the unlikely event that the M-1s could not hold the line.

The decoded message was reporting the attack on a guided missile frigate, the *Blaine*. This had already been noted from satellite photos, but when Gray rang the president back, he used it as ammunition for the Pentagon's overall argument. "What we're saying, Mr. President, is that the stakes, not just for the Koreans but for the United States, are enormous here. On top of Vietnam any hint of defeat in Korea could undercut confidence not only among our allies but in our, ah—"

"You mean my administration?" Mayne cut in.

"To put it bluntly, sir, yes."

"Bluntly is what I get paid for, General, but I've been talking this over with Harry Schuman. The fact that we can't hold so far with forty thousand U.S. troops and the ROK forces might be just as bad a signal to send. If confidence has been undermined, then it's been undermined—I don't want to send any more boys in there if we're going to lose the place anyway."

"All I can say, Mr. President, is that General Cahill and ROK command concur with the JCS assessment. It's a very tenuous situation, sir."

Harry Schuman scribbled a note and pushed it across the desk as Mayne asked General Gray what precisely had happened to American air cover. As he answered, Gray could tell from the echoing quality of his voice that the president had put him on "conference," Harry Schuman sitting in. As far as Gray was concerned, Schuman should stay in his office in Foggy Bottom for all the help he was to the military. Always prevaricating.

"General," put in Schuman, peering over his bifocals as they switched to visual conferencing, "it's my understanding that we have the Seventh Fleet moving in off—" he turned to look at the stand map of Korea next to the president "—Pohang?"

"Yes, sir, but their planes aren't in range yet for air cover

over the Uijongbu corridor. In any case, we'll have to clear the area of MiGs before our navy choppers and what few A-10s we have left can go tank hunting up there. And the point is, sir, no matter how much air cover we can provide, the battle's ultimately going to be decided on the ground. Our tanks and theirs'll be too close to—"

"All right, General. But keep me informed half-hourly. Sooner if necessary. Meanwhile I'll authorize reinforcements from Japan. What I want to avoid right now is sending any troops from the United States. Everyone knows Japan's our reserve base for Southeast Asia—so we can do that without causing undue panic."

"Fine, Mr. President, but if you'll permit me, sir, I need to know whether the reinforcements can be *deployed* immediately." Mayne glanced across at Schuman to see if he had any advice.

"General," put in Schuman, "it'll be twenty-four hours before they reach Korea. I think it'd be a good idea to leave that question open for now."

"Sir," answered the general, "we're going to need marines or airborne if we're to secure a landing zone. NKA infiltrators have effectively cut off—"

"Pusan hasn't fallen," put in Mayne.

"No, sir, but it's a long way south."

"I know, two hundred and fifty miles," answered Mayne.

"Bridges over the Naktong are cut, sir."

Mayne nodded, glancing up from behind his desk at the large stand map of Korea. "Well, General, you people try to secure the airfields around Pusan. NKA can't get across from Seoul very quickly. Cahill's at least seen to that."

General Gray stiffened, sensing an undercurrent of condemnation of Cahill's action. "In any event," continued the president, "you people have a lot of faith in the M-1s. They could still turn this thing around, couldn't they?"

"Yes, sir, I think they will. I merely wanted to know about deployment of reserves in case—"

"Let's see what Cahill's armor can do south of Uijongbu first."

"Very well, Mr. President."

Mayne flicked the phone off "conference" and, his hands forming a cathedral, leaned back in the leather chair, looking thoughtfully across at Schuman. "What I want to know when this is over, Harry, is how the hell were so many North Korean agents allowed to infiltrate the South? My God, they're blowing everything up left, right, and center."

"They've had a lot of years to plan it, Mr. President."

Mayne's cathedral collapsed. "Well, blame can wait. What we need now is for those tanks to stop the NKA dead in their tracks."

Schuman was about to comment favorably on the pun but thought better of it.

Entering the Pentagon crises room, several of the Joint Chiefs' aides noticed it had been freshly carpeted, golf-green. The military psychologists, Gray told them, had found that the old deep maroon tended to depress people. With Mayne and his defense cutbacks, there'd been enough of that. Besides, Gray had always kept a putting iron and one-hole mat in his office to unwind after the conferences. Used to look like hell on maroon. Several of the aides allowed themselves a smile at the general's joke, but it was difficult to do with the wall map of Korea showing red arrows well below what had once been the DMZ.

When all the chiefs had arrived, Gray informed them that despite his conviction that the M-1s Cahill had unleashed from the revetment areas beyond Seoul would turn the battle, it would also be necessary to start moving the thirty thousand reserves across the Sea of Japan. It was a logistics problem that airlift alone, even if they'd had landing strips, could not solve. They would have to use troop ships.

"They've got no subs to speak of," an aide said encouragingly.

"No," interjected Admiral Horton, chief of naval operations. "They didn't have any Nanuchkas either—until they attacked the *Blaine*."

"Nor any Skimmers to speak of," put in another aide. "That could be air-launched."

"We're not sure it was hit by an AS missile," replied the admiral. "Could have been by one of those MiG-29s they didn't have."

There was a heavy silence, but Horton wasn't about to back off from his criticism of Gray's heavy reliance on the National Security Agency. With the same determination that the admiral had argued against putting out frigates as a naval group screen without air cover, he believed just as strongly that the military could not gather intelligence by electronic means alone. "What I don't understand is how you people missed the MiGs. We've got a billion-dollar satellite up there that's supposed to be able to read *Pravda* and you don't see any fighters?"

"I doubt we missed them, Admiral," replied the NSA representative, James Halpern. "My guess is they were probably ones we already know about but were flown down from Manchuria or across the Yellow Sea from Shanghai. Training's probably done over southern China."

"*Probably* won't do it for us, Jim," said the admiral. "You either know where they are or you don't."

There was another awkward silence. Gray moved to a more positive note. "At least we have the reserves to move. I didn't think we'd get that much from the president, to be quite honest."

"Why not?" asked Admiral Horton. His directness was unnerving to the other combined chiefs of staff. "He can do that. Problem is getting Congress to go along."

"I don't think there'll be any problem there, Admiral," General Gray assured him. "This isn't a Gulf of Tonkin Resolution. It's invasion, plain and simple."

"Maybe, but I'll sleep a lot better when I know your tank boys have stopped them, General."

"So will I," conceded Gray. "What about the *Blaine*?" It was a little tit for tat.

"Blindsided," replied the admiral. "The carrier *Salt Lake City* was in contact with her right up until she was hit. Another frigate is going to assist. If the *Blaine*'s still afloat."

"You keeping Senator Leyland posted?"

"Yes. Soon as we know anything more definite, we'll inform him."

Later, after he was satisfied everything that could be done by the Pentagon was being done, General Gray rose, indicating the conference was over, but as they filed out he asked Air Force General Allet about the extent of the sabotage in South Korea against the airfields. Gray already knew the answer, but it was a pretext to get Allet off to the side. "Bill, we've been caught with our pants down on this one. Everyone here's been so goddamned worried about Europe and the Mideast—" He paused.

"Point is, now we're in it, we ought not to paint too rosy a picture about any short-term victory."

William Allet looked up in surprise. "You mean you don't think the M-1s can buy us the time we need?"

"Oh sure, they'll knock the crap out of those damned tin cans. Point is, I think we need to give those North Korean bastards a lesson they won't soon forget. Senator Leyland's of the same mind."

Allet nodded, but General Gray didn't know whether that meant acquiescence or mere acknowledgment.

CHAPTER TWENTY

BRENTWOOD HEARD THE screams of men trapped in the tangled debris of the red-hot bulkhead, the jagged hole amidships causing the *Blaine* to heel dangerously to starboard as she reduced speed from flank speed of thirty-five knots to slow ahead at five, still taking water. He ordered the port side compartments flooded, trying to bring the frigate back onto an even keel, lifting the punched-in starboard side above the waterline. Over the sound of the men the Phalanx Gatling gun kept up its murderous fire, aided by the *Blaine*'s remaining helicopter attacking the retreating patrol boats as they sped southeast now that they'd crippled the American frigate.

The *Blaine*'s designers, sacrificing weight for high speed and greater maneuverability, had installed state-of-the-art electronics in the warship, but with its armor plate less than an inch thick, the result was a gaping wound in her side roughly twelve feet in diameter, fires raging inside the aluminum superstructure, reaching temperatures unheard of in the slower, heavier ships of old—the decks becoming so hot that fire hoses and the firefighters' boots began to burn. Twenty-three men were killed outright by the explosion of the Exocet against the ship's side, another nine dying from burns and hemorrhaging before the asbestos-clothed rescue crews could get anywhere near them. Four men simply disappeared at the point of impact. But by far the most damage was caused by the toxicity of the fumes as the ultramodern materials in the superstructure, everything from tabletop plastics in the mess to plastic-coated wiring, melted, their deadly poisonous gases spreading in a hot fog throughout the ship, the *Blaine*'s position now visible on SATINT photos as a white blob, so dense that normal infrared was having difficulty penetrating it. Some of the crew had time to don gas masks, but except in the more heavily armored and protected combat information center immediately below the bridge, the curling, tumbling smoke pouring through aisles and ventilation shafts of the ship killed another fifty of the crew. Within five minutes half the CIC watch became nauseated to the point they could no longer properly monitor the radars, which for the most part were either malfunctioning or useless anyway, the Exocet's impact having severed communications within the gutted ship that was now moving ghostlike in the fog. Messages had to be sent by runners trying to circumnavigate the inferno in the belly of the ship, where white jets from the fire hoses crisscrossed, the water immediately vaporizing as it struck superheated metal.

The second missile, eight minutes after the first, hit her port rail, its high orange explosion erupting in front of the bridge, where Brentwood was issuing orders for damage control. The fireball enveloped the bridge, sections of the bulletproof glass bulging like some grotesque metallic monster, the sudden heat driving the exploding glass into the bridge, molten shards instantly killing the helmsman and decapitating the OOD. Brentwood saw a blinding white slab, his body crashing into the deck, hands pressing hard against his face, his fingers awash with the warm blood, the second officer, his clothes lacerated, his left arm shattered, eyebrows burned, stumbling, grabbing the battery-operated megaphone, yelling at Brentwood that they should "abandon ship."

"Very—" began Brentwood. He nodded and blacked out, his face appearing to the second officer as if parts of it had melted, flaps of skin dangling loosely from what had been a face. Hearing the order to abandon ship, the tactical warfare officer one deck below in CIC inserted the key, opening the Clark Meyhew automatic destruct charge, setting the fuse on the Mark II decoder for fifteen minutes.

Only a yeoman, first-class, who had been on the chopper deck astern, assisting the helicopter to land on the fifteen-degree-kiltered deck, noticed, while racing forward along the relatively undamaged walkway, that although the port quarter was a shambles of twisted railing, the decking about it buckled from the missile's impact, the multimillion-dollar ship, though limping badly, was still afloat. He shouted at one man who was tightening his life jacket on the starboard rail not to jump, that their sister ship in the carrier screen, the *Des Moines*, must be coming for them. For a second, the yeoman couldn't see the figure he was yelling at, a gunner's mate still wearing his asbestos hood, looking strangely like a polar explorer, lost momentarily in the churning white smoke that was boiling up from the well deck. The wind shifted for a second, smoke clearing, but the mate was gone. Looking over the side, the yeoman could see several oil-slicked men striking out in an Australian crawl, others breaststroking, the noises of their cries remarkably like wounded seals, heading as best they could through the black scum of the debris toward one of the half dozen Beaufort life rafts that were now bobbing unconcernedly on the oil-smooth chop like huge, contented orange-glow igloos.

In the Uijongbu corridor it was still raining late in the afternoon, the monsoon seeming to set in for the rest of the wet season. As forty M-1 tanks fanned out into overwatch positions, two wingmen or flank tanks overseeing the advance of a middle tank, which in turn would pair up with one of the remaining two and watch over the advance of the third tank, Clemens took delight in watching them spreading out to encircle the PT-76s. The latter were now withdrawing across the paddies, heading for the foothills of the mountain range as fast as they could. Clemens cursed, being unable to move in for the kill in his disabled tank, but did his best on his radio from his defilade position on the reverse side of the hill to guide some of the M-1s through the thick, choking smoke that the fleeing North Korean tanks and infantry were laying down across the darkening green of the countryside and around the enormous flooded ditch that the NKA had blown out of the ground to serve as an antitank defense. Through the noise of the battle and that of his own tank's motors Clemens could hear the American commander brusquely alerting all tanks to go around the ditch that Clemens had warned was not just another paddy. The commander also ordered the tank crews to ready their fording equipment in the event that any of the other normally waist-deep paddies, flooded by the monsoon, proved too deep to negotiate on tracks alone.

Within minutes of General Kim ordering the PT-76s to withdraw as fast as possible, the M-1s were in trouble in the paddies, as Kim knew they would be. It was the culmination of a plan he had fought so long and hard for in Pyongyang, arguing against all the norms of military logic for a monsoon offensive rather than a winter or spring attack, when the ground would still be hard enough to allow for a rapid armored advance. He knew, as did everyone else, that the Abrams M-1 was quite simply the Cadillac of tanks, that next to them the PT-76s appeared as "Model Ts," in his aide's description, the old 76s tough and reliable in their own way but with nothing like the speed, armor, or shooting power of the American tanks. The Americans had long known this, and what Kim had banked on was the Americans' very willingness to engage the PT-76s. He knew the M-1s, like the 76s, were also capable of fording waterways with the huge flotation skirting, but he also knew something else, something that came from a simple mathematical equation in the art of modern war.

The Abrams M-1 had been designed to go up against the equally heavy Russian T-90s and T-80s, a strategy that had not bothered itself with the old "tin can" PT-76 and so was not prepared for the difference between the ground pressure exerted by the fifty-four-ton M-1 and the ground pressure exerted by the fourteen-ton PT-76s. The difference, the M-1 exerting a ground pressure of 210 pounds per square inch as opposed to the mere 150 for the PT-76s, meant that each fifty-four-ton M-1 now had to spend vital time once it was stuck in the soft flooded ooze of the paddies while its flotation skirting was unfolded. In those few minutes the much lighter PT-76s were still free to maneuver. This single fact proved deadly for the Americans as Kim closed the trap.

In the mutual slaughter that Kim had predicted to his "dear and respected leader," the M-1s' laser range finders and 105-millimeter cannons put out an awesome rate of fire, crippling the PT-76s. The latter, having served Kim's purpose of drawing the Americans into the paddies, were now sacrifical lambs as crack NKA infantry, able to set their aim on the temporarily stationary M-1s, delivered the knockout blows with everything from the old but reliable inventory of wire-guided Russian Saggers to sophisticated one-shot, one-time RPG 18s and 22s and the AT-5 Spandrels, whose hit probability at five hundred meters exceeded 74 percent. Contrary to what those outside the military believed, Kim was well aware of the fact that tanks, even the most modern ones, advancing without infantry support were much more vulnerable than was generally supposed and were one of the easiest targets to destroy, providing they stopped long enough for you to sight them in. The M-1s were able to spot some of the RPG 18 launchers because of their back-blast, which in the paddies created an aerosol spray of water behind the launchers. Even so, the M-1s, the giants in the contest of armor against armor, were for these few minutes at their most vulnerable, standing still in deep, muddy water, many of them knocked out of action by the inflatable skirting catching fire and in effect baking the four-man crews, especially the drivers, whose positions forward of the turret offered them less protection, and many of whom were machine-gunned as they tried to escape fume- and flame-engulfed tanks. Many of the gunners on the left flank, from Clemens's company, most of whom hailed from the San Diego area, were among the first hit. Some of them drowned as they fell badly wounded, "flopping around in the water like winged ducks," as Kim wrote in his glowing report to Pyongyang.

Few of the M-1s were holed by the NKA's antitank weapons, the crews killed by spalling, the ricocheting of equipment knocked loose inside the tank by impact, creating a deadly hail of shrapnel whizzing around inside the tank like a scythe amid chaff, many of the Americans blinded from shattered periscopes, others burned alive despite the automatic extinguishers, which could do little if any of the M-1s' fifty-five rounds of high-explosive and armor-piercing shells were hit. A dozen or so of the American commanders who opened the hatch to snatch a quick look at the all-round situation were dead within seconds from the NKA's hail of infantry and sniper fire, the Americans slumping over the hatch, from which they had to be either dumped quickly overboard before the hatch could be closed or dragged back down, bleeding, often blocking vital steering and gunnery controls.

Kim's infantry paid heavily, as well as the PT-76s, thirty-four of his tanks destroyed outright by the deadly accurate M-1s, which, for as long as they lasted in their stationary, or near stationary, positions, delivered as much as they got. But there were many more PT-76s than M-1s, and soon, under the combination of the more mobile PT-76s and heavy infantry antitank rocket and missile fire, all the American armor, except for one platoon of three tanks, was knocked out of action.

Kim knew that if the Americans were allowed time to retake the ground by air strike, then their logistical genius for quick repair would come into play, and tanks that had now been taken out of the battle because crews had either been killed or their M-1s had lost a track could be in action again in six or seven hours. Kim had never forgotten his father's stories of how North Korean commanders had stood, mouths agape at the incredible speed with which American Seabees had cleared rough-timbered ground and laid down mesh airstrips within twelve hours. Accordingly, Kim sent in mop-up squads of demolition experts, some of whom were cut down by the few remaining M-1 crews manning the commanders' and the loaders' machine guns before they, too, were silenced. The NKA squads placed plastique on the tanks with times fuses set for forty minutes, giving all NKA squads time to clear the area.

The Uijongbu corridor exploded in a roaring cacophony of black-orange balls spewing into the dark and monsoon-riven sky as the plastique and rounds aboard the tanks exploded. Kim would have liked to use the tanks, but the NKA didn't have the supporting ground crews for each tank that the American army had now instituted, or enough electronic experts to service the sophisticated equipment on board. Only two M-1s were spared so that when the day came that the South had been crushed, the tanks could be forwarded to Kaesong, where they would be airlifted by the big Russian Condor transport to Moscow for thorough dissection by the Russian army.

Seoul was Kim's, and as tens of thousands of NKA regulars, reservists, and artillery closed in the ring of steel about the capital, it was an ashen-faced Cahill whose officers had given him firsthand observation reports of the carnage in the city above—"like tossing grenades into packed rooms," one of them had said. Cahill was now confronted with the inevitable. By midnight Korean time the situation was so hopeless that, amid the shambles and smoke of what had been one of the most sophisticated electronic communication centers in Asia, the American general asked permission from Washington to surrender Seoul into enemy hands.

General Gray replied that any delaying action would help the reserves on the way from Japan, while SATINT suggested the NKA were pausing to ford the Han across pontoon bridges. Approximately a half hour later Cahill received more reports that thousands of NKA regulars, wearing normal peasant and worker garb, had, hours before the last bridges were blown, infiltrated the columns of refugees and were now causing further chaos and thwarting South Korean artillery batteries trying to reach the Han from the South.

At 1:00 A.M. the U.S.-ROK command, in effect Cahill, was given an ultimatum: seven hours to surrender or "answer for all the consequences."

To the rest of the world now watching, to America holding its breath, the seven hours granted Cahill to make up his mind—Washington's mind—was puzzling. In Europe it was being touted by commentators, especially in the Eastern bloc, as a "generous" act of humanitarianism in modern war; obviously Kim was giving time for as many civilians to escape over the few remaining bridges as possible. In fact, Kim had given Cahill seven hours for one reason only: By that time it would be dawn, the usual time for the U.S. and ROK flags to be raised, so that Kim and his troops, as all the world looked on, could be photographed in full light as the American and South Korean flags were taken down. He could see no reason why Washington would prolong his artillery's systematic destruction of the city. He was correct.

The flags were hauled down, furled, and General Cahill, but not General Lee, was instructed to hand the American flag personally to General Kim. Without looking at it, Kim peremptorily handed it to an aide, who set the Stars and Stripes on fire, lifting it high on a bayonet, waving it joyously for the cheering NKA army and the world to see.

CHAPTER TWENTY-ONE

"IF WE DO nothing, it will be interpreted as weakness. Confidence in our treaty obligations in Europe would be seriously under—"

"I agree, Comrade," said the premier. "But our Far Eastern Fleet has already put to sea. Is that not message enough?"

"No," answered the defense minister, a stocky Georgian leaning back in one of the leather-backed chairs in the premier's extraordinarily long, rectangular office, trying to compose himself. It was no use; he was leaning forward again, his impatience and anxiety evidenced by his nicotine-stained fingers fiddling unconsciously with a thick glass ashtray. He looked across at his comrade from the ministry of the interior, then left at the premier in his seat at the top of the T-shaped conference table, and spread his hands on the green baize in a gesture of accommodation. "What I wish to point out, Comrade Premier, is that sending our Eastern Fleet out of Vladivostok is standard procedure whenever the American Seventh Fleet enters the Sea of Japan. The Americans know this. Our Warsaw allies know this. It does not send a strong enough message, in my opinion." His fingers tapped hard on the table. "We need to take a step which will send a strong message that the Sea of Japan cannot be used as an American lake, as—as an operational area from which to bombard North Korea. As it was last time." The defense minister turned and pointed at the Pacific Rim map. "It would be the same as if we sailed into California waters. Can you imagine?"

"I understand this, Comrade," put in Premier Suzlov, "but I think a strong enough message is being sent to everyone with the deployment of our Eastern Fleet. The public in the Warsaw Pact countries don't know it is standard practice. They will see it as a *very* strong measure on our part. So will the American public. The important thing, Comrades, is that *we* know the Pentagon understands it is normal procedure.''

"But the Pentagon suspects us of being in collusion with North Korea," answered the defense minister, undeterred.

Suzlov looked about the table for an informal poll. "Our dear comrades in Pyongyang have made a grave error by invading South Korea. They did not consult us. Nor inform us of the time—''

"They never consult us," grumbled the marshall of the air force. "They are a law unto themselves.''

"Law?" skoffed Admiral Doldich. "They have no law. They are ruled by an overfed ego in a Mao suit. He has placed us in a very dangerous position, Comrades. If we do not support him, we are seen as deserting a socialist brother, an ally. If we do support him, we risk widening the conflict. We have enough trouble now with every republic from here to Turkey clamoring for semiautonomy.''

"Which means *autonomy*," chimed in the minister for defense. "This is precisely why, Comrade Premier, I urge a tougher stand here. Comrade Doldich is quite correct that everyone in our fifteen republics is watching us. If we falter, if Moscow fails, appears weak-kneed on this, it will only fuel divisive elements from Estonia to Mongolia. To make matters worse, Beijing is being foolish again about their borders.'' The minister meant Damansky Island and the far eastern border along Outer Mongolia, which fronted the Soviet Union's back door and laid claim to more than a million Soviet troops permanently stationed there. Particular knowledge of this area lay with Kiril Marchenko, a colonel in the service of the STAVKA, the general headquarters of the Soviet Supreme High Command, one of the specialist advisers invited on occasion to Politburo meetings. It was to him that Suzlov now turned for his opinion. The STAVKA suspected that Doldich's dispatch of the Soviet Far Eastern Fleet had less to do with sending messages to the U.S. Seventh Fleet and more to do with Doldich wanting to have the fleet "at sea" rather than risk having it bottled up, the fate of the German surface fleet in both world wars. But it had been STAVKA's intelligence reports suggesting the USSR show its muscle to

China that had caught the premier's attention. He favored giving a clear signal to Beijing that, despite the post-Gorbachev rapprochement between the two countries, Moscow would not yield to "unrealistic" demands for more territory on the far eastern border. If China became obstreperous, which it seemed to be doing, as it had done in the sixties and seventies, then the Russian navy was there, ready to pulverize anything that moved against the vital Soviet security zone between the Manchurian border and the USSR's huge year-round ice-free port of Vladivostok.

Marchenko was about to begin speaking when the defense minister put up his hand, his rows of "Afghanistan service" ribbons catching the early morning light from the wood-paneled beige walls and the portrait of Lenin directly behind him. "If I can return for a moment to the matter of our European allies. If we do not take stronger measures, if we do not come quickly to the aid of our North Korean comrades . . ." He held up his hand like a traffic cop to stop oncoming objections until he was finished. "Despite their rashness, if we, as the leaders of the socialist world, do not come to their aid, our Warsaw Pact comrades have every reason to doubt our commitment to *them* in the future—to socialist solidarity all over the world. And if, God forbid, there is ever an attack against our Soviet Motherland and we need them to aid *us, and quickly*, then our reactions now to our comrades in North Korea will decide the issue." No one stirred or even raised an eyebrow at the defense minister's use of "God forbid"—Gorbachev had used it all the time during his years in office. "The question for them, Comrades," continued the minister, "will be, why should we fight to protect the Soviet Union's borders? Why should *we* be the buffer zone in Eastern Europe if the Soviet Union does not move to aid one of our brother socialist countries in North Korea?"

Colonel Marchenko made eye contact with the most powerful man in the Soviet Union, though he knew that as colonel he was not entitled to speak at such a high-level meeting unless specifically invited to do so. STAVKA not only supported the premier's position regarding the Eastern Fleet and China but also shared the defense minister's concern. Marchenko's intelligence sources had painted an even more frightening picture of the spread of underground dissident movements within the republics. Gorbachev and his precious *perestroika* had lifted the lid from the can of worms. Encouraged by his call for reforms, the underground, indeed everybody, from the decadent rock and roll stars—"Bye-bye, Miss American Pie"—to literary professors quoting the leftist traitor Orwell, had started yelling about freedom. Freedom for what? Marchenko wanted to know. Freedom to be degenerate—sex, drugs, and mayhem in the streets? The KGB had over three million files on "latent dissidents" who had surfaced in the Gorbachev years, in addition to those already known as "dubious characters," all of them just waiting and watching like delinquents for the first sign of the parents' discipline weakening. But even this was as nothing compared to the threat posed by the clique of "delinquent generals" on overcrowded Taiwan. Agents who often sent in conflicting assessments were as one in their reports that the Taiwan clique were positively drooling at the prospect of a Soviet-U.S. conflict and/or Chinese preoccupation in a Sino-Soviet conflict over Far Eastern borders. And the present North/South Korean War was a prime time, one that would not come again, for Taiwan's American-equipped navy, vastly superior to Mainland China's, to retake some of the offshore islands.

"Yes, Colonel?" said the premier encouragingly, struck by the officer's fearlessness in the presence of so much top brass.

"Mr. Premier," began Marchenko, "I suggest that the Far Eastern Fleet be authorized to carry out AIRTAC exercises in the South China Sea, supported by amphibious elements from our base at Cam Rahn Bay. This would be enough of a departure from normal Far Eastern Fleet maneuvers to give a clear signal to the Americans but still be part of the usual exercises so that it should not unduly alarm Washington. It would also be far enough south of Korea not to be viewed by the Americans as Moscow interfering with Korea. It would also mean our fleet would be strategically positioned between the offshore China islands and Taiwan to signal the opportunist generals in Taipei, very clearly, that we will not tolerate an attack upon Mainland Chinese territory which would further destabilize our Far Eastern border situation. This would also put us in Beijing's favor and may reduce tensions on the Far Eastern border itself."

Premier Suzlov nodded, pondering the suggestion. "A double play?" he said, using an American baseball expression he had picked up years before during his stint as KGB head of station in Washington. Suzlov turned to the Politburo. "Well—?" his gaze resting on the ranking minister of defense. "Ilya?"

"Yes," said the Georgian, "but we must still be seen supporting Pyongyang. Not to do so will only encourage the Americans to become further involved. They already have reservists en route from Japan. The problem will only—"

"A volunteer force," cut in Marchenko. "From all socialist countries. We could fly them out from Berlin." The STAVKA colonel realized this time he'd not only spoken out of turn but jumped the gun, butting in on the defense minister's response. But, as the British said, "In for a penny, in for a pound." *"Dobrovol'tsy"*—"Volunteers," he continued, "from all socialist countries would absolve Moscow of any direct intervention, yet we would all be aiding our North Korean comrades." Suzlov looked quickly about the room as Marchenko added, "Socialist solidarity." It was heartfelt for the colonel, no mere slogan. He hated America—a "mongrel mix of races," he called it, his father, an adviser in North Vietnam, killed by the American Division outside Khe Sahn.

The defense minister turned toward Suzlov, unable to decide whether the upstart colonel deserved a damned quick put-down or a promotion. Never would have happened before Gorbachev, of course. Gorbachev was the one who had encouraged this kind of "spontaneous" exhibitionism the colonel was indulging in. Still, the defense minister saw the colonel's point was supporting his. "Where," the defense minister asked Marchenko, "would you get these *volunteers* from?"

"Cuba. East Germany—" Marchenko paused. "We were still able to get volunteers for SPETSNAZ for Afghanistan—"

"Afghanistan was a mistake," said the air marshal. It was the party line; it had also cost the air marshal many pilots and put pay once and for all to the idea, as the Americans had learned in Vietnam, that air superiority alone could decide a war. Unless you dropped atomic bombs.

"Yes," agreed Premier Suzlov, "but the colonel is correct. We did get volunteers despite the unpopularity of the war. And for SPETSNAZ." He was referring to the toughest, most hazardous duty of all, the special forces.

"And," said the colonel, pausing, a little more cautious, "we could offer some kind of inducement. Recognition by the state.'' He meant bonuses, of course. Cash, coupons for the specialty stores normally open only to the Party.

Suzlov saw the heads nodding. Lenin would have understood perfectly. He was no fool, this colonel.

CHAPTER TWENTY-TWO

THE MOMENT NATO HQ in Brussels had heard that the North Korean army had crossed the DMZ, all units along the NATO front, from Jutland in the north to Austria in the south, went from normal ''alert'' status to ''military vigilance.'' If the Soviet–Warsaw Pact armies were to invade the West, there were three ''most probable'' points of entry. The first was the Fulda Gap. Here the end of the barbed wire East German funnel thrust into West Germany between Mount Fichtel Gebirge in the south and the Harz Mountains in the north, where Hitler had used thousands of SS-controlled slave laborers to build the V-2 rockets in the deep underground tunnels near Nordhausen. The second probable point of attack was in the far south near Burghausen on the Austrian–West German border, where tanks could race through the Hof Corridor, utilizing the plain about the Danube northeast of Munich. The third most probable point of entry was in the far north along the Elbe on the North German Plain. Here attacking forces would almost certainly try to isolate Jutland while racing to Hamburg and Bremerhaven to cut off the vital ports needed by NATO for the massive U.S. reinforcements, which after one week of war would have to start pouring in if NATO was to have any hope of stopping the Soviet–Warsaw Pact juggernaut.

It was clearly understood, as laid out in the UN Charter, that an attack on any of the sixteen NATO signatories would be considered by the United States as an attack upon it. It was possible, of course, that the Soviet–Warsaw Pact forces could attack all three points at once. This was considered highly unlikely, however, as even with the Soviets' overwhelming numerical advantage of forty thousand main battle tanks against NATO's twenty thousand, six combat soldiers to NATO's one, it would mean the S-WP splitting their forces against the numerically inferior but mechanically superior NATO armor.

In Fulda Gap, fifty kilometers northeast of Fulda on the central German front, stood "Tower Alfa," NATO's forwardmost observation post along the "trace," the fifty-yard-wide DMZ that stretched for 550 miles, separating the two Germanys. Here the Americans of the U.S. Fifth Army's Blackstone Regiment had the responsibility of guarding the fifteen miles of the arrowhead-shaped sector.

Not far back from the tower, in the platoon hut, they had been through the routine quite literally a thousand times, snatching up rifles on the double, the white and turquoise walls of the hut a blur as they raced out to man the machine-gun posts and the M-1s, never knowing when the choking "horn" sounding the jump from normal "alert" to "military vigilance" to "reinforced alert" and finally to "general alert" meant a drill or the real thing. But the American's were always keen no matter how many times they'd been through it, as mindful as Hans Meir, the Wermacht liaison officer serving with them, that if the Russian T-90s came bursting through the Gap, the men at Alfa would be at the point of maximum danger, that a world war might stop or proceed, depending on the swiftness and the bravery of their response. Meir had never forgotten his grandfather telling him how if just one officer, one man, had stood firm on the bridge across the Rhine in that fateful summer of '36, World War II might never have begun, Hitler having secretly ordered his men not to proceed if resistance was met. After that, Meir's grandfather had explained, it was all downhill to Munich.

Being stationed at Alfa, therefore, was an awesome responsibility for young men, and their commander, Lieutenant General Sutherland, never tired of telling them that their greatest danger lay not in the "bean count," the two-to-one numerical superiority of the Russians and Warsaw Pack armies in men, tanks, and aircraft, a count used to get U.S. congressmen to vote for better, more sophisticated weapons, but in the very dullness of the Alfa routine, the general constantly lecturing them on the need to be ready. When they had heard of the NKA invasion, however, and the continuing debacle of American arms in South Korea, General Sutherland said no more. The sight of their flag burning on jubilant East German television was enough.

For Hans Meir the threat was more pronounced, for while his parents lived in West Germany, in Frankfurt, only seventy miles southwest of the Fulda Gap, his sister, her husband, and two children whom he had never met still lived in East Berlin, a hundred miles northeast of Fulda. Meir knew that if war ever did break out, the vast sea of the Soviet-WP armies would immediately close the lonely, narrow one-hundred-mile-long corridor that ran from West Germany to the "island" of West Berlin. Hunkered down in the most forward machine-gun post by the lonely, cream-colored watchtower overlooking the pleasant, rolling countryside beyond Fulda Gap, Hans Meir prayed, though he was not sure whether he believed in God or not, that the *Korea-Sache*—"Korean business," as they were calling it— would soon be over, that tempers would cool, logic applied. As darkness descended over the "trace," he saw the lights of East German farms come on like tiny stars in another universe.

CHAPTER TWENTY-THREE

A LONG, SHINY black Zil swept out of Sheremetyevo Airport in a gray dawn, heading south through the Green Belt before reaching Moscow's outer ring, passing through the flickering sunlight of the Garden Ring Road, arriving a half hour later in Dzerzhinsky Square outside the Children's World. The route was a small variation in Director Chernko's routine, for normally the head of the six-hundred-thousand-man KGB would have been driven either to the front of, or behind, the ocher-colored facade of Number 2 Dzerzhinsky Square. Chernko preferred the older seven-story All-Russian Insurance Company building fronting on Lubyanka Prison rather than the ugly, modern headquarters in the suburbs. Chernko did not intend to make any mistakes, and small changes in his arrival routine were part of his plan to keep any potential assassin off balance. He knew that to confound madmen wanting to kill you was a difficult thing even in the USSR, but there was no point in helping them. Besides, it was a good example to set for those of his agents still in training. Variation in standard procedure was the most difficult thing to imbue a good agent with. After all, they'd been brought up in a world of *apparatchiks*, "bureaucrats," for whom conformity to the rigid system was safety. And there was another good reason for obeying strict procedure: Orthodoxy, in terms of chain of command and basic trade craft, was essential if the First Directorate's one hundred thousand agents abroad were to function with any discipline. But now, summoned home for an "extraordinary" meeting of the Politburo when the Korean War broke while he was attending a high-level USSR-U.S. "peace study group" in Zurich, Chernko knew that if his supposition about what the Politburo meeting would be about was correct, then the agents he wanted now would be the most unorthodox, the most willing to adapt to quickly changing circumstances. Indeed, the first order the tall, ascetic-looking Chernko gave his aide, upon reaching his seventh-floor office, was to bring the files of all operatives in CANUS—Canada and the United States—who had been "disciplined" in the last two years for exceeding their authority and/or violating "operational procedure." As the major went to "Records" on the fourth floor, Chernko pressed the buzzer for tea. The Swiss could make superb coffee, but their tea—it was as weak as a congressman's principles. Chernko liked his tea "so dark," he had told his staff, that "I can fill my fountain pen with it." When it arrived on a silver salver, he took a cube of hard sugar, placed it between tobacco-stained teeth, and sipped the steaming brew, glad to be home, looking out over the square.

The breezes blowing softly down from the Lenin Hills to the west carried a smell that for Chernko was as nostalgic as watching the shivering leaves of beech in the last golden light. It brought a sense of sadness, of time lost, never to be recovered, of long summer days when he and the other privileged *nachalstvo*—the "establishment"—enjoyed their cool dachas at Uspenskoye by the Moscow River. Yet as well as nostalgia for the past, the smell of autumn carried with it the feeling of hope—that the Americans would see all the signals; the idea of the volunteer force released by the Politburo and quickly picked up by the Western media; the presence of the Eastern Fleet in the China Sea; the maneuvers off Cam Rahn Bay. "Amphibious maneuvers," the phrase that had been suggested by Colonel Marchenko, had precisely the right tone, telling the Americans that if the Soviet Union wanted to land on the Korean peninsula in force, it could.

Chernko's aide returned with the list of agents. There were seventy-three who'd been cited for infractions, the most common being excessively high expense accounts.

"Back when the American President Carter proposed a tax on the two-martini lunch," Chernko told the major, "it raised more hackles with our Washington head of station than with the capitalists on Wall Street."

The major said he was surprised by the number of infractions.

"It's the American air," Chernko said, half-jokingly. "It encourages rebellion against rules." The major gave a noncommittal nod. Sipping his tea, Chernko ignored the cases of inflated expense accounts and odd "unadvised" liaisons with street women. Finally he had ticked off fourteen names, eleven men, three women, who had been singled out for *uprostit' delo*—"cutting corners"—a phrase that meant that in these cases the agent had, upon encountering unexpected difficulties, violated strict precautionary measures, a series of obsessionally followed multiple checks against entrapment or enemy surveillance.

"These are the ones," Chernko said. "Arrange special meets," he instructed the major. "No code transmits. No pouch. By hand. Vancouver the point of entry. Advisories to Toronto and New York."

"Yes, sir. Special couriers?"

"No. We'll use our two best *Izvestia* people to take the instructions. Journalistic cover." Chernko thought for a minute, picking up another cube of sugar. "Travel pieces—'This Land Is Your Land—East To West.' That sort of thing." He saw the major didn't get the play on words of the popular American folk tune. He toyed with the sugar cube as he looked down the list, tapping the yellow paper. "Vancouver is sister city to Odessa, isn't it?"

The major was embarrassed—he didn't know. "Ah—we have a consulate there."

"Yes, it is," said Chernko, answering his own question. "Odessa and Vancouver in the fall. A nice travel piece."

CSIS, the Canadian Security Intelligence Service, had Montreal and Toronto pretty well bottled up, and the CIA, of course, had New York well covered. But Vancouver was known within the First Directorate as not only an easy entry point to the United States but also as one of the most beautiful cities in North America, which would make the journalist cover more convincing.

"How many agents will be going through?" asked the major. "I'll start preparing the paperwork."

Chernko drank his tea until the sugar cube crumbled under the pressure. "None," he answered the major. "No, we don't want to start putting in new people now, Major." He gave an enigmatic smile. "It would upset the CIA!" He glanced up at the clock below the mounted emblem of the gold shield and sword. "I have to meet with the premier in half an hour. Two cars."

"Yes, sir," answered the major, picking up the phone to order the cars while deducing that if delivery of the Directorate's message necessitated the Directorate using two of their top journalist agents in *Izvestia* and not even entrusting it to top secret code traffic between Moscow and the Soviet Embassy in Washington, then the message must be one of the most important Moscow center had sent in years.

It was.

As the Zil picked up speed across the square, the twin flashing red lights on either side of the Spassky Gate changed to green, the guards coming to attention and saluting. Chernko looked across at the major.

"Major, do you remember Rust?"

The major thought hard—was he one of the agents in place, a sleeper in America—or was it Canada?

"The German hooligan," said Chernko. "He landed a plane in Red Square. *Here!*"

"Yes, of course."

"About this time of morning," added Chernko as the battered white Volga sedan they were riding in behind the decoy Zil limousine slowed before entering the Kremlin's sulfur-colored inner sanctum. From here Chernko could see the red star high above the bell tower of Ivan the Great. Chernko allowed himself a smile of satisfaction. "Of course, that hooligan performed a great service to the revolution."

The major, who prided himself on a photographic memory, was having a bad day. "How is that, Comrade Director?"

"He allowed Gorbachev to fire the minister of defense and his clique, all the old *bednyagi*—"farts"—in the party, the ones resistant to change." Chernko saw that another Zil was entering the courtyard. It was Admiral Doldich. All stars and flags. A wonder there wasn't a brass band.

"And," Chernko continued, "the hooligan helped Comrade Gorbachev break down opposition to *perestroika* and opposition to detente with the United States. New openness. Gorbachev and Reagan. A breath of fresh air, you see. Just what we needed."

The major was nodding; it had indeed been the *schastlivoe vremya*—"happy time"—for the KGB. "So that now," Chernko said as the door opened, "we are ready." He nodded to Doldich. "Morning, Admiral."

"Comrade Director," responded the admiral. The navy man would not have made a good agent, Chernko thought—his expectant look, his need for help, was written all over his face. As they walked to the premier's office the admiral announced forlornly, "The fools have attacked an American frigate."

"I know," replied Chernko. "Satellite pictures show it is still burning. Is *your* fleet steaming south?"

The admiral looked up warily at Chernko. "*Your* agents can tell you an American frigate is afire, but they miss my whole fleet. How is that?"

Chernko started to smile but saw the admiral wasn't in a joking mood. "Of course we did not miss it," he explained. "But I mean have you reduced speed?"

The admiral glanced about as they continued walking toward the premier's office. "Yes."

"Will you stop it?"

"No. The Americans would see that as indecisiveness."

"Quite correct," commented Chernko. Foolhardiness was

one danger—indecisiveness the other. "Yes," he assured the admiral. "The best thing is to steer a middle course."

The admiral frowned with concern. "This is very difficult to do. With Japan on your port side and Korea on your starboard, there's not much room."

"No."

One of the premier's aides rushed by them, a sheaf of cables in a folder.

"I've never seen them move so fast," said Chernko.

"Now what's happened?" worried the admiral aloud.

"The damned Cubans," interjected a voice from behind. It was the air marshal. "With that loudmouth Castro gone, I thought they'd settle down. But no. I tell you, their biggest export is trouble."

CHAPTER TWENTY-FOUR

IN SEOUL, THIRTY-THREE thousand American soldiers, in the biggest mass surrender since Corregidor in 1942, made the short but humiliating march across the NKA's pontoon bridge, built in the shadow of the old Chamshil Iron Bridge, toward the shell-pocked Olympic Stadium, where they were separated according to regiment. Fifty-two senior officers were weeded out as quickly as possible, the NKA interrogators insisting, upon pain of death, that the officers sign confessions of "criminal imperialistic aggression against the Democratic Republic of Korea."

Only three signed. Seventeen of the others, described over the loudspeakers as "recalcitrant warmongers," were shot in the dressing rooms below the baseball stadium, their bodies dumped on the diamond from a captured U.S. Army truck that roared about the field of six thousand prisoners, the blood of the murdered men dark on the poorly lit artificial turf.

One of the bedraggled and shell-shocked soldiers, a U.S. private from the Eleventh Division, clutching a military blanket about him, watched each of the bodies fall limply onto the tread-torn diamond, wondering aloud why they had not killed all of the officers who had refused to sign.

"Why seventeen?" he asked numbly of no one in particular.

"To scare us," answered an ROK lieutenant who had torn his intelligence corps patches off moments before he was captured. "To show the others what happens if they hold out. To show us what will happen if we resist."

The American looked around the dimly lit stadium, few of the lights working after the artillery barrage, "There are ten times as many prisoners here as guards."

"But they've got the guns," said the ROK soldier.

One of the bodies they saw was a major, his collar stud standing out even in the poor light. He looked as if he were smiling, but it was an illusion—a death mask of pain—tortured before they shot him. The private noticed for the first time that the white U.S. stars on the truck's doors had been smeared red. When it drove out of the stadium, there was a silence heavy with the stench of pickled cabbage and excrement. The NKA had forbidden use of any of the toilets behind the stands, and most of the wounded were now crowded into the north corner beyond the diamond, where they had to defecate into a shallow trench.

Soon the loudspeakers were ordering all intelligence officers to report to the stands.

"Fuck you!" the ROK lieutenant shouted in perfect English, and got a weak round of applause.

CHAPTER TWENTY-FIVE

HEADING OUT INTO the mid-Atlantic, the U.S. nuclear submarine USS *Roosevelt* received its digitized "burst" message via its trailing VLF aerial. The message lasted less than two-hundredths of a second—too short a time to alert an enemy sub to vector. The *Roosevelt*, one of the Sea Wolf class II dual-purpose (Hunter/Killer and ballistic missile) subs, normally patrolled deep at a thousand feet below the surface, but in order to be on station to receive the scheduled burst message through its very low frequency aerial, the sub's captain, forty-three-year-old Robert Brentwood, had had to bring the *Roosevelt* to 150 feet. The message told Robert Brentwood and his crew two things: They were "spot on" their prearranged patrol route, and the "balloon had gone up" in Korea. This meant that they, like all NATO units, were now on the second stage of the four-step-alert ladder. But as to exactly what had happened in Korea, they did not know, nor, under the CNO's standing orders, was it necessary for them to know what had gone on over seven thousand miles away. Besides, the coded messages, in order to keep the location of the American subs secret from the Russians, had to be kept as short as possible.

"Your brother's out there somewhere with the Seventh Fleet, isn't he, Captain?" asked the executive officer, Peter Zeldman.

"Last I heard," answered Brentwood. "GM frigate."

"Your kid brother might end up in action before you do, Skipper."

"Hopefully, Pete, none of us'll end up in action," said Robert thoughtfully, adding, "My guess is it's some border incident on the DMZ. Probably blow over in a few days."

Zeldman wished he hadn't said anything about the skipper's brother—he'd merely meant it as a bit of conversation, something that was in short supply on a six-month patrol back and forth across the Atlantic in various attack and defensive patterns between Norfolk, Virginia, and Holy Loch.

The immediate question for the crew was, how long would it be before they were in Scotland? Robert Brentwood knew that, next to being home in the United States, the high point for most of them was refit and supply at the Scottish sub base. While the repairs at Holy Loch were usually minor, not requiring a stay of more than a day or two, sometimes they had more shore leave than expected if he decided a new coat of anechoic, or "sponge," paint, as the crews called it, was needed. The red paint absorbed active sonar pulses sent out by sub chasers and so denied them any return echo or at least diluted the echo so much that it was too weak to be of any use to the enemy. Painting the bottom of the 360-foot-long, 42-foot-wide sub that served both as an attack submarine and a Trident II balistic missile carrier was a job that took two weeks in the dry dock, affording the crews, alternating shifts, at least a week's liberty, even longer if the sub's barnacle-encrusted exterior had to be scraped down and new primer applied. However, with *Roosevelt* on second-stage alert, Executive Officer Zeldman had a hunch none of them would be seeing Holy Loch for a while.

He put the question to Brentwood, who said, "We'll see."

"Skipper's not exactly bursting with information, is he?" said the third officer.

"He's got his reasons," answered Zeldman, not permitting himself to be drawn into taking sides. But the lieutenant did have a point.

Zeldman couldn't quite figure Brentwood out either. He was one of those men, Zeldman thought, who seemed to be born old. It wasn't that he looked old, despite slightly graying hair, but rather that he had an unflappable manner and deep-set brown eyes with a penetrating quality about them that constantly made you feel he knew what you were going to say before you said it. Luckily, however, unlike some people, who thought they knew everything, Robert Brentwood wasn't the least bit arrogant or impatient with others, the kind of man, Zeldman concluded, a boy would be lucky to have as a father. But Zeldman doubted if the skipper would ever be one. He wasn't married, unlike most of the crew, or even engaged.

Zeldman had put it down to Brentwood's obsession with his work. He was first and foremost a submariner—everything else took second place, so much so that Zeldman had become convinced that though Robert Brentwood came across as a strong, silent type, he must be driven deep down by a burning ambition to surpass his father's achievement as admiral. Zeldman was quite wrong about this—Robert Brentwood had no intention of trying to surpass his father's reputation. He merely loved submarines. Always had. Years before, when Lana and David had asked him how he could possibly sign on for "coffin duty" in what was essentially nine titanium-alloy spheres welded together and covered in a superstructure that at times looked stronger than it really was, Robert Brentwood had no good answer. All he could tell them was that he did not share their fear of death by sudden implosion. Death was inevitable; the way you went, he told them, was beside the point. And what Zeldman had at first taken to be his quiet, all-knowing air was nothing more than shyness.

Brentwood was conscious that his reticence was often misinterpreted by the men on the *Tennessee* as aloofness, a kind of carefully constructed emotional shield between captain and crew so that orders were less likely to be questioned in times of crisis. With women it was different, for he discovered that most of them intuitively understood that his shyness was not a deliberately erected barrier or a lack of feeling but rather the quietude of a man confident in himself and his work, a man most of them liked, but one whose most passionate affair was with the sea. A pretty young Englishwoman in her late twenties, during one of the welcome-ashore parties arranged by the British Admiralty at Holy Loch, had once unabashedly told Robert Brentwood over her half pint of Guiness, "I'd like to go to bed with you."

"Thank you," he'd replied.

"It's so, then?"

"No. Sorry."

The woman had appeared stunned. "Why ever not?"

"Not before marriage."

The girl, "a smashing blond bit," as his English host had described her to Brentwood before her arrival, had spat out half her Guiness in astonished gaiety at the American's joke. When she saw he was serious, surprise quickly gave way to anger. "You're not a sailor." It was delivered with all the finesse of a depth charge, meant to shake the very bulwark of his masculinity.

"I can assure you I am, ma'am."

"Don't 'ma'am' me! Are you queer?"

"No, ma'am."

"Jesus—you're worse than Bing Crosby," she said, pushing a napkin at him to wipe off the chocolatey stain of Guiness.

Brentwood had taken the train to London a few days later for a bit of sight-seeing and had ordered a biography of Bing Crosby at Marriage's bookstore.

"Back list, sir," the manager informed him without bothering to consult *Books in Print*. "We could have it in five days."

Brentwood told him he wouldn't be able to pick it up for about three months but offered payment.

"That's not necessary, sir," the manager informed him. Brentwood ordered the book anyway, mad at himself for having mentioned how long he'd be away, the length of his next patrol, a breech of security for which Brentwood roundly chastised himself.

A few days later, when he returned to the ship, then in the floating dry dock, he asked Zeldman one morning, "You know anything about Bing Crosby?"

The executive officer had shrugged. "Big shot in the forties and fifties, I think. A crooner."

"What's that?"

"A singer."

"What kind of singer?"

"Not sure. My Grandpa used to talk about him. Love songs, far as I can recall. Pretty slow."

"Was he ever married?"

"Think so," said Zeldman, who'd been pouring the coffee in the galley at the time, his voice rising above the hammering echoing through the *Roosevelt* as yard "birds" replaced one of the hatches on the six multiwarhead missile bays aft of the sail which the sub carried along with the eight attack torpedo tubes forward.

Standing in the sail as they'd slid slowly out of Holy Loch for another NATO Atlantic patrol, the second officer had asked Zeldman, "Ex, you ever heard of a Bing Crosby?"

"For crying out loud," said Zeldman, "what's with this Bing Crosby? Skipper was on about that yesterday."

"Yeah?" said the second officer. "He was asking me this morning."

From that point on, Captain Robert Brentwood became known as "Bing" among his crew, and the boat's unofficial theme song was "Moon River," crew members occasionally providing impromptu renditions during off-hours. Zeldman figured that if the skipper heard some of the associated jokes, his hair would turn completely gray overnight.

CHAPTER TWENTY-SIX

LIN KUANG STOOD alone in the forest darkness high above Taiwan's Taroko Gorge, the sound of the rushing river far below melding with the swishing sound of the wind in the pines, and here he pledged to Matsu, goddess of the sea, by all that was holy to him that he was ready for battle. The waiting was over, the invasion fleet ready, as it had been since his grandfather's day, since that terrible day in 1949 when the Communists had driven the Kuomintang to the sea, when, thanks to the mercy of Matsu, the fleet of the two million Chinese Nationalists under Chiang Kai-shek had arrived safely on Taiwan and made it their own.

Lin Kuang had never been "home" to the mainland, but one night, as a junior officer years before, he had been the captain of one of the Nationalist navy's motor torpedo boats that had landed a raiding party during the intermittent fighting over Quemoy Island just off the mainland. The purpose of the raid was ostensibly to gather and update intelligence reports on the Communist ports' defenses. But the real purpose was to keep every soldier in the KMT practiced and ready, to prevent them from ever sinking into the passive acceptance, shown by the rest of Taiwan's eighteen million, of believing that Formosa, as Taiwan used to be called, was home. The KMT ruled Taiwan and built up enormous financial reserves to the point that it was now, after Japan, the richest Asian nation. But, with the legendary Chinese patience, the KMT had never lost the belief that the Communists in Beijing were but temporary usurpers, one day to be thrown out. Unlike so many Westerners, who accepted the legitimacy of Communist China and had now turned their back on Taiwan, Lin Kuang and his fellow officers never doubted that one day it would be their day to return. Superbly trained, equipped, through their billions of surplus dollars, with the latest in technological advances in the West, the KMT knew that the Communist hold on Mainland China was in large measure an illusion; the Chinese Communist party, constituting only 10 percent of the population, could no longer hold it together. There were hundreds of millions in the far-flung regions of China who had never been anywhere near Beijing, for whom Beijing might as well have been on the moon, as there were many millions in the far-flung republics of the Soviet Union whose loyalty to Moscow was as tenuous as a failing marriage, the parents unable to control the children who sensed the great divide and wanted to go their own way. Promise the far-flung republics their own country, said the KMT, and the center was yours.

Lin Kuang closed his eyes, breathed in the cool, damp air, happy that autumn was upon them. It was a time for change—when men would need to brace themselves for the stormy seas of winter. In his mind's eye he was no longer inhaling the air of Taiwan but that of Hangzhou, where his father's fathers had been born and raised, the city of which Marco Polo had once said, "In Heaven there is Paradise, on earth . . . Hangzhou." There on the West Lake, serene amid a garden of gladioli, lawn, and fish ponds, was Mao's villa. It was Lin Kuang's dream to be the one to retake Hangzhou, to personally raze the villa to the ground.

In Washington State, Mount Ranier's volcanic cone was visible from Seattle's Northwest University over sixty miles away, the mountain's peak shrouded in a rosy hue of pollution and midday sun, while fourteen miles west of the city a Trident nuclear submarine glided gracefully out of the upper reaches of Hood Canal. It passed the spidery wire webs of the onshore degaussing stations, which would wipe the sub clean of any telltale magnetic signature that might otherwise be picked up by a ship nearby, especially the Soviet trawlers that often lay listening north of the Bangor base beyond the Strait of Juan de Fuca, which ran between Washington's Olympic Peninsula and Vancouver Island. Sometimes, as was the case this day, a Coast Guard cutter plowed ahead of the Trident, making lots of noise and running interference against hydrophone arrays that might be trailing behind a Russian trawler as the sub headed into the vastness of the blue Pacific.

David Brentwood's father had often taken him out to see the ships leaving the east coast, and David always found it a calming experience, which was why he'd driven out to the placid waters of the canal. The fight with Melissa was still officially on, but he'd hoped that if he could cool down by the time he got back that evening, then she would have simmered down, too. He needed her, especially now, for on the circuitous route down through Tacoma and up across to Bremerton, the "classic" rock and roll he'd been listening to on the Buick's radio had been interrupted by a news flash that an American frigate had been attacked in the Sea of Japan, but as yet the Pentagon hadn't given out the name. Somehow David knew in his gut that it was the *Blaine*. By the time he'd reached Bangor, another radio flash had cut into the program. It *was* the USS *Blaine*. No information on casualties. Then forty minutes later a Pentagon announcement saying the ship had apparently been hit twice. The National Security Council and combined Chiefs of Staff were meeting with the president.

David felt terribly guilty. He had grown up in a naval family; his grandfather had fought at Midway. But instead of following his brothers into the navy, he had bitterly disappointed his father by joining the army reserves at Northwestern instead of the naval reserves. Now he suddenly felt somehow responsible, that somehow he should be in Ray's place or in Robert's place aboard the Atlantic Fleet sub, wherever it was. His father had always told him that he didn't care what "line of work" David got into after college, "so long as you're happy." That at least was the official "liberal" stance of ex-Admiral John Brentwood. He had never indicated any disappointment about David joining the army reserve, yet David felt it whenever his father was talking to someone about "the boys." Likewise David held back from talking about his father's honor-clad career, for much as he admired his father, David always felt pressured to perform as well as his two navy brothers. One reason he felt he couldn't, but which he had never confided to anyone, not even Melissa, for fear of them thinking him weak, was that he had a terror of the sea itself. From a distance he could admire and enjoy it as one admired people on the high trapeze, but the very idea of dying at sea, of being entombed forever in the great dark abyss, sent a cold shiver through his bowels. His father, obviously without meaning to terrify him and more as a simple point of information, had once told him when David was a young boy that the Marianas Trench in the Pacific was as deep as Everest was high. To David it became an *idée fixe*, a phobia that no doubt, like all phobias, would merely have sounded silly to someone else, but one that for a young man from a distinguished naval family was nothing short of cowardice. The thought of the closed-in darkness, the enormous pressures that could "crumble bones to dust"—that was another of his father's favorites—began to haunt David, to obsess him so much, he'd gone to the library at his primary school and, with all the guilt of a pornographer, had looked up whether his father was right about the Marianas Trench. Surely nothing could be as deep as Mount Everest was high. His hands trembled as he flicked over the *M*s in the encyclopedia, heart racing, ready to shut it immediately should anyone approach him. The Marianas Trench *wasn't* as deep as Everest was high—it was *deeper*, by another *five thousand feet*.

Later David's rational side did battle with his irrational shadows—after all, to die was to die. But logic held no ground in the battle between primeval fear of being buried in blackness and the calm logic that argued that however you died was in the end immaterial. The memory of seeing his grandmother's body borne away on a wet and windy fall day in New York, the sounds of horns honking impatiently and uncaring from behind the hearse as, rain-polished, it pulled off into the cemetery, had stayed with him. Watching the coffin lowered, hearing the run of the ropes and the thump of the clay to close you in forever, he'd decided there and then he would be cremated when it came his time. But did the soul still rise or did it die, too, in the funeral pyre?

Driving back to Northwestern, anxious about whether or not Ray had been injured or killed, he heard a Pentagon "spokesperson" come on the radio informing the press that "at this point in time we cannot categorically say whether the missile was fired by another vessel or a plane." The woman droned on with more "points in time" instead of "presently," and it all added up to the Pentagon wasn't sure what the hell had happened. David watched the long, black sub, now no more than the size of a small branch, floating out on the clean and vibrant blue, taking what he could from its deceptive serenity. As much as he'd feared the sea, he also felt a strange communion with it at times, an attraction of opposites. David thought of his mother, pained at the thought of her pain, on the other side of the country, and it plunged him into a dilemma. Should he go back that evening to be with his folks? His father, of course, would never admit it: "Not for me, son, you understand. But it'd do wonders for your mom. Thrown her for a loop, David." Well, Dad, Mom. She handles loops pretty well. Why don't you just say, "Davy, *I* need you"?

Or should he wait a few days first until the Pentagon knew for sure what had happened, who was hurt? Driving over the Seattle overpass, David thought of how Melissa would be waiting for him now, full of sympathy and feminine comfort. God, he could play it to the hilt if he wanted, stoic expression, the Brentwood tradition. Just as quickly he was ashamed of even thinking of using it to his advantage. As he thought of her, he felt himself getting hard. Was it normal? His brother thousands of miles away, the *Blaine* and Ray in God knows what shape, and here his kid brother at home was so damned horny that the mere thought of having a woman could override his concern for his brother.

David could see her now. She was slipping off her jeans but nothing else—yet parading for him in the semidarkness of his room. He could feel her hand cupping him, squeezing, bringing him to her in one long, even pull. . . .

A light changed to red and he hit the brakes. Next to him a big Mack semitruck shuddered, its raw power barely held in check. The driver, chewing gum, looked down at him, shaking his head.

When David got to her dorm, it was four in the afternoon. There was a note for him folded and taped to the doorknob. "Davy—it's dreadful. I just heard. Be back from seminar five-thirty. Wait for me. Love you, Lissa."

He went down to the dorm's lounge room and wandered over to the pop machine before flicking on the TV.

"Dave!"

He turned around to see it was Stacy—only guy he knew who wore a bow tie to class. He had a short neck, too—looked ridiculous. And loaded with library books for effect.

"You get the message?" asked Stacy.

"Melissa's?" Davy asked, annoyed that Stacy knew ahead of him. Had probably read it, too.

"No," answered Stacy. "Your dorm."

"Haven't been there yet," answered David.

"Oh—reserves are being called up. Fort Lewis. Your name's on the list."

David felt a rush in his gut. "You sure?"

"Brentwood. D.—that's you, old buddy. Hey—listen, I'm sorry about your brother."

Damn Stacy—why the hell didn't he give you the messages one at a time and in order, for God's sake? David wanted to ask him what exactly he'd heard about the reserves but hesitated—Stacy thrived on the drama. "All right, so what did you hear?"

"You mean you haven't heard?"

"Jesus Christ, Stacy—what's going on?"

"It was hit. Pretty bad, looks like. CBS is running an in-depth report at seven."

Sometimes David didn't know whether Stacy was just plain dumb—"in-depth"—or was just too "gee whiz" to realize how insensitive he was to others. *"In-depth!"*—Christ. Some file footage probably with a cutesy lead-in: *"Blaine*'s shame." Or how about "Bam in the Sea of Japan"—that would be par for the course these days. David's anger was now turning to consternation and turmoil. The upgraded Hazard Perry–class frigate, as his father proudly told anyone who'd listen, was one of the world's most sophisticated warships, its Phalanx radar-weapons system capable of tracking and destroying multiple targets simultaneously. How was it possible that the *Blaine*. . .

For a moment he thought his brother's frigate might have been blindsided, one of the defense systems turned off as it had been on the *Stark*. David just as quickly dismissed the idea from his mind. Hell, that mistake on the *Stark* had been imprinted on the mind of every cadet out of Annapolis. His brother Ray had even made up some kind of sign about it, or so his dad had told them: "Don't forget the *Stark*" or something. It was also clear now that Stacy hadn't been trying to be a smart-ass with his "in-depth" pun and that he, David, had simply been overreacting. Then Stacy gave him a slip of paper, the number at Fort Lewis that all the college reservists were supposed to call. Was Stacy helping him or helping Stacy? Very considerate of him to have it all ready like this. Almost as if Stacy was in a hurry to see him off campus.

"I guess this is the downside," commented Stacy as they approached the quad, cypress trees glistening with rain from the night before.

"Of what?" pressed David, trying to be civil despite the conflicting responsibilities and choices coming down on him: his Mom, Fort Lewis, Melissa—a possible extension might be granted from the army until fall term's end—but how would it look, with his brother . . . .

"Downside of the army paying your tuition," Stacy explained enthusiastically. "You know, tit for tat."

"Yeah," said David, trying to hide the fact that Stacy was getting to him, if that's what he was up to.

"You could get a deferment probably," opined Stacy. "I know a guy in commerce. He's a corker at writing out requests for deferment. He did it for a—"

"Corker?" asked David. The little shit was definitely trying to get to him.

"Sorry," said Stacy with an air of Ivy League superiority, except that he was on the opposite side of the country. "Corker's a British expression," explained Stacy. "You know, means first-class. Top of the line. My roommate's a Brit. I pick up things like that, I guess."

"What things? Brits or the way they talk?"

"Way they talk, old boy. Don't you know?"

Barf. Brentwood didn't know which would be worse, putting

up with this crap for another term or fighting gooks. Melissa had got after him one time for calling them that. As he reached his dorm, gladly saying good-bye, or "Toorooloo!" as Stacy had put it, David Brentwood knew there was only a slight chance he could apply for a deferment from the call-up of reserves. He started to get mad with his brothers, just as he had as a youngster, always having felt he had to "measure up" to them. What the hell had Ray been doing out there anyway? Daydreaming? Walking slowly up the cement stairs, the dark, shaded mouth of the dorm swallowing him like some leviathan of the deep, he recalled they'd said something on the radio about patrol boats having attacked the ship, but had they been Russian or North Korean? Or did they say *South Korean*? He was confused. Everything seemed to be coming apart at the seams.

CHAPTER TWENTY-SEVEN

BECAUSE HE HAD been summoned to report immediately to the *Chungang Chongbo-bu*—South Korea's CIA—headquarters during the hours before Seoul's fall, Chin Sung was unable to attend the funeral of Lee Sok Jo, his young colleague who had been with him as the students had rioted outside the Secret Gardens.

Chin had been ordered to proceed on what he was told would be the most important mission of his career. He was flown out of Seoul by one of the few remaining helicopters to Pusan and then to Tokyo. The Apache chopper ferrying him across the Sea of Japan barely made it, being mistaken for a split second by an F-15 out of Japan for a Soviet Mi-28 Havoc, the ROK pilot ferrying Chin frantically firing off packs of flares lest the heat-seeking missile be fired by the Americans by mistake.

From Pusan, Chin had flown to Tokyo. The immaculate khaki-uniformed and white-gloved Japanese police were out in force at Narita, the airport having become a target once again for the Japanese Red Guard terrorist faction, who, encouraged by North Korea's bold move, were now issuing more than their usual weekly number of bomb threats.

The sight of the Japanese police was at once reassuring to Chin and disturbing. In his grandfather's day the Japanese had been the ever-present enemy, not only from across the East Sea but as occupying troops whom it had taken Korea over thirty-five years to evict; their loyalty to the emperor and their legendary cleanliness and cruelty were inextricably linked in Chin's mind to all the images of childhood hatred. But now, as allies of the United States, Japan and the ROK were in the same boat, one that, as South Korea's ashen-faced president told Chin, was rapidly sinking, doomed if the Americans could not stop the NKA. Washington, he confided in Chin, as if Chin were hearing it for the first time, had lost all stomach for another fight in Asia. After the humiliation of Vietnam, said the ROK president, the American public simply would not tolerate another Asian "adventure."

Ninety minutes after he had left the president and was en route to Pusan, Chin heard that Seoul had fallen. The shock of it caused him to walk about Narita's crowded rotunda-shaped waiting tower all but oblivious to what was going on around him. No one seemed to be talking about it, most of the passengers Canadians, Americans, and Australians on stopovers out of Shanghai and Communist Hong Kong. Didn't they realize what was happening? If Korea went, the Western world would not have a single foothold on mainland Asia.

Chin sat down, listening carefully to the public address system, worried he might have missed the first call for his flight to Europe. Everything was strange: the disinfectantlike odor of the waiting room, the buzz of tourists preoccupied, like him, with the TV arrival and departure monitors, the fresh fruit juice counters, the doll-like complexion of the young girls in white behind the counters—above all, the absence of familiar smells. Nothing was familiar, nothing reassuring.

"Heard about Seoul, mate?" he heard an Australian asking a friend. It reminded him of the old beer ads by Crocodile Dundee.

"Nah," said the other Australian. "Last I heard, the Commies'd surrounded it."

"Surrounded it? They've taken it, mate. Lock, stock, and barrel. Yanks threw in the towel—hour or so ago."

"What? The whole shebang?"

"Not yet. Only Seoul so far. But it's just a matter of time, I reckon." The Australian made a joke about the South Koreans having four gears in their tanks, one forward, three for reverse. The other Aussie indicated Chin sitting near him, the Korean Airlines bag at his feet, its loop handle around one ankle.

"Sorry, mate," said the Australian to Chin. "Me and my bloody big mouth. Just skylarkin', sport."

"Skylarking?" Chin had never heard the expression before, but it was an old joke, usually made by Americans about the ROK. It hurt. And though the two Australians had no way of knowing it, in Chin's mind their comments had confirmed the necessity of his mission. Even if the U.S. reserves from Japan arrived, they were not battle-tried. They might not even be able to land their convoy if blocked or otherwise engaged by the Soviet Eastern Fleet steaming south, or if they were attacked by one of the NKA's diesel submarines. President Rah had been right—if the Republic of Korea was to be saved from communism, from utter defeat, drastic actions were called for.

The announcement came over the PA for the Lufthansa flight to West Berlin. It was surprisingly clear-voiced, unencumbered by the usual hollow echoes of most other airports he'd been in, and Chin took it as a good sign.

Aboard the plane Chin waited till takeoff, watching the passengers watching the cabin attendants watching the video of what to do in an emergency. He knew what to do—it was getting there that would be the problem. They had a safe house in Kreuzberg, three miles south of the old Reichstag and two miles southeast of the Church of Reconciliation. *Reconciliation.* Normally it would have elicited a wry comment from him to Lee Sok Jo. But Lee Sok Jo was dead, and besides, Chin was too tired.

Before the movie, in which a gray-haired Tom Cruise was playing a dignified old general, forced out of office for opposition to a new space-decked beam weapon, Chin went to the toilet and unbuttoned his shirt, taking a sheaf of deutsch marks from the money belt. They didn't want him to use any traveler's checks, and signatures that either the West or East German intelligence services could trace. One of the deutsch marks had been torn in half; the meet in Berlin would take place only when the other half of the bank note was received by Chin. The fact that the note had been hastily torn, ripped rather than neatly cut with scissors, was only a minor detail and made do difference to Chin, but the tear spoke volumes in terms of the urgency with which the operation had been set in place. It was, thought Chin, as if the person in the *Chungang Chongbo-bu* who had torn the deutsch mark had done it with trembling hands.

When he returned to his seat, Chin sat back and tried to watch the movie. People were laughing, but Chin couldn't hear it properly, the earphones crackling. Besides, his sinuses were acting up again. He pulled out his Dristan bottle and took a sniff. At thirty-five thousand feet the sinuses cleared, but he knew from his previous trips abroad that the trouble would start as soon as they began to descend, the pain like a red-hot needle being pushed through the bone directly above the bridge of your nose.

The thing that worried him most was that the success of the NKA's special forces sabotage meant that the NKA network had been extensive and well trained, so that it was naive for him to think that Pusan, a major gateway to the West, had not been carefully watched by NKA operatives. The question was, had they had time to make him and/or follow him onto the same flight?

He looked around. The plane seemed full, and they would have had to bump someone or buy out someone else's ticket to get space on such short notice. Of course, it wasn't essential that they start at this end; they could merely fax instructions through to Berlin and follow him out from Templehof. If it was someone on the plane, it would be someone, Chin thought, very much like him, no luggage to declare so as not to be delayed in customs. They'd be carrying an overnight bag—maybe a camera. Touristy-looking.

CHAPTER TWENTY-EIGHT

COLONEL DOUGLAS FREEMAN dreamed of war. The thought of commanding vast armies on a European front as battle moved back and forth across the surface of the earth so filled his imagination, the scenes of glory that visited him so powerful, that at times he couldn't sleep. Then, walking quietly down to the main floor of his house, or rather the army's house, which overlooked Monterey Beach, he would stare out into the darkness of the sea, beyond the line of fluorescent waves, at once convinced that his destiny was nowhere near fulfilled yet anxious as to what form it might take and when. In his basement den, TV earpiece in so as not to disturb his wife, Doreen, sleeping above, Freeman would replay the videos of all the wars; from the reconstructed sites of the Peloponnesian War to videos of Victory at Sea, the Great War, the long-lost battles of Indochina, and the most recently released footage from Britain's war office of the Communist insurgency in Malaysia and the Falklands War. Everyone else who Freeman knew in the Armored Corps was busily writing tanks off, the debacle of the M-1s in the Uijongbu corridor supporting them. But Freeman believed the problem was not the team but the tactics. They were still fighting World War II; that was the trouble. The hardest thing, he knew, as did Guderian, Liddell Hart, and Patton, was to get a man to change his habit. Simple things—ask someone who loves coffee to drink water, a smoker to quit, to move anyone out of a mode that, for all its inconveniences and ill side effects, he's grown familiar with, and you might as well talk to a rock. And after forty forget it. Freeman knew colonels no older than he was at fifty-five who, finding themselves in a crowded field for promotion, had simply stopped trying, accepting that colonel was as far as they'd go. Freeman wanted to "break out."

He heard a faint scratching upstairs—their cat, Alexander, clawing at the kitchen door, and Doreen getting up to let him out. Freeman preferred dogs—cats could be so damned uppity—but he and Doreen had compromised, the Persian doglike in his loyalty.

Freeman expected loyalty, unconditionally, and it was something he and his wife had given each other. For the most part it had meant a happy marriage, with Doreen independent enough not to let the service life from here to there get her down. And now, with their two girls out of the nest and settled down, life was easier than it had been for years. But Freeman knew he would have few, if any, regrets if suddenly called upon to serve in Europe if war broke out. While other colonels he knew were planning their retirement, talking about secure portfolios with the bright young men from Harvard business school, Freeman spent most of his time in his basement, poring over his mock-ups of European battlefields, demolishing the army's main battle tank and armored vehicle tactics. History repeated itself, and yet it did not. Like a football field, the terrain remained the same, but each match was different, no matter if you'd played the game a thousand times before. The important thing, he kept telling himself, was to keep yourself fit and ready.

He heard the slap of the morning paper on his porch and went upstairs. Still in his robe, he glanced at the headlines and began muttering. Korea was a rout. Air strikes from the Seventh Fleet were unable to inflict any decisive damage on the Communists so long as the North Korean troops continued to mingle with the panic-stricken streams of refugees. He switched on the TV— "Good Morning America." The NATO line was on full alert, Soviet-WP forces flexing muscle along the line but both sides informing the other of any "unusually" heavy movement above battalion or squadron strength. But the State Department spokeswoman was assuring everyone that "despite the worsening situation in the Korean peninsula," the State Department and the White House believed "that there is no danger of an outbreak of hostilities elsewhere." The NATO and Soviet-Warsaw Pact alert, it was explained, "is fairly standard procedure in times of such tension."

Freeman shook his head sardonically at the State Department's use of the "Korean peninsula" instead of plain "Korea." "Peninsula" made it seem not only far away and out of sight but a minor inconvenience; merely a wart on the body politic. Freeman sat there as the flickering TV images filled the darkened room like the flashes of distant artillery, more conscious now of the crashing of the sea no more than a hundred yards away, suddenly depressed by it all, wondering how many other men and women in their time had sat through their imagination's lonely vigilance in the night and dreamed of far-off glory, of things yet undone, of leading their country out of dark hazard. He was thinking again of football, of how the "T" formation that had revolutionized the game had come directly from Coach Shaughnessy's study of Guderian's Panzers' tactics during the infamous blitzkrieg of 1940 that had overrun France, thought to be the greatest military power of her day, in less than ten days. Shaughnessy was careful not to tell his players or anyone else he'd been studying the Nazis, but he started drawing "funny" diagrams on the board and then suddenly it happened. The Chicago Bears devastated the Washington Redskins' defense. Bears seventy-three, Washington zip. Then Shaughnessy took the T formation to Stanford, derided as the all-time losers in the Pacific Coast conference. Another blitzkrieg. A lightning run which shot Stanford from the doghouse of the Pacific Coast conference to whipping Nebraska twenty-one to thirteen.

But Freeman knew, as Shaughnessy had, that tactics are always changing. Shaughnessy's T formations were no longer as effective in the modern world, and just as the football coach had studied the German general, it was now time, Freeman believed, for the general to look at the new game of football. Increased sophistication in communication, allowing instantaneous instructions from coach to player, was akin to the state-of-the-art electronic communications between the tank commander and his echelons. There was less time to make a decision, and the only way to counteract it was to buy time with more sophisticated deception. He had thought about it long and hard and now on impulse took out a piece of blank $8\frac{1}{2}$-by-11 typing paper, wrote down his assessment—a battle plan—folded it meticulously, slipped it into an envelope, and went upstairs.

Doreen was putting on the toast. She was used to him being up so early, but this morning he was scratching his wrist. He was chafing at the bit.

"You okay?" she asked, switching on the coffee grinder. It sounded like a loose bearing he'd once heard rattling around in an M-1's gearshift.

"I'm okay," he answered. "How much extra is special delivery?"

"Another dollar. You're better off using fax," she advised.

"No," he said. "It's personal."

"Who is she?"

"Larry Oakes. Two-star general. In the Pentagon."

"What's he got that I haven't?"

He slapped her on the bottom with the envelope. "Clout!"

"Can I ask what it's about?"

"Europe," answered Freeman. "Possible attack plan for the Russian C in C."

"You'll look a bit foolish if you're wrong."

"Can't win it—"

"If you're not in it," she finished for him.

"It's worth a try."

"They won't attack NATO," she said matter-of-factly.

"Maybe." He put the envelope beneath his car keys.

"And what would happen to me?" she asked.

"Well," he said, kissing her on the cheek, "it's all hypothesis."

Alexander was scratching at the door.

In Washington, D.C., it was hot and muggy, thunderstorms moving in across the river from Virginia. The president's coffee had gone cold by the time his national security adviser, Harry Schuman, and Joint Chiefs in the White House situation room had filled him in, suggesting different plays, now waiting for his decision. Senator Leyland wasn't there, but everyone in the United States was hearing what he thought about the situation.

Outside the Capitol, where there were so many flashbulbs, TV lights, and microphones that it looked as if they were making a movie, Leyland's call was for "decisive action . . . not a time for pussyfooting . . . not simply America's honor we're talking about here but her security." Security was getting high marks in the polls, but Leyland wasn't saying just where the line should be drawn: Pusan? Subic Bay? Wake Island? Midway? Honolulu?

"How about San Diego?" said Trainor, watching the senator's TV performance.

"If we pull everybody out," argued General Gray, "it can only be interpreted for what it is—a humiliating defeat. People have never forgotten the sight of us scrambling off the top of that embassy in Saigon. And I might add, Mr. President, pushing off so many who had been loyal to us. We can't desert the ROK."

"General," Mayne pointed out, his tone growing tougher by the minute, "if it's already a 'humiliating defeat,' we might be wise to cut our losses."

"I think, Mr. President, we have to stand and fight."

"That's what Cahill was supposed to do on the DMZ, not—" The president waved his hand at the crisis map, a sea of red dots spreading like measles over South Korea, a sprinkling of blue in a rough semicircle in the southeastern corner of the country, its outer perimeter an arc stretching south of Pohang on the east coast through Taegu in the center seventy miles inland and down onto Yosu on the south coast. Pusan, the vital southeastern port, was halfway between Yosu and Pohang.

"Report from NATO?" asked Mayne.

"No unusual movement," reported General Gray. "My hunch is that Moscow's as worried about this as we are, Mr. President."

"I think you're right, General, but I don't want any trigger-happy private pulling the trigger. Anyway, to be on the safe side, I've put a call through to Suzlov."

Admiral Horton's bushy eyebrows lifted. "I wouldn't trust Suzlov as far as I could kick him. Their fleet's already off Manchuria. I don't believe that 'maneuvers' line they're giving us for a moment. There's a lot of traffic in and out of Cam Rahn Bay."

"They advised us of that, Admiral," Mayne pointed out.

"Trojan horse," the admiral responded. "Probably carrying enough troops aboard that battle group to reinforce Pyongyang. Three or four hours, they could be unloading at Wonsan." He could see the president didn't know where it was. He moved the pointer up along the east coast of North Korea. "Only sixty miles from the DMZ. They come down that coast road, we'll have another front we have to contend with."

"If they turn up at Wonsan," said Mayne, "we'll ask them to stop."

Sometimes the admiral simply despaired. If the president's advisers had kept their boss half as well informed about naval matters as they did the latest Gallup polls, the whole country would be better off. The entire Soviet order of battle was clearly evident in the satellite, and he reminded the president of this. "There's everything in there from the nine-thousand-ton Mike subs to the Alfa and Yankee classes as well. And—"

"Then if we can see them, Admiral, the Russians must want us to know where they are. Doesn't that tell you something? Our satellite photos tell us they're only proceeding at five to ten knots, Admiral. Hardly battle speed, is it?"

"I haven't been advised of this."

"We got the call before you arrived," interjected Trainor. "From NSA." The admiral was embarrassed, and Trainor could see there was going to be some ass kicking in Naval Intelligence when the admiral got back to the Pentagon. Then suddenly the embarrassment vanished from the admiral's face as he cast a cold, professional eye over the chart, surveying not only Korea but the entire northeast Asian operational area, from Japan to the China Sea. He quickly calculated the Soviet Fleet's position. Why had they slowed? Surely this would upset their amphibious maneuvers timetable, a highly complex combined-services operation that had slim time margins at the best of times. "My God," he said, turning to the president. "That places them on the thirty-ninth parallel, one thirty-one longitude. Off Wonsan." Mayne looked across at General Gray. "Order the reinforcements en route from Japan to disembark at Pusan as quickly as possible."

"Yes, sir."

CHAPTER TWENTY-NINE

"RESCUE BOATS ALL out, sir. Three returning."

"Very well," answered the skipper of the USS frigate *Des Moines*, binoculars steadied against the flying bridge stanchion, watching through the haze as the two LAMP helos dropped vermilion marker flares about the Beaufort rafts now two miles or so from the *Blaine*. The crippled frigate was listing hard to starboard, white smoke still pouring from her. Now and then a crimson streak of fire could be seen erupting from deep inside her.

The skipper of the *Des Moines* had to make the decision whether or not to use the choppers as additional rescue vehicles or to release them to complement his radar as part of his protective screen. All about him he could hear radio chatter between the *Des Moines* rescue boats, including the ship-to-shore launch.

"Helos to resume screen," he ordered. For the price of a few more men lost to exposure in the oil-slicked sea, the young skipper decided he would prefer to use the choppers as airborne eyes. Something had gone drastically wrong aboard his sister ship, the *Blaine*, and until he knew why, he'd rather be reprimanded for overcaution—protecting his ship—than have to explain to a naval board of inquiry, as Brentwood would have to if he pulled through, how it was that his men had abandoned the multimillion-dollar ship when she was still afloat. And though it was sheer chance, it wasn't going to do Ray Brentwood, badly wounded as he was, any good to have to explain how it was that he and his second officer were the first to be pulled out of the chuck by the *Des Moines*. Already there'd been strict orders that upon return to port, there was to be no comment made to anyone until Admiral Horton, chief of naval operations, had directed otherwise.

Most of the men picked up were already dead, drowned, despite their life jackets, by the oil breathed in, some asphyxiated by the lack of oxygen in the oil slick fires that were still burning around the smoking ship.

"Sonar contact. Range three thousand yards."

"Bearing?" asked the skipper.

"Zero five niner."

"Hard right rudder," instructed the skipper. He flicked onto the CIC channel, pushing the "squelch" button, drowning out the rescue craft traffic.

"What have we got, Tom?"

From the CIC below the bridge there was a pause, the computer racing through engine signature matchup.

"No match." Which meant it might be hostile but was definitely *not* friendly. The young skipper was thinking fast; the signal was increasing, its echo louder. "It should be closing," said the skipper to the electronics warfare officer down in the CIC. "How about passive mode? Any noise from it?"

"Negative."

"Very well. Torpedoes ready?"

"Ready."

"Stand by to fire."

"Stand by to fire."

"Lookouts sharp!"

The starboard lookout shouted an alarm, but it turned out to be one of the white igloo-shaped Beauforts, bobbing up and down.

"Contact fading," reported the radar operator. It was an awful decision—the damned thing was thirty fathoms below the surface. A new sub with no known signature waiting? The computers weren't sure, absorbing a lot of clutter from the dozen or so small rescue boats and clutter from the sewn-in metal reflectors on the Beauforts.

"Size?" asked the skipper.

"Can't say," reported the EWO. "Echo returns, but it's fuzzy."

The skipper decided he couldn't take a chance.

"Range?"

"Twelve hundred yards."

"Bearing?"

"Zero six three."

"Stand by to fire one and four."

"Ready to fire one and four."

"Bearing?"

"Zero six three. Holding steady."

"Range?"

"Eleven hundred yards and closing."

"Fire one and four."

Two MK-48s, the most sophisticated torpedoes yet made, were running through the chop in excess of sixty feet a second, diving, homing in on the target.

"Contact fading."

"He's hightailing," came a voice in the background, but the target wasn't running; it was the angle of the ship turning that only made it seem that way.

Two miles ahead, the sea rose out of itself, shattered white and brown-speckled, the *whoomp* of the second torpedo quickly following the first—the sound of both reaching the frigate seconds later. The brown spots in the white mushroom were Beaufort life rafts, tumbling down into the vortex of the collapsing column, its bubbling base dirty with oil, bodies, and pieces of metal, possibly from the *Blaine* as well as from the target blown skyward, several of the *Blaine*'s crew spilling out of the falling Beauforts, disappearing along with the glints of metal.

The contact vanished from the screen, the skipper not knowing whether he'd sunk an NKA diesel sub, though this surely would have made more noise, or whether what he had torpedoed was a hunk of metal having sunk and been suspended in the heavier salinity layers and for which he had killed over forty more of the *Blaine*'s already shipwrecked crew, those close to the explosions of the torpedoes having their spines and necks snapped by the concussion.

It was his prerogative and his conscience. No board of inquiry would fault him. Parents would write and say they understood, which would be the most terrible burden of all.

As the Lufthansa began its descent into West Berlin's Tempelhof Airport, Chin felt the thickening pressure growing above his eyes. He had already taken an Ornade capsule but didn't want to use any more. He wouldn't be driving, which the label warned against, but he would need a lot of concentration. Nothing less than the fate of his country hung in the balance. In the seat pocket in front of him he saw a copy of *Paris Match*, a headline about the Communists for Peace volunteer force. Chin was sure the force wouldn't be composed of merely anyone who wanted to fight "U.S. imperialist aggression." It would be the cream of the crop, special forces from the SPETSNAZ air/marine commandos most probably, including the pith-helmeted Vietnamese, the latter particularly courted by Moscow as the outstanding brothers in the fraternity of socialist states—the ones who had "humiliated the Americans in Vietnam." Even so, it wouldn't be the numbers, initially only a few thousand or so, from different Communist Bloc countries that would leave East Berlin that would prove to be strategically important in the eyes of the world, but rather their entry in the war on the side of the NKA.

Even now, as affluent young West Germans cruised down West Berlin's neon-sparkling Kurfürstendamm less than two miles from the eastern sector which, though it had ostensibly been socially and commercially integrated with the West, retained its own political demarcation proscenium, radio and television were reporting that there were already mass rallies in East Berlin's Alexander Platz. The vast square was jam-packed with everyone from athletes marching in from the Sportforum in the suburb of Weissensee to workers from as far away as Karl Marx Stadt in Dresden over a hundred miles south, bussed-in crowds spilling out onto the *Unter den Linden* and down past the reinstated statue of Frederick the Great. While bands played the national anthems of each country, from Cuba to the dozen or so African nations, hardly any of which the East Germans knew much about, the crowds kept growing and surging amid calls for socialist solidarity in the face of American aggression. The East Germans were clearly taking the idea of the Communist volunteer force much more seriously than Moscow.

Then the "Internationale" was struck up by the band of the *Bereitschafts Polizei*, the blue-uniformed civil police, and enormous splotches of yellow, red, and black, the long-time colors of the GDR, and the red flags of the revolution struck a vivid contrast to the white uniforms of the athletes and the hazy blue sky.

The East Berlin parades could be seen by hundreds of West German residents, mainly Turkish *Gastarbeiters*, or "guest workers," looking through the holes made by souvenir hunters and down over the remnants of the Wall from apartment balconies in Kreuzberg, the suburb an island of foreigners within the island of West Berlin. It was here Chin now headed within twenty minutes of landing at Tempelhof. The cab driver, like the older *Insulaners*, "islanders," of West Berlin, had heard all the noise before, whenever the Communists wanted to whip up an anti-West rally.

Kreuzberg was a suburb which had always been avoided by most West Germans, not because of its proximity to the old Wall but because for most West Germans, Kreuzberg belonged to the Turks who had come in their thousands after the Wall had gone up in '61 and who, though they liked deutsch marks, did not like what they saw as West German decadence. The Turks were not so much unfriendly as separate. It suited them and it suited the KCIA, for in the netherworld of the only West German city exempted from the drafts, in order to attract businessmen and young workers, Kreuzberg had also become a haven for dropouts and squatters. It was good cover.

As they entered the outskirts of Kreuzberg, Chin could still hear bands less than a mile away and the tinny sounds of loudspeakers. It seemed as if someone was being denounced and then a band would strike up, but because of a wall of trees, he could not actually see what was going on, and this put him on edge.

When he arrived at the row house, a three-storied "redbrick" place out of the nineteenth century, he saw that the gardens about it had run wild, unattended for years, many of the cobblestones in the street visible beneath the worn-out scabs of bitumen through which green vegetation was poking. Very unGerman this, he knew, but one way that frustrated landlords had of trying to force squatters out. All it did was attract more.

Appearing twisted and bent, the woman walking toward him as he looked through the bubbled glass was an elderly asthmatic, and it took her a long time to answer his knocking. Even before she reached the door, Chin could smell strong, pungent Turkish coffee and sausage. A battered-looking "Golf" van came round the corner, stopped, and unloaded a group of Turks, who were talking and laughing at the day's end as they dispersed down the street, most of them smoking and taking no notice of him. Several doors opened, children spilling out onto the road, greeting fathers who, putting lunch boxes down on the pavement, lifted their children high, twirling them in the air. They seemed oblivious to the racket from beyond the Wall.

Chin looked up for the late sun to judge how much time they'd have, but it was blocked by the Wall, which in this part of Kreuzberg had not been battered by the souvenir hunters of Gorbachev's heyday. Then he wondered whether the rapidity of the fading light was due to the fact that the Wall that rose straight up from the small backyard of the house was blocking it or whether someone in Seoul had slipped up in the panic, looking at the wrong month on the calendar and so getting the wrong sunset time. For all his training, this uncertainty panicked him for a moment, but then the door opened at last. As he entered, bowing graciously to the Turkish woman, he saw the other agent emerging from the bathroom, head bowing, apologizing profusely. When the old woman had passed them back into her kitchen, they matched the deutsch mark, though it was hardly necessary, as Chin recognized him as one of the agents whom he'd worked with years before, when assigned to Bonn. Behind the Wall they could still hear massed bands playing in the distance as they got into a gray BMW.

Six miles south along the line of what used to be the graffiti-scrawled Wall they came to a cream-colored, nondescript apartment building in the suburb of Neukölln, near East Berlin's Schönefeld Airport. From the black-tarred top of the apartment they would be able to see the airport proper, the two giant Condors, Soviet-made transporters, sitting side by side a hundred feet from each other, not far from the main terminal, their tail planes much higher than Chin remembered from any of the recognition charts he'd had to memorize. The sheer brutishness of their size, their lower half a blue wave pattern, the top a mottled khaki, made them frightening even from a distance. Now more bands could be heard; they were obviously marching south from Alexander Platz, the sound of the bands drowning out the usual putt-putting sound of East German cars, Tribants mostly, which during his earlier posting had always sounded to him like the two-stroke motorbikes that weaved in and out of the traffic, a law unto themselves. He could smell a mixture of high octane in the late afternoon air as the fumes from planes landing and taking off from Schönefeld washed over into the west on a brisk east wind.

There was a tremendous roar from the southeast, and turning, he saw it was the 6:00 P.M. Aeroflot flight from Moscow. Through the smell of the kerosene fumes he could detect a faint but much more pleasant odor of fried liverwurst. Everything was at once so foreign yet so familiar. Here the DMZ had been known as the death strip. The mixture of the familiar and unknown was unsettling in Chin's tired and nerve-racked condition, the Wall having once created the deadly illusion of safety in the West. But he knew that the Wall had served exactly the same purpose as the DMZ in Korea: locking in the peoples of the Communist blocs, prisoners of the totalitarian state, the same state that was at this very moment ripping his country apart, sweeping down from the North like the barbarians of old had swept against the Great Wall of China.

Then it happened. The other agent had assured him that everyone necessary from the trade legation in Berlin was ready. But now he learned that one of the men, the RSO, or resident station officer, had had to back out at the last moment because he felt someone, either from the West German police or from the Staatssichereitsdienst, the reinvigorated East German state security service that was supposed to have disappeared in the post-Gorbachev euphoria, was watching him.

The agent Chin had made the meet with realized that Chin was not listening to the explanation. Instead, the older man had stopped, hands in his light gray summer raincoat, looking up and down the street. "Those trees," he said, "I'd forgotten—"

"What?" asked the other agent apprehensively. "Oh, yes, ginkgo trees—the same as home."

"But they are not indigenous to Germany," said Chin. "I remember it struck me the first time I was here. In—"

The other agent didn't care about trees and, as delicately yet persuasively as he could, ushered Chin inside the apartment block. "I don't know about the trees," he told Chin. Chin could see he didn't care, but the older man was smiling. It was an omen. How was it, how could it be, that the same beautifully exquisite fan-shaped leaves, native to his homeland, were here—across the other side of the world in Berlin? Still green and full of light, bravely growing and vibrant in the dying rays of the sun that were now flickering down through leaves intertwined with—or were they being strangled by?—the indestructible barbed wire that was still there. Through the branches of the ginkgo trees, he saw scrawled, "Freiheit!" Freedom. The *um* and the *yang*. From the turmoil of opposites would come the eternal calm. Such omens were no accidents, he believed—they only underscored the unity of all and the unhealthy futility of drawing borders through men's souls. His country was on the edge of the precipice—the time had come for the clash of the opposites. From inevitable turmoil would come the eternal calm. "When do they start to load?" he asked.

"Eight o'clock," answered the other agent.

Chin glanced at his watch. "Then we have two hours to fill." The other man said nothing.

The five of them were in a room on the eighth floor, on the eastern side of the apartment. "Huh," one of the agents grunted. "After Gorbachev they were going to be one big happy family."

"That was before Suzlov," said another agent. "Everything changes and everything stays the same."

One of the three men who were introduced to Chin began brewing coffee, another passing around cigarettes from a gold-plated case. Chin hesitated. He'd tried too many times to give it up.

"American," said the junior man, an army captain in mufti.

Chin took one, sat back, and felt himself relax, inhaling the smoke deeply, letting it settle in his lungs before blowing it out in a bluish-gray jet. He felt strangely at peace; everything would now be simply a matter of operational procedure, and the men under his command could tell from his face there was no turning back. Chin told them of the five U.S. and ROK servicemen who had been beheaded at Panmunjom. Seoul HQ, he said, had told him that they had a video of it, not a very good one, taken by one of the last units to get out of the DMZ but that Seoul HQ wouldn't release it to the media. "Upset the folks at home," Chin said. "What about the airport security tower?" he asked.

The younger one drew a sketch on the back of a copy of *Abendzeitung*. "We'll take them out—or at least keep them occupied," said the agent.

"Have you got suppressors?" Chin asked. "We don't want any noise here putting us off."

"Yes, suppressors, of course. Two rifles." The agent identified the marksman, who inclined his head respectfully. Chin told him to destroy the sketch, then said nothing more. There was a long silence, the evening light growing dimmer.

Chin could sense the question in the air from all four of them but calmly kept smoking, composing himself. The marksman was looking over at the younger agent and the other two with the suitcases. "We were wondering, sir," began the younger agent tentatively, "should we not do it now?"

Chin looked around at them, taking his time. "Are you afraid? I don't want people who are afraid."

"No, no," hastened the younger agent with forced unconcern. "I merely thought that while there is light—"

"There is no point in half measures, son," Chin responded. He paused, the end of his cigarette glowing more brightly now. No one put on any of the lights in the apartment. Nearing the end of the cigarette, Chin asked for another one, accepted it graciously, tapping it on the gold case, thinking for a moment about his young colleague who'd been trampled outside the Secret Garden. Chun looked up.

"Are there any other questions?"

There were none.

"Lee," he signified to the young agent. "You should stay here to stop anyone at the door. Just in case. I will go up with the others." Chin nodded toward the other three men.

"Sir?" asked Lee. "With great respect, sir, I would wish to be with my colleagues. We have served together in Berlin and—"

Chin shrugged easily. "Very well. I will stay here." He looked around, seeing it would be no problem to put the sofa across the door. But in truth he expected no one. They had been very careful, they all said, leaving the embassy and all their homes in West Berlin at different times, using different routes. And Chun was now sure no one had followed him from Tempelhof.

When Chin glanced at his watch again, it was 7:30. "Go up in ten minutes," he said quietly. "Has the roof lock been freed?"

"Yes, sir," answered one of the men. "Everything has been prepared."

CHAPTER THIRTY

IN THE WARS of the early and mid twentieth century, the naval yards of Bremerton, west of Seattle, had seen many a ship launched and not return, but it was never something you got used to, most navy wives carrying it about with them until the day their husbands retired. Beth Brentwood, as had her mother-in-law and many navy wives before her, thought about it every night when she put the children to sleep and went to an empty bed. Sometimes she would take one of his shirts to bed with her.

She had prepared herself for the knock on the door, one of the things Ray's mother had told her she must do as a service wife. It was no good trying not to think about it, Catherine Brentwood had told her—only fools told you not to think about death. We moved toward it since the day we were born. To imagine it, said her mother-in-law, and how it might come to your loved ones, could sometimes help dampen the fear. Once you'd imagined it in all its obscenity, it wouldn't have such a hold on you. Besides, Beth knew, navy wives were expected to have the right stuff, too. And so, dutifully, Beth had imagined Ray's death, or at least receiving the news of it. A naval officer, probably another captain, smart dark blue uniform with the pristine white cap and gold braid. A gentle knock on the door. Their eyes would meet and the first thing would be to get four-year-old Johnny, who right now could talk the leg off a chair, away from the door. Jeannie, eight next week, would realize, with her sensitivity, that something dreadful was up, and Beth knew she could appeal to her maternal instincts to look after Johnny while the naval officer, cap off, smiled down at the two children, waiting. Then when the children were out of the room . . . "Sorry, ma'am, but I have to tell you the USS *Blaine*—"

But it wasn't like that at all. Jeannie had been bratty about cleaning up her room. "Daddy wouldn't make me!"

"Oh yes he would."

"No he wouldn't."

It had quickly degenerated into Jeannie charging, "You don't love me!" and Beth answering, "Ungrateful brat!" and telling her, "No!" she certainly could not have Melanie home to play after school. Jeannie had obeyed, bringing home not Melanie but Judy, the daughter of one of the *Blaine*'s petty officers. Beth had had to make a polite but firm stand and tell Judy that she was sorry but that Jeannie hadn't been behaving and had been told she couldn't have any friends to play after school.

"That's okay, Mrs. Brentwood," Judy answered, bubbling. " 'Bye, Jeannie."

There'd been another fight and Beth had sent Jeannie to her room, after which Jeannie had immediately sent young Johnny on a reconnaissance patrol. "Let Jeannie come out, Mom. Pleeeease."

"No!" Beth screamed, frightening Johnny.

When the doorbell rang she felt like a quivering mass of guilt, anger, and sheer frustration, and on top of it all the dishwasher that was supposed to have been fixed still didn't work.

The sight of the woman in navy uniform drained Beth of all color. She forgot to invite her in—heard Jeannie's "Can I come out now?" and the woman saying something about "normally the Pentagon . . . but in light of the news reports . . . as soon as they have further information . . ." Beth finally rallied enough to ask the woman in, but she said no, asking whether there was anything else she could do.

"Mom, can I *come out*?" Beth's hand was on her forehead, trying to sort it all out. She asked the Wave whether Ray's parents had been notified.

"No, ma'am. Our policy is to first notify . . ."

"When will you know for sure?"

"Mom, *CAN I COME OUT!!*"

"Be quiet, Jeannie! Sorry—"

"We should know something more in a few hours, Mrs. Brentwood. I'll be sure to call you."

When Beth walked back from the door, Johnny saw something was wrong.

"That lady an officer, Mom?" he asked timidly.

"Yes." Beth was stirring the fiercely boiling water, macaroni packet unopened.

"She talking 'bout Daddy?"

"What?" Beth looked down at Johnny. He was holding the macaroni packet. "Can I open it, Mom? I'm strong."

"Sure," she said. Johnny's cherubic face grinned, then turned red as he tried valiantly to puncture the push-open tab. "I help you when Daddy's away, Mom."

Beth bent down and hugged him. "Jeannie," she said, "you can come out now."

Jeannie came out pouting but not daring a repeat offense so soon.

"Come over here, honey," said Beth, her voice soft, unhurried. When she pulled Jeannie to her, the three of them clung in an embrace, and it struck her how small a physical space a family actually occupied, a feeling of being infinitessimally small.

Of the *Blaine*'s 192-man crew, only 61 were rescued, 11 of these dying within the first two hours of pickup and 14 still in shock with third-degree burns, including Ray Brentwood, whose face was so badly disfigured that the *Des Moines*'s sick bay petty officer, after giving Brentwood a shot of morphine, had left him till last, assuming he was a goner, the time better spent on some of the others.

The burn victims, the worst of whom were Brentwood and the lookout who'd been on the port side when the second missile slammed into the port side railing, were ferried to the *Salt Lake City*. Ray Brentwood's nose was nothing more than a skewed lump of flesh, a molten pulp now cooled, so that he was forced to breathe through his mouth, creating an alternately groaning and dry whisking sound, which, for all the compassion they had in the sick bay, several of the attendants found difficult to tolerate and so sent someone else to do the awful job of dabbing cool saline solution to wash away the globs of congealed blood from raw flesh that had once been the patient's face. Brentwood's eyes, having been partially protected by sunglasses but more effectively by the instinctive closing of the eyelids during the flash of the explosion, seemed not to be affected, but if he survived, this would have to be confirmed by later and more detailed tests once they got to Tokyo.

Now and then he tried to say something, but no one could understand what it was and put the garbled and repulsively snorting sounds down to pain- and narcotic-induced delirium.

In the Tokyo U.S. Army hospital, beneath the junglelike oppressiveness of an oxygen tent, Brentwood felt a searing burn, so intense he kept blacking out beneath the onslaught of pain, the merest eddy of air tearing across his raw flesh like a white-hot rake. What the voice inside was trying to ask was, what had become of his men and of his ship? But no answers came—only enormous and elusive shadows bending over the tent, the roar of his own breath like that of a doomed animal, the rushing of what he thought was his own blood unbearable but which was in fact his urine, something wrong with the catheter, filling the bed, the warm, acidic solution burning his already burned legs, his sense of smell gone. He was screaming.

With all the will he could muster, Ray Brentwood prayed for death, and now the jungle of vines, IV tubes swinging about him, went completely black, shrinking to a distant point of light, then a gentle blue, the color of the Exocet's exhaust growing larger until it began to fade, flickering away, replaced by a long, intensely white tunnel and within him a sense of rising above the earth, his screams unheard.

CHAPTER THIRTY-ONE

THE MEETING REQUESTED by the U.S. ambassador in Beijing had begun at 10:00 A.M. precisely, being late for an appointment being considered a grave insult to the Chinese. The white lace antimacassars draped over the arms and headrests of the old-fashioned maroon velvet lounge chair struck the American ambassador as typically Chinese, at once quaint and sensible, like the covered teacups on the table between him and the Mao-suited chairman of the Chinese Communist Party. The ambassador noticed that there was a new spittoon, this one a lighter gray than the last, which sat at the chairman's feet, looking for all the world like a child's potty. The lounge chair was so comfortable, it made it difficult for the American ambassador, for anyone, to make a strong point, for in order to drive home the argument, most people found it necessary to first do battle with the chair that enfolded them, yet sitting on the very edge of the chair gave one the undignified attitude of an eager schoolboy in front of the chairman, a position contrary to the dignified pose striven for by the representative of one of the two most powerful nations on earth. President Mayne's instructions had been clear: to assure the Chinese that the presence of the American carrier *Salt Lake City* in the Sea of Japan was not meant in any way as a threat to the People's Republic but was there merely to assist American action in South Korea—as it was permitted to do under the terms of the UN Charter, of which the People's Republic was a member.

The Chinese chairman was smoking as he spoke to the interpreter, a thin, young man in rimless glasses who, even as he was receiving his insructions, was looking at the American.

"The chairman understands, Mr. Ambassador, but there must be no interference north of the thirty-eighth parallel. On this matter we are resolved."

The American ambassador acknowledged the point but did remind the chairman that it was not South Korea that had "initiated engagement" and that the U.S. government might find it necessary to operate above the thirty-eighth parallel if it involved securing the safety of South Korean and/or United States citizens.

"The chairman disputes your claim insofar as it was clearly the repeated provocations and aggression of the Seoul government which precipitated hostilities. As a good neighbor of the Democratic Republic of Korea, the People's Republic of China may have no alternative but to aid the Koreans in freeing their country of hostile and antidemocratic elements."

"But not in the South?" asked the ambassador. The chairman began the long warm-up preparatory to spitting, sounding as if he were in the grip of a severe case of catarrh. The chairman spat at the spittoon, the ambassador's eyes scrupulously avoiding witness of the act by watching the interpreter, who was busy informing him at the chairman's request that the Chinese People's Republic did not recognize such designations as "North" or "South" Korea, that there was only one country and this was the Democratic People's Republic of Korea. The Seoul "gangsters" were merely usurpers, in Beijing's view. The Chinese People's Republic had no quarrel with the United States and wished friendly relations with Washington, but the chairman, on behalf of the Politburo, would reserve China's historical right to intervene should the integrity of the People's Republic be threatened. Which meant that, as in 1950, Beijing would not tolerate any American action north of the Yalu along the Manchurian–Soviet–North Korean border.

As the ambassador left the Great Hall of the People, Tiananmen Square was bathed in pale sunlight, the soft and dirty haze over the city pierced here and there by high-rises, but in the main Beijing remained a flat city on the North China Plain, its pace still remarkably unhurried compared to Western cities. The endless rivers of bicycles, "Flying Pigeons" jostling with less expensive models for position, never ceased to entertain the ambassador as it was here that the ingenuity of the Chinese was so everywhere apparent, their ability to use bicycles to haul everything from sofas to enormously high stacks of crated chickens, and even twenty-foot-long heavy steel rods for concrete reinforcing—as impressive an example as you could find of a people anywhere in the world coping with day-to-day problems that were all but inconceivable in the West.

As the Cadillac drew up outside the Great Hall, the ambassador saw construction workers walking by with woven cane safety helmets. They were not a people to go to war with.

When the chauffeur opened the back door of the black Cadillac that to passersby looked remarkably like any other limousine except for the thickness of the bullet-proof windows, what they could not see was the heavy armor plate between door panels, beneath the chassis, and sandwiched in along the roof interior. With all the added weight it meant that the car could get only ten miles to the gallon, but the ambassador didn't give a damn about the cost, for he knew that if push came to shove, the car he was in might be the only thing capable of getting him to Beijing Airport in a hurry and alive.

As usual, scores of Chinese stopped what they were doing and stared at the "Big Nose." Beyond the stone flag and heroically cast soldiers that celebrated the revolution, the lines passing through Mao's mausoleum were ushered through with an efficiency one normally didn't see in China. It was just as well, for the vast crowds would have led to chaos if they had been left on their own to visit the mausoleum and to see Mao under glass. The ambassador sometimes wondered whether Mao's head, the only part of him visible, was a wax fake. Whatever, he was still venerated by most of the Chinese despite the killing and madness of what was euphemistically called the "Cultural Revolution," when the Red Guards had run amuck, putting people to death arbitrarily, sending intellectuals to work as peasants, and putting the country twenty years further behind the West. Someone at the embassy had worked it out that with China's population of over a billion and the line outside the mausoleum the usual four abreast, the line would never end. Their sheer numbers and the fact they had the A-bomb made them a formidable force, and the American ambassador recalled the chilling words of General Lin Biao so many years ago when MacArthur had threatened to cross the North Korean border with China and drop nuclear bombs on the staging areas in Manchuria. Upon hearing this, Lin Biao shrugged. "So we lose a million or two."

* * *

Next into the lounge chair was the Russian ambassador, Leonid Guzenko, conveying cordial greetings from Moscow and reassuring the chairman that Premier Suzlov wished Beijing to know that the dispatch of the Soviet Far Eastern Fleet from Vladivostok had been effected solely in order to send a "clear" message to Washington that interference in Northeast Asia would not be tolerated.

The Chinese premier spat again and said that "*no* foreign navy" was welcome off the Chinese mainland. Which, the Russian ambassador replied, was precisely why Admiral Golchin had been given explicit orders to keep not only outside China's twelve-mile territorial sea zone but beyond the two-hundred-mile extended economic zone as well.

The chairman spoke quickly to his interpreter and then began clearing his throat again.

"The chairman says that it is his understanding that Soviet aircraft are capable of flying more than two hundred miles from the Minsk."

Guzenko fell silent for a moment. To answer yes might be interpreted as lending credence to the traditional Chinese mistrust of the Soviet Union. To answer no, however, would be absurd, for quite clearly the Chinese knew Soviet aircraft had a strike range well beyond two hundred miles. He remained silent.

The premier spat, then turned rather stiffly, the ambassador noticed, to his left to take the lid off his teacup, disturbing the antimacassar. The Russian followed suit; it was a pleasant, mellow-tasting Long Jing green tea from the hills of Hangzhou. A Chinese aide entered and spoke softly to the interpreter. The premier listened while offering the ambassador more tea. The Russian accepted. He had been in China for five years, and it was as if he had just arrived. You were never sure how to read the signs, but you always knew that beneath the surface current there was a subcurrent and beneath that a contrary current and beneath that . . . They thought themselves better than everyone else.

The premier was staring ahead, smoking, asking about the response East Germany had received, as Moscow's proxy, to the call for an international Communist volunteer force to aid China's "friend and ally," the Democratic Republic of North Korea.

"It has been excellent, Comrade Chairman," Guzenko replied truthfully. "In fact, we had to discourage the Cubans." He smiled. "They wanted to send *three* regiments." The premier nodded, seemingly approving both sides of the argument. "Your restraint," he said wheezily, "is proper." It was said in the tone of a headmaster addressing one of his junior staff. "There is no point in pulling the tiger by the tail."

The Russian seemed pleased and he added jokingly, "Even if it is a paper tiger, Premier." The chairman showed no emotion as he spoke, but his interpreter nodded sharply.

"Many mistakes," began the interpreter, "have been made. The tiger has claws. And in the desert."

The Russian turned to his own interpreter now, not knowing what to make of it. The Russian interpreter was also unsure of the chairman's phrasing but hazarded a guess. "I think, Mr. Ambassador, the chairman is referring to our intervention in Afghanistan—the rebels backed by the Americans."

The ambassador was sitting forward on the lounge chair looking distinctly uncomfortable, as if he had gas. "The chairman is quite correct. I think we have all learned our lesson about 'adventurism.'"

The Chinese interpreter wasn't sure but informed the chairman this might be a reference to China's Vietnamese war about the same time as the Russians were in Afghanistan.

"We are concerned," the interpreter abruptly told Guzenko, "about the Kuomintang."

"I am sure they will behave themselves," said the Russian in the spirit of family members talking hopefully about a deviant relative.

"And if they do not?"

"Moscow's position on this has long been clear," answered the ambassador. "If the Taiwanese stick their nose in where it is not wanted, we would insist that they withdraw immediately."

"What would you insist with, Comrade?"

Guzenko was surprised at the sudden shift. Why were they talking about a policy mutually agreed upon long ago by the two Communist giants? Taiwan was an "outlaw," as much a nuisance to Washington as any wayward republic in the USSR, as Tibet was to the Chinese People's Republic. There was only one way to deal with them.

"We would insist by force of arms," said the Russian ambassador. It was the one thing the Russian could state unequivocally without any clearance from Moscow.

The chairman rose. "Then we can rely on you?"

"Certainly."

"I wish to release this pledge of support publicly," said the chairman.

"Yes," Guzenko answered. There was no problem in this. Standard party policy. Taiwan had no legitimacy at all either here in Beijing, Moscow, or anywhere else. It had long been relegated to the backwater of history—ejected even by the UN years ago.

"Thank you," said the chairman, nodding his head and extending his hand. The Russian struggled awkwardly from the lounge chair, making a mess of the lace antimacassar. "Will Beijing contribute to the volunteer force?" he asked quickly.

"We will see," said the chairman through his interpreter.

Satisfied the Chinese didn't suspect that Moscow had dispatched the Far Eastern Fleet to remind the Chinese who was boss in Northeast Asia, the Soviet ambassador nevertheless left the Great Hall of the People dissatisfied with the meeting. Everything had been going fairly well, he thought, until Taiwan.

The orange light on the Zil's phone console was blinking relentlessly. "Yes?"

"Mr. Ambassador?" It was his military attaché, a bright young man but with a high, piercing voice that irritated the ambassador.

"Yes?"

"The Kuomintang navy with transports is entering the Taiwan Strait on the Chinese side."

The ambassador issued an oath for which he used to be soundly whipped as a boy. "Their course?" he demanded.

"Looks like Weitou."

The ambassador repeated the oath. Weitou was on the Chinese mainland at the mouth of the Jinlong Jiang and within the range of the two offshore islands of Jinmen and Xiamen, formerly called Quemoy and Amoy. Both islands were heavily fortified and belonged to Taiwan. The ambassador felt ill. Soon as word got out, and the chairman would make sure it did, that through its ambassador Moscow had promised military intervention should the KMT attack the People's Republic, the Soviet Politburo would be furious. He thumped the armrest. Damn the Chinese! They must have known the KMT were in the Taiwan Strait as they were talking to him. A trap, and he'd fallen for it. But who would have thought Taipei would—

"And sir—?" came the military attaché's piercing voice.

"Yes, yes," snapped Guzenko. "Go on."

"Ah—Mrs. Guzenko would like some Lucky Strikes."

The ambassador slumped back into the plush leather. "Friendship Store," he ordered the driver, who was beaming broadly—it was a chance to stock up on Western goods with foreigners' exchange yuan.

The ambassador knew that no promise with China would be honored unless the USSR was directly threatened—Moscow had enough problems of her own. But if the Kuomintang attacked Quemoy or one of the other offshore islands and Moscow didn't supply military support, the ambassador's career would be over. The driver was beeping the horn constantly and the Chinese moved—when they were ready.

CHAPTER THIRTY-TWO

A HUNDRED FIFTY miles south of Seoul on the littered and burning outskirts of Taegu, long spirals of thick, blackish smoke—rubber tires burning—there was a no-man's-land, not yet reached by the NKA but abandoned by the retreating ROK forces trying to consolidate the perimeter a few miles south of the city, the distant thunder of the NKA artillery unrelenting. James Law, a World Press photographer, was tired and disgusted with himself. Despite his best efforts he always seemed to be a half hour behind the action. The only people moving through now were terrified refugees, women crying, exhausted, some with babies strapped with blankets to their back, moving in a kind of shuffling half run, energy long gone but still going out of sheer terror. But the world had seen countless women and babies in countless wars. He called over two boys who, like so many others he'd already seen, didn't seem to belong anywhere, as if appearing out of nowhere, scavenging through the rubble for food, clothes, anything they might barter away, including themselves. They were crouching over the body of an American GI, stripping it bare. As he approached, they started to run, but he held up his hands, patting the air to calm them. "You boys speak English?" They both looked at each other, frightened and suspicious. Finally one of them nodded. "Hello—how are you?"

"I'm fine," said Law, smiling. "Listen—" He peeled two ten-dollar notes from his billfold. "You like?"

"Sure."

"Okay—" Law looked about, indicating to them that he wanted something on which to write. They produced a few scraps of paper.

"No, no," he said. "Big. Over there—boxes. You savvy?"

"Hello?"

"Here, I'll show you. The packing box." He stamped it flat and made a writing motion. One of the boys pointed to Law's shirt pocket.

"No, no," said Law. "No pens. Too small. Something big—super-duper." One of the boys started to run toward one of the fires and came back with a block of soft charcoal.

"Now you're talkin'," said Law. "Good boy. Now here, hand me the carton." He ripped it in half. Taking the most ragged part and using the piece of charcoal, he printed in a childish hand, "WE HATE AMERICANS. YANKEESS GO HOME." He poured some water from the dead man's canteen into a Kleenex and squeezed it under one of the boys' eyes. He took ten shots with the ASA400 film, shooting half on f16 and half on f8 as backup, and three with the Polaroid, one of which was very good because it showed the boys really scowling, their eyes full of hate, getting the American's corpse in nearby and violating the old World War II photographers' taboo by making sure the puffy face was plainly visible—flyblown and blood-congealed. Law heard the crack of a rifle, and the next minute shots were whizzing nearby. ROK or NKA, he didn't care. He got into the jeep and took off, the two boys running into what had once been a bakery shop, its counter covered in glass and spilled flour that looked like snow.

When Law reached the Pusan office, he sauntered in, announcing to the army clearance officer that "I just got a shot you wouldn't believe. Christ, make you sick." He showed the officer one of the Polaroids, which he knew he could use to jump the fax line if the wire service dispatcher deemed it good enough.

"Jesus," said the dispatcher. "This won't get past the censor."

"Christ, it's not a military installation. We're in a war. What are we talking here? Another Vietnam cover-up? That's the way it is, Sam—that's the way it is."

The censor passed it, Law checking to make sure the photo credit was well within the fax-sized paper.

Within an hour it was *the* photograph of the Korean War; two Korean youngsters, crying in their sorrow, shouting their hatred for the United States, and pleading for it to stop the war—to get the hell out of their country.

In Moscow it came in over the wire from the Soviet Embassy in Washington and was rerouted to 2 Dzerzhinsky Square. These days Chernko was staying there, a cot set up in the annex to his fourth-floor office. He took the photograph to the major, who awoke, startled, from an early morning nap after having been up all night.

"Good," he said. "Very—" Then he saw the photo credit: James Law. *"Gospodi!"*—"My God!"

"You see, Major," said Chernko. "Now it pays off, eh? All the training. The waiting."

"Yes," agreed the major. He had difficulty recalling Law's face, as he had been one of the early illegals they had shipped over in Gorbachev's time.

The director was walking away, holding the photograph high. "Power of the press."

In Washington protestors had already started to gather about the White House, the photo galvanizing opposition to the war.

CHAPTER THIRTY-THREE

IN EAST BERLIN the loading of the two giant four-engined Russian Condors, the world's largest military air transports, was delayed. Each plane, assigned to carry 345 members of the Communist freedom volunteer force from East Berlin to Pyongyang via Khabarovsk, was waiting for the Cubans to arrive. The East German commander was annoyed, but the members of his "shock troop" company, though in full battle packs, had no complaint. The shock troops, who had been waiting for more than an hour in the huge, overheated cavern of the Condors, had been told that some of the Cubans sent to help the Communist volunteer force aid North Korea were women. They had also been told by their political officer they must not call any of the women "señorita" or "señora"; this was something you heard only in the decadent American Western films so beloved by equally decadent West Germans. Nowadays, they were instructed, they must address the women as "Comrade"— "Compañera."

But as they filed into the transports from the long flight from Havana to East Germany, the Cuban women, over twenty of them, aroused intense curiosity among the East German soldiers. For the East German troops, used to the hard, athletic beauty of their women, the Cuban women's beautifully developed bodies were more supple in appearance. The compañeras' swarthy Latin color, their golden faces beaded by sweat, their combat bras damp through their dark green T-shirts, were an intriguing and welcome sight. Several of them had to press up closely against the men in the tightly packed aircraft, but not one of the men complained. It would be a long flight.

"Sicherheitsgürtel anfassen!"—"Fasten seat belts for takeoff."

Despite some eager helpers, not one compañera needed assistance, most of the 120 Cubans wearing red A patches on their black berets, signifying they had served in the African campaigns of "fraternal assistance," from Angola to the Sudan. As the pilots in the two Condors began the preflight checks, the seven hundred troops in the Condors were ordered to put in their earplugs as the engines went into their distinctive high-vibration scream. For a few token Bulgarian assault troops who had never been flown into action before, the noise in the huge military plane, devoid of the insulation normally accorded passenger aircraft, was frightening.

Despite the noise and the earplugs, Dieter Meir, a tall, blond East German, the cousin of Hans Meir at Outpost Alfa, managed to introduce himself to the Cuban woman next to him.

"Ich heisse Juanita"—"My name is Juanita," she said, putting out her hand. Seeing his surprise at her knowing German, she explained, shouting close to him in competition with the engines, "We have many experts from Germany."

Meir nodded rather than saying anything, as now conversation was impossible; the huge Condor's engines were in takeoff pitch as its nose wheels traced the white floodlit semicircles about Schönefeld terminal, beginning its lumbering and surprisingly bumpy roll toward the main tarmac. The second Condor was taxiing three hundred yards behind as they moved to the runway best suited for takeoff in the "Berliner Luft," the legendary and invigorating wind that blew down across the old Prussian plain across West Berlin's Grunewald Forest and toward the East German farms beyond Köpenick's Forest.

When Juanita had said "Germany" rather than "East Germany" to Dieter Meir, it struck a responsive chord, for he hoped that one day, in the heart, it truly would be Germany again—not just when the politicians declared it was, but united in spirit, in the same way that he and the Cuban, from thousands of miles away, had come together in their common ideology—to help unite another country. And yet he sometimes wondered if it would ever really happen in his own country, families, like his own, separated too long, still split asunder, the lingering legacy of having been apart for so many years beyond the 500-kilometer anti-Fascist barrier.

Inside the terminal the *Bereitschaftspolizei*, "police band," played a stirring rendition of "Freiheit and Peace," at once entertaining the waiting passengers for domestic flights and being recorded by Soviet and other Eastern bloc radio and television networks to mark the historical occasion on which the socialist world had, after so much inner turmoil, risen as one in defense of the North Korean workers' democracy in their struggle against the American assault. How typically stupid the Americans had been, to assume that after the workers' democratic movements had erupted in the Gorbachev years somehow everyone would suddenly throw away the good socialist things with the bad and declare themselves lovers of capitalism—to trade socialist evils for the evils of capitalism, as if there was nothing in between.

Aboard the Condors none of the more than seven hundred East German, Cuban, Bulgarian, and Romanian troops could hear the ceremonies or speeches, but there was excitement in the two huge transports. Most of the troops were under twenty-five and needed to prove themselves. On the other hand, Gen. Hans Demmler, commander in chief of the Communist volunteer force, could hear the pomp and circumstance over the earphones, plugged into the aircraft's circuits, but he took no notice of it. His hands were full, hoping that as the different segments had not time to train together for the task at hand, they would fight well as self-contained units, not that he had any choice, for full integration would be impossible with the language barriers, despite liaison officers. Right now he was checking that each AK-74 was "tipped" to protect the barrel and front sight from any damage during the flight. Although the troops had been told the flight plan called for a landing at Khabarovsk, if possible they would refuel in the air, as Moscow deemed this would be an impressive logistical display for the Americans.

The engines screamed in protest as the flaps were tested, then the brakes released, as the first Condor began the long run, gathering speed on the south runway.

The shadowy figures of the two men on top of the eight-story Kreuzberg apartment block were all but invisible, one of them leaning against the northeastern corner of the water tank for support, head bowed, right hand as one with the gripstock of the Stinger. It was not the Stinger of Afghan guerrilla fame, which in fact had often refused to fire, but the much improved

Stinger-POST, incorporating the passive optical scanning function so that the taxiing Condor completely filled the aiming circle, the edge a little fuzzy due to diffused heat waves at the plane's extremities that were rising and falling in the Stinger's sight like a mirage. The Condor was equipped with exhaust "baffles" or shields to minimize hot exhaust trail, but as the fully loaded plane rose and banked hard right, both port and starboard engine exhausts were clear to the naked eye, let alone the Stinger. The agent squeezed the grip, the Stinger's backblast illuminating the rooftop momentarily, scorching the agent's arm. The Condor, in effect a climbing fuel tank with soldiers aboard, exploded like napalm, reminding Chin of the American *Challenger* explosion years before, one of the Condor's engines, part of the wing still attached, cartwheeling, disintegrating in a fizzing halo of scarlet, ripping open the night. The main body of the Condor fell as if lead-weighted onto the tarmac like a zeppelin tight with hydrogen. As it burst into flame, the first fifty rows of tiny soldiers could be seen, heads black as burnt matches, then disappearing in the orange-white inferno that raced hundreds of meters down the runway.

Suddenly the second Condor, in the process of braking, tires sparking, exploded, its four nose wheels collapsing, fuselage sliding along the tarmac like the chin of some huge dinosaur on black ice, plowing into the inferno of the first plane. In the near daylight intensity of the scene, as fire engines screamed, the Koreans atop the building three-quarters of a mile away could plainly see some Communist volunteer troops in the second Condor sliding down open ramps, most of them afire, others coming down into a fuel slick that ignited seconds later, the ribbing of the fuselage now stretched against the skin of the aircraft in an X ray of the plane, the roar like a thousand Christmas trees going up. There were no screams, or at least none that could be heard over those of the sirens. A tongue of flame like a ceremonial dragon's shot out and around the rear cargo door of the second Condor, incinerating the remainder of the fleeing troops. Then the Koreans could hear a rattling sound, becoming louder as searchlights streamed out into the night like wildly flung straws darting over the West German apartments across the bulldozed remnants of the Wall. The rattle was the sound of machine-gun fire.

"Let's go!" said Chin.

* * *

There was shooting all the way from Treptow, East Berlin's most southerly suburb around the East German airport, to Pankow, East Berlin's most northwesterly suburb sixteen miles away. Most of the shooting, however, was concentrated in Kreuzberg, from where the missiles had been fired.

Outside the apartment the big crowds that had been racing down from the back-street *Kneipen*—pubs—toward the wire fence along the River Spree now started to retreat as rifle and machine-gun fire increased. In the flickering light of a flare, Chin saw an old man in pajamas and robe standing in the street below the apartment complex gazing skyward. A woman darted out from beneath the ginkgo trees, shouting and dragging him away toward an old section of dilapidated Wall. It wasn't certain who had started the small-arms fire, and Chin suspected that some of it was probably the ammunition going off in the Condors. In any event, now both sides were at it.

"All here?" Chin asked in the darkness, the sky to the east glowing a soft pink.

Each man responded. They went down to the gray BMW. "All in?" Chin asked again.

"Yes."

"Where are the cases?"

"The trunk."

Chin told the agent driving to pull up on the first bridge over the Teltow Canal on their way back to the trade legation offices. But there were too many people about, mainly police cars, and so they went on to another bridge, where Chin got out and ditched the two missile launchers. "Switch off," he told the driver. He listened for a moment, thinking he'd heard footsteps nearby, but now all he could hear was the cracking and pop of the small-arms fire. "Now, look," Chin said, stamping his feet; it was damper and colder by the canal. "You've all done a first-rate job. I want you to know—well—" He paused, never having been one for sentimentality but wanting to convey to his men how satisfied he was with their performance. The four of them had volunteered even though they didn't know all the details of the job. "*Komawŏ*"—"Thanks," he said.

He shot the driver first, the man immediately behind the driver second, and then the third man, trying to scramble out the back door. The man in the center was begging for mercy, his hands up high. "Don't, sir—"

Chin stood back a little and shot him, trying not to get anything on his coat.

In a chain reaction all along NATO's line, phones were ringing off the hook. Fighting had broken out in Berlin, and reinforced alert had become "general alert." At Outpost Alfa, two hundred miles southeast of Berlin, Hans Meir, who, with an American, was manning the observation tower, the scene of Hitler's Nazis crossing the Rhine Bridge clear in his mind, knew what to do. He was no less than Samson at the pass, only instead of the jawbone of an ass with which to beat back any breach of the "trace," he had a direct line to Fifth Army Corp's 120-millimeter artillery batteries behind him, ready to pulverize the Soviet and East German echelons, groups of five tanks, if they dared try to punch their way through. So far it was a quiet country night, broken only by the sound of an owl. Then Meir heard at least half a dozen dull thumps some miles to the east of him.

"What was that?" asked his American partner. There was a faint pinkish glow, the size of a tennis ball from where they were. Hans Meir didn't hesitate. Lifting the phone, he rang Alfa HQ. He couldn't get through, the line sounding like frying eggs.

In the main Alfa hut, a mile behind in a pine wood, they were trying to contact the Alfa tower, but all they heard was static on the line. It wasn't a line in the old-fashioned sense of radio cables but rather the elaborate fiber-optics links that were on the blink. The fiber-optic system was thought to be impregnable against an EMP, electromagnetic impulse, the kind of shock blasts given off from a nuclear explosion, scrambling all sensitive electronic equipment from televisions to defense computers.

At NATO's headquarters in Brussels, calls were coming in from pay phones from as far south as Burghausen in Bavaria to Fehmarn in Schleswig-Holstein in the north. All computers and fiber-optic radio links between Bonn and NATO HQ, as well as all those between armor/infantry, artillery, and air forces, including ground tremor sensors, had simply gone mad, spewing out nonsense that was only adding to the general air of confusion.

In the rain-slashed darkness around Alfa One, Klaus Meir

and his army buddy, Johnny Malvinsky, were watching the gap when the American, his Bronx accent markedly different from Meir's college-taught English, said something to Meir, who thought it sounded like "ruf."

"Ruf?" Meir said.

"A ruf," Malvinsky seemed to say. "It's moving."

"Where?"

Malvinsky handed him the infrared binoculars. "Two o'clock. By the rise near the barn."

The American was right, but it wasn't a roof, at least not a full-sized one, more like that of a small toolshed moving, the infrared's scope picking up spotty thermals.

"What magnification is this?" asked Meir. They were still trying to call Alfa HQ since the order had come, taking them from "military vigilance" to "general alert."

"Usual setting," said Malvinsky. "I didn't change it."

At Alfa HQ the infantry troops had come through the door of the metal hut as if there were a fire, the lift-up counter slamming back, held down by the sergeant as he kept intoning, "Let's go! Let's go!" The M-1s were coughing once, then there was a deep, throaty purr, surprisingly quiet to infantry used to the old M-60 Pattons. No one questioned whether it was a drill or not. They'd get their butts kicked if they didn't do it within the required time, and suffer restricted furlough as well. Outside as they moved beneath the pines into the Bradley armored personnel carriers, thirteen men apiece, the column started off for the trace near the tower.

Facing the Fulda Gap from the east, Meir's and Malvinsky's opposites, members of a squad from the East German Fourth and Soviet Thirty-ninth motorized rifle divisions, saw something moving along the plain through their infrared.

"Looks like an outhouse moving," the East German told his Russian liaison officer. The Russian agreed, but the thermal images were spotty because of the light rain that had now started. In any event, the shape of whatever they were watching was much smaller than that of a tank. Besides, in regular radio "line," rather than fiber-optic, contact with their headquarters, they would have been informed of any unusual traffic.

The farmer driving the tractor, like many others all along the front between East and West Germany, was one of those who had agreed to work farmland in and around the DMZ provided they were given special *Risikoentgeld*—"danger subsidies." But the subsidy, the East German farmer had long since told his wife, did not cover nuclear war, thank you very much. Working patiently and only at night, he had fitted the tractor with a rude protective shield of steel slabs either side and over the roof, his wife and two children sitting behind him now on seats he had expressly built, along with a huge white flag that would flap as he unfurled it at the Fulda Gap.

"And how," his wife had asked him, "do we get across the strip?" She meant the mines, many of which had been left in place after the West and Gorby's Moscow had ended. He told her he had mounted a roller plow on the front with loose chains to act as flails. Nothing new about the idea. Old as the hills.

"It's crazy," she told him.

"Woman," he'd replied, "when the shooting starts, there will not be time to discuss what we might do. But where do you prefer to take your chances? In the East or the West?" His argument was sound enough as it went, and a rainy night was made to order. Still there were two things wrong. The first was a loose bearing on the tractor's right wheel, making the squeak of the tractor even more pronounced than usual, and the white flag, which wouldn't have made any difference to the East Germans anyway, was so sodden by the rain that it didn't flap at all but merely hugged the pole like some thin, frightened ghost.

In the East German's infrared scope, whatever was moving disappeared for a moment behind a wood but then came back into view as a frosty outline, nearing the fence. In Tower Alfa, Hans Meir tried to make his decision. This wasn't the bridge leading to the Rhineland in '36 where Grandfather said you could see the Nazis clearly coming at you. But, of course, using camouflage is precisely what the Russians would do—trying to confuse Alfa just long enough.

"Den Mann halten!"—"Stop him!" ordered the Russian liaison officer. "It's an American mine clearer." The East German fired. Meir didn't see the antitank missile streaking toward the trace, only a white streak across the infrared scope—

"Back-blast!" said Malvinsky. "Ten o'clock."

"Got it in sight?" asked Meir.

"In sight," answered Malvinsky, the AT launcher on his shoulder.

"Fire!" Malvinsky fired at the East German back-blast position, bits and pieces, probably signed pine branches, still visible like pricks of orange snow melting on the glass of his infrared scope. Nearer the trace was a rolling fireball followed by pieces of flaming debris from the tractor. There was no human noise, the tractor stopped.

The East German fired back at Alfa's Tower, but by then Meir and Malvinsky were halfway down the staircase and into the jeep, pulling back to the next position in the pines on their side of the strip. The East German rocket missed the tower, crashing into the pines above them, setting the trees afire and throwing long shadows across their dugout. Meir fired three scarlet emergency flares. As they burst high above Alfa One HQ, the Soviet headquarters at Zossen-Wünsdorf, using the old-fashioned radio lines, ordered "Operation Home Rule" to begin.

Twenty Soviet–Warsaw Pact divisions, infantry and four thousand tanks, began to move, preceded by hundreds of strike aircraft, primarily SU-24/Fencers for ground support and MiG-29s with air-to-air Alamo and antiradiation, antiradar air-to-surface missiles, with Russian NR-30-millimeter tank-destroying cannon, all converging on the Fulda Gap. Most of the Soviet aircraft, while visible on U.S. and European satellite pictures, were not visible to the knocked-out fiber-optical radar systems on NATO's central front, north and south of the Fulda Gap. Here the Soviets' AFOMs, anti-fiber-optic measures, had been most effective. Nevertheless, in response to the firing in Berlin and the flares at Outpost Alfa, NATO, as part of its "forward defense, flexible response" strategy, sent hundreds of M-1s toward the gap while concealed 155-millimeter and 203-millimeter artillery guns behind Fulda began pounding the gap through which the T-80 echelons were expected to come pouring. Heavily camouflaged tanks of the U.S. Eleventh Armored, dug in behind preselected revetment areas in defilade positions, waited should the Soviet–Warsaw Pact tanks burst through the saturating fire of the artillery. In their deep, second forward observation hole, Hans Meir and Malvinsky hunkered down by the infrared periscope, ground shaking all about them, trees trembling as the deadly artillery rained down only a quarter mile from them.

"This is it, Fritz," said Malvinsky.

"Ja." Meir pulled a walkie-talkie from the OP's alert kit.

They had been taught and they understood that for them there could be no retreat; their only job, the most important of all as NATO front line OPs, was to hold as long as they could, to give situation reports vital to the artillery and tanks. To buy time. As Malvinsky stayed glued to the periscope, seeing the earth erupting before him in jagged white plumes on the infrared, he could hear Meir cranking up the minefield charge box for their sector. "Are they coming through yet?" he asked Malvinsky.

"Jes-us!" replied the American. "Have a look at this!"

General Sutherland had told them the Soviet–Warsaw Pact would most likely come through in echelons of fives, two wingmen, one either side of the three center tanks, to protect the flank. But to see his infrared's circle blocked solid with the thermal waves of so many, to actually see more tanks than he had ever seen in his life coming straight for him, was something that no amount of live ammunition training or anything else had prepared them for. "Settle down," said Meir, as much to himself as to Malvinsky. "Are they in our sector yet?"

"Hundred meters to go," said Malvinsky. "Hold it."

The forest was erupting, trees splitting all about them, the scream of hot metal shards carving up the air. Malvinsky swung the infrared scope around, and for as far as the eye could see there were thick, billowing clouds of white and the steady *pomp, pomp, pomp* of smoke grenades spewing from the forward tanks, whose earsplitting frenzied sound was now joined by the NATO mines as they began detonating. Meir heard the distinctive screech of tracks coming off roller wheels and the heavy metallic thumps as the mine-disabled tanks were shoved aside by those behind them, who now continued converging on the gap.

Within minutes a squadron of NATO A-10s came in low over the trees, wings rocking in their tight subsonic turns, almost like a car too powerful for its driver. Despite the rain and darkness, the A-10s performed superbly as they flew below the Russian MiGs and Fencers, their telltale "bug-eyed" twin jets, well back, and the high lizard-patterned khaki-green camouflage visible in flare light. The A-10 Thunderbolts showed astonishing virtuosity, seeming almost to hover momentarily, their noses down, the forty-two-hundred-round-per-minute, 30-millimeter cannon spitting out long orange tracer. Wherever the cannon fire struck the Russian tanks' reactive armor, the 30-millimeter bullets disintegrated the explosive-reactive armor, destroying the cannon's bullets before they could even penetrate the tank metal proper. But wherever the rain of orange cannon fire found the thinner-skinned turrets, it was enough to stop the tank, the air inside the vehicle a whizzing cloud of white-hot razors, many of the tanks exploding as the A-10s' fire not only passed through the turret but superheated the tank's own supply of ammunition by the loader. As well, most of the big American 155- and 203-millimeter self-propelled guns behind Alfa One were able to change position before the answering Soviet–Warsaw Pact batteries got a fix on them and were now tearing into the more than five hundred T-72s and T-80s that were the first to reach the Fulda Gap.

The Russian tanks now within range, the dug-in M-1s and M-60s were waiting until the last minute for their best shots to stop the Russian tanks that had survived the deluge of artillery and the assault of the U.S.-Luftwaffe Tenth Tactical Fighter Wing. At times the NATO fighters managed to penetrate the thick Russian MiG cover, below which Soviet Fencers were providing ground support for their tanks, their wings no longer swept back but in the straight lateral position as they came in low, seeking out the American A-10s.

Meir and Malvinsky knew that sooner or later they'd be overrun, no place to hide, caught between the steamrollers of the two opposing armies. Through screaming bursts of shell fire, Malvinsky was reporting tanks coming through in battalion strength, fifty at a time now that gaps in the minefield had been "ribboned out," Russian and East German infantry having quite literally laid out fluorescent tape to mark the safe entry points through the strip. If that happened, disaster would result, with the Soviet–Warsaw Pact echelons pouring through, then splitting off into arrow formations, one right, northward towards Kassel, the other left, south to Frankfurt-am-Main, driving deep wedges into NATO's central front.

The crisis for NATO at the moment, as Meir knew first in his forward position, was that of the fiber-optical-guided missiles. Long lauded as much cheaper than the old standard missiles, which had to carry their own expensive independent control system, the fiber-optical missiles were guided by images relayed back to a central fire control. But the fiber-optical missiles had now been neutered by the fuel/air explosions, one of which Meir and Malvinsky had seen earlier that night as a pinkish glow to the east and which had sent out pressure waves twice as powerful as that emitted by a two-kiloton nuclear bomb, the waves ineffectual against the old-fashioned clunky Russian missiles.

A motorcycle messenger sent out from Fifth Army headquarters to Alfa One was told about the fiber-optics screw-up and, hitting 120 miles an hour on the Autobahn, had taken the message back to Fifth Army HQ, which had still not heard anything directly from NATO's commander in chief of the thirteen-hundred-kilometer-long central front. Nevertheless, despite the dire situation, Fifth Army's General Willison was refusing to unleash his "stochastic" robot mines, designed to be set loose to roam for up to four days, attaching themselves to the magnetic field of a tank, exploding with fifteen kilograms of explosive. The trouble with this plan, Willison realized, was that if the NATO forces couldn't stop the enemy tanks at Fulda Gap and NATO tanks had to go into the Gap, then the stochastic mines would be just as much a menace to his U.S. M-1s, German Leopards, and British Challenger tanks.

The Fulda Gap was now a caldron of flying steel and volcanic earth as the Russians' spearhead column littered the ground, many of its tanks still burning, crews dead or dying aboard, hulls of others ripped apart, but still the Russians kept coming. With a six-to-one-man ratio over NATO, the gap might still be breached, MiGs and F-15s thundering overhead in the night, contesting the space above the potential breakthrough point.

All Meir and Malvinsky, eyes red with fatigue and fear, could do was to keep changing dugout positions as much as they could, trying not to expose themselves to either enemy or "friendly" fire as they watched tracer arcing out from the M-1s, machine guns finding the range, the M-1s' lasers put out of action either by the tanks' own reactive armor packs exploding or direct Russian fire, over a hundred of the M-1s destroyed by East German 125-millimeter armor-piercing discarding sabot rounds, capable of piercing the M-1s' twenty-centimeter steel.

It was now twenty minutes after two on the morning of September 3 when, because of the massive blackout of communication along NATO's line, NATO's European commander in

Brussels, General Koch, had no alternative but to formally release, by pay telephones (those not plugged into the fiber-optic system) and dispatch riders, all sector commanders to proceed on their own initiative. He emphasized the main tenet of NATO's "forward defense, flexible response" strategy—that the enemy should be engaged as far forward as possible to buy the desperate one week needed for the American reserves to enter the war—the nuclear option being the strategy of last resort and only permissible under the express orders of the NATO council.

For Koch, there were only two decisions he could have made: one, to do as he had done, or secondly, to defer to the NATO council. But in the time it would have taken him to convene the council, he was ordering the recall of "dual-based" troops from the United States, that is, those troops who, on paper, were in Europe but were only at half strength, an economic measure left over from the habit of bleeding the United States' European garrisons to put reserves into Vietnam. Koch knew the decision he made to recall the dual-based troops was in effect a decision that might force the U.S. president's hand. Unlike his NATO brief, Koch did not have the authority to move any U.S. unit higher than a forty-thousand-man corps. His request for the dual-based troops, a request that he knew would be known as quickly by Moscow as Washington, would be one that in effect would widen the war, but if not made, would make it impossible for the United States to reinforce Europe in time. Besides, if he didn't issue the order, the Russians would see this as a weakness, and what might have been an intention simply to gobble up territory along the near front would expand anyway, encouraging the S-WP forces to press on farther into Western Europe, knowing the farther they went, the less likely it was that NATO's nuclear option would be invoked.

Koch fully understood what it would mean for President Mayne, but historians could argue about it—if there was anything left to write about, which there wouldn't be if the Russians broke through. The president could refuse the request, of course, but this also would be seen by Moscow as a lack of resolve, another Munich sellout, and would only encourage Moscow to grab even more territory.

When Press Secretary Trainor got the request, he was at once shaken and relieved. He and the president had been holding a decidedly gloomy discussion in the Oval Office with the Joint Chiefs about the political necessity of ascertaining whether or not the missiles that hit the *Blaine* were fired by NKA patrol boats or had come in via low air attack, launched by a Russian aircraft out of Cam Rhan Bay or from the southbound Soviet Far Eastern Fleet. But now the request from SACEUR—Supreme Allied Commander Europe—endorsed by CINC south and CINC north, rendered the question about who fired the missiles academic.

"The North Koreans have tripped the whole goddamned thing off anyway," said Trainor.

A call came in from Premier Suzlov. He was demanding that Mayne order NATO to cease its fighting and surrender all territory gained by the Soviet–Warsaw Pact countries.

"Mr. Premier," answered the president, "I don't want this. You don't want this. Call off your people in Korea."

"I have no people in Korea. *You* have people in Korea."

"I mean call off Pyongyang."

"We have no people in Pyongyang. It would be interference in the internal affairs—"

"Then," said Mayne calmly, "call off the 'fraternal assistance' you are giving East Germany, Poland, Czechoslovakia—"

"We did not begin this, Mr. President. Your act of terrorist aggression in East Berlin—"

"That might well have been terrorist, Mr. Premier, but it was not an act of any government in the NATO alliance. Of this I can assure you."

"Assure me? You can assure me of nothing. However, if you contain your NATO armies, I will agree to—"

"Mr. Premier?"

The line went dead.

"What the hell—" began Mayne.

The NSA electronic experts overseeing the White House and situation room communications punched all the right buttons, including time-of-conversation-cessation tone for number coding into the computer. Most likely explanation, they informed the president, "connections purposely cut."

The president looked up, astounded. "By whom?"

The Joint Chiefs of Staff knew and Trainor knew they knew, but coming from them, it could look almost self-serving. It was for a moment as if each of the Joint Chiefs, aware that history was being made, did not want to come out on the wrong side of it.

"Some of Suzlov's generals see this as their chance," Trainor proffered quietly, adding in an even quieter but more chilling tone, "As the British would say, 'in for a penny, in for a pound.' "

"What in hell does that mean?" snapped the president.

"Go for broke," Trainor answered quietly. "The NATO forces might be reeling. They won't get another chance like this for a hundred years."

Harry Schuman, sitting next to Admiral Horton, was nodding in agreement. "Hard-liners have been fretting ever since Gorbachev's reductions. I think Mr. Trainor is correct in his assessment."

Mayne was rubbing his forehead. "All right, General Gray. What'll it take?"

"Rollover, sir." He meant the NATO policy of "Atlantic necessity," of the U.S., British, and other NATO navies, but primarily the U.S. and British, having to accept enormous losses, simply roll over them, to get the reinforcements of men, matériel, and food to reinforce Europe if they were to have any hope of pushing the Russians back.

"Mr. President," said Admiral Horton, "in the first hundred and eighty days we're looking at a minimum of six thousand cargo ships. Each ship making six round trips. Means a minimum of thirty-four cargo ships a day—excluding naval battle groups for escort and carrier air cover for the convoys."

"Can we do it?"

"We do it or we lose Europe."

"Do it," said Mayne.

At that moment Trainor knew that from here on in the government of the United States would function from the bombproof shelter of the White House situation room and that Senator Leyland had just lost his bid for the presidency.

On Capitol Hill, the entire place under the heaviest security ever seen in Congress, President Mayne, for purposes of national as well as European morale, made his address from the House of Representatives, as Roosevelt had done, to show Europe that both Democrats and Republicans supported him. As he mounted the podium, the silence was palpable as the speaker invited the president to address the Congress.

"This day, as you know, war has broken out in Europe. Soviet and Warsaw Pact armies have attacked the NATO alliance through the very Iron Curtain that for years past has been the front line between the forces of freedom and those of oppression. And once again the United States has been called upon to stem the tide of tyranny, a tide given its force by those in Moscow who in their unshakable Communist determination wish to rule not only the Soviet Union but the rest of the world. I have asked Premier Suzlov several times this day through the offices of the Soviet Embassy to agree to a cease-fire by midnight tonight Washington time. Should I not receive the answer we want—that hostilities shall cease at that time, that both NATO and the Soviet–Warsaw Pact units will withdraw to the positions they occupied before hostilities began—then we must understand that a state of war exists between the United States and the Soviet Union and her allies.

"I ask you all to join with me, to pray for peace but to stand ready for war. We, like all our allies, hold our breath for all mankind, for all those on the edge of the abyss. But should it come to this, we are resolved to fight if we have to, with all the resources at our disposal. Let us be calm, but let us be firm, firm in the conviction of those Americans who have gone before us, for those Americans who went to do battle with the evils of Hitler and all those of his ilk who would make us slaves and extinguish the flame of freedom.

"Rest assured that the United States will do all in its power to bring the hostilities to a quick and peaceful end. But if our overtures of peace are rebuffed, then the fate of our children is at hand and will reside in our determination, as Americans, to stand up to a bully in the only way we know how: in the words of another American, 'to give him a thrashing he'll never forget.' God bless you all."

Trainor was stunned, as was a good part of the Congress, by both the brevity and starkness of the president's oration. Suddenly the Congress erupted in applause as the president walked from the podium, surrounded by a standing ovation. The people, Trainor saw, had done what Americans had always done, rallied about the presidency in times of national peril.

The Secret Service contingent was unable to hold the members of Congress back as they crowded about to shake the president's hand, but Mayne, his demeanor calm, his stride purposeful, walked up to Senator Leyland and extended his hand. For that moment in the nation's history, following the speech, beamed all over America, time seemed to stand still and there were no Democrats or Republicans, blacks or whites, there were only Americans.

Trainor's beaming smile was not simply that of a PR man's victory but one of genuine affection for his boss; the Hemingway paraphrase about thrashing a bully was a master stroke, he thought—so long as no one pointed out that Hemingway had committed suicide.

By 4:17 A.M. the Fulda Gap was choked with armor, the destruction by the U.S. Fifth Army's artillery so unabatedly concentrated that east of the Gap, the plain looked like some vast scrap-yard strewn with the steaming hulls of over a thousand Soviet–Warsaw Pact tanks, the majority of these being obsolete T-62s, of which the Russians alone had twenty-three thousand in reserve and which were unofficially known among the Moscow general staff as *dryan*—"fodder." But the tanks had kept coming, and at times, with the air-ripping scream of wire-guided and fire-and-forget antitank missiles, blanket artillery fire and the cacophony of assorted small arms, barrels threatening to jam from unprecedented sustained action, U.S. and West German troops manning tank graders had to be called in to clear a way through as more and more NATO units arrived.

The A-10 Thunderbolts, or rather those who managed to penetrate the increasingly accurate Russian SAM screen, continued to buoy Meir's spirit. Though his hearing had long gone in the deep fallback bunker at Alfa Two, he still managed to glimpse with awe the twin-engined jets screaming in just above tree level, their seven-barreled Avenger Gatling guns ripping into the oncoming S-WP armor, the stream of the thirty-millimeter cannon fire stuttering into the tanks at over seventy rounds a second, many of the tanks exploding, illuminating others nearby, which became the Thunderbolts' next targets.

The antitank missiles were doing well, but it was the A-10s that, despite their losses, continued to be the best antitank weapon NATO had in the field, augmenting the fire of the hidden M-1s, whose 120-millimeters kept thumping away from the woods, although many of the American tanks, over 270, had been destroyed by the Soviet–Warsaw Pact onslaught.

As the carnage continued, the biggest surprise to the NATO generals was that the highly touted and sophisticated Swedish BILL antitank missile system, proven so effective at homing in above and destroying the thinner turret armor, was being foiled by the less-sophisticated Soviet T-72 tanks with reactive armor packs *on* the turret as well as around it and by foil-spewing ejectors that scrambled the incoming missiles' homing radar. The BILL top-attack warheads, however, extracted a savage price in S-WP infantry—the Russian BMPs', personnel carriers', much thinner armor and tracks easily penetrated by the missiles. But still the Russians and East Germans kept coming, astonishing the forward infantry companies of U.S. Fifth Army by using infantry in several instances to clear a minefield before the tanks by running through it, an old Russian tactic from the days of Stalingrad.

Then at 5:00 A.M., 4:00 P.M. Washington time, Russia's C in C of the western military theater, or TVD, Major General Agursky, received reports that brigades from West Germany's First Armored and Second Mechanized Divisions were being diverted to reinforce the NATO semicircle of armor around Fulda Gap and that the British First Army, in position south of the Elbe, was moving farther south to shore up the areas depleted by the West Germans heading to Fulda. Agursky gave orders for the second phase of Operation Home Rule to begin.

The Soviet Ninth Armored Division, leading an attack of thirty Soviet–Warsaw Pact divisions, fifteen armored, fifteen mechanized infantry—over 430,000 troops in all—struck and broke through in the far northern sector of NATO's front, twenty-four miles east of the Elbe, racing for Hamburg and Bremerhaven, the prime designated ports for U.S. resupply of NATO. NATO headquarters had never been sure of Denmark's willingness to intervene against such a Russian–Warsaw Pact right hook south of Denmark's border, and eventually NATO's high command was proven correct, for while the West German Sixth Mechanized Division in Schleswig-Holstein attacked bravely and without hesitation, aided by elements of the eighty-four thousand First German Corps and northern elements of British First Army, the Danish Parliament debated the advisability of becoming involved. By the time they'd decided to send a stern note to Moscow, the Soviet–Warsaw Pact blitzkrieg, spearheaded by five thousand Soviet T-90s, was racing through the dawn toward the vital ports of Hamburg and Bremerhaven, the T-90s' 135-millimeter laser-guided cannons blasting everything before them. The most strategic bridges over the Elbe were quickly in the hands of the Soviet 207th Airborne, who started a row at Soviet-WP HQ by taking all the glory for having secured the bridges when it was in fact SPETSNAZ, special force teams, already in place in the west, who had secured the bridge crossings by thwarting NATO demolition teams in the first place.

At Fulda Gap, one of the T-90 five-tank-platoon leaders was Lt. Sergei Marchenko, twenty-four, younger son of Kiril Marchenko and who, because fate had decreed that he be less than five feet six inches in height, had automatically been conscripted to the Soviet Tank Corps, the 2.10-meter-high T-90 being the lowest silhouette of any tank in the world. But what Sergei Marchenko had really wanted to do was fly. What his father wanted him to do was to obey orders so that in time, *if* he acquitted himself well as a tank commander, he might qualify for transfer. Right now, however, all that Marchenko was concerned about was staying alive—the inside of his T-90, devoid of the air-conditioning of the American M-1s, extraordinarily hot due not only to the heat generated by the tank engine and the motors that moved the ten-ton turret but by the smoke and heat generated by the scores of other tanks in the 270-tank division moving toward the Gap. Many of the tanks were now more than fifty meters apart, violating the normal "twenty-five meter" rule of Russian armor. It was an on-the-spot decision by the corps commander to get his tanks as far from one another as possible in an effort to avoid the heavy punishment being meted out by the Thunderbolts and the legendary concentrated artillery fire of the Americans, many of whose guns, as in Fulda Gap, had long been pre-positioned to saturate grids within grids. Apart from the heat in the tank, Marchenko, after two hours of combat, found the noise of the turret grating and squeaking, the crash of the automatic loader, the high whine of electric motors, and above all the radio traffic, was so overwhelming in its disorientation, he removed his headset. It was a court-martial offense during full attack, but whoever wrote the rules in Moscow didn't drive a tank.

Lowering his dust goggles, he stood up, one arm resting against the 12.70 machine gun, as he watched another wave of armor and armored personnel carriers barely visible forging through the dust and smoke, the whole division under orders that once they were through the Gap, they were to wheel south in an effort to engage American and German reinforcements on the Fulda front. What they needed now, thought Sergei, even more than air-conditioning, was for the Soviet MiGs to deal a death blow to the F-15s and F-16s and to clear the skies of damned American Apache helicopters that were so good at ducking down in gaps between the woods and ambushing two to three tanks at a time before they were blown out of the air. Unfortunately the Soviet Havoc helicopters at a distance could easily be mistaken for an American Apache, and some of these had reportedly been hit by mistake. Above the smoke and crash of the ground battle, over sixteen hundred NATO and Soviet–Warsaw Pact aircraft, from high-performance jets to subsonic ground-support aircraft, were engaged in a fierce battle, the NATO air force's biggest problem being that, although they could launch over three sorties per plane in the first ten hours compared to the Russians' two, their heat-seeking missiles had in fact taken out over thirty of their own aircraft in the high-tech confusion over Fulda Gap, where Murphy's Law operated with even more devastating effect than in peacetime, when laser beams fixing on the wrong target simply resulted in an embarrassed pilot. In the skies over Fulda it meant the death of a pilot, and while the West could replace fighters at a faster rate than the Soviet–Warsaw Pact alliance, the Russians' reserve of pilots had been carefully built up to a three-to-one superiority over NATO.

As Marchenko looked above him, glimpsing dogfights through the wafting smoke cover, he wondered who was getting the best of it. He wasn't even sure what was happening in *his* sector, let alone what part he played in the master plan. All he knew was what any other commander, tank crewman, or infantryman knew. Their local action, no matter how small, was merely part of the master plan hatched by some genius out of the Frunze Military Academy after Gorbachev had so stupidly signed the INF, ridding NATO of all its medium-range missiles and so forcing NATO to face only two possibilities in Western Europe: either a modern conventional war now under way or all-out, long-range nuclear holocaust in which no one would be the winner.

Marchenko's troop of five tanks had stopped for refueling when they got the message "German armor ahead." Marchenko felt his stomach tighten—half fear, half excitement. If you beat the Germans, you'd done something. The Americans were tough, but this was German home soil. "What are they?" Marchenko asked. "M-1s or Pattons?"

"Leopards."

"Ones or twos?"

His wingman signaled that they had finished refueling.

"What's the difference?" asked the sergeant over the radio from the starboard tank. "A Leopard's a Leopard."

"Oh," said Sergei with mocking nonchalance. "No difference. You clod. The Mark Two's reach is an extra thousand meters. A slight advantage over the One, wouldn't you agree?"

That was the major trouble, thought Sergei—never quite knowing what you were up against. His was only a small piece in a great puzzle of war. Now, if you were a fighter pilot—then at least you'd have more freedom of movement than in one of the nine hundred tanks now wheeling en masse to engage the West Germans, the biggest problem for the next few miles being a collision in the damn smoke.

Then news came that another lightning strike had been unleashed by Major General Agursky, this time a left hook around the Czechoslovakian-Austrian corner near the Bohemian Forest across the river in Inn—fanning out on the ancient Danube plain and streaming into Bavaria.

"Who told you that?" shouted Sergei.

"American armed forces network."

Sergei smiled, despite the dust-thick air that was making it almost impossible to breathe. Why did the Americans tell everybody? He couldn't understand it. They had to tell everyone everything. They were telling the world NATO was on the run, reeling two hundred miles to the north, the S-WP forces smashing through the Dutch 415th Armored and now two hundred miles south through the Austrian border, splitting the NATO defenses into three sectors. Clearly Agursky was set on the right course, seeking to divide the NATO pockets, then pulverize them into submission before America's enormous production potential to resupply could come into play. And the farther the S-WP penetrated into West Germany, the less likely it became that the Americans could even consider the nuclear alternative. The radio crackled, informing Marchenko that the Leopard tanks were Mark Ones.

"God is good," Sergei said, and got a belly laugh from the driver, who was so wound up that the thought of his T-90 having a thousand-meter advantage over the Leopard One seemed to him nothing less than a gift.

"I'd still rather be up north," put in the driver, "if the American radio is right."

"Why?" asked Sergei.

"I'd rather be fighting those Dutch hippies," said the driver. "The Krauts are a different matter, Comrade."

"Ah," said Sergei, dismissing the odds, "we'll shit all over—" The tank swerved violently to the left to avoid a forty-five-degree antitank slab. There was a tremendous *thwack* and Sergei saw a fine red mist, then his gunner's head rolling by his feet, the man's torso bubbling with blood.

In the south on the Donau, or Danube, plain, the weather over the Bohemian Forest was closing in, hampering NATO's Fourth Allied Tactical Air Force, so that to its Luftwaffe commander, General Heiss, it seemed that even God was against NATO.

Before Congress had even heard of and ratified the president's declaration of war, the British convoy, under British naval escort, was already under way, the U.S. Navy to take up escort duty twelve hundred miles north of Newfoundland, and in Manhattan, Lana La Roche, née Brentwood, had carefully aligned the arrows on the child-proof safety top of her vial of sleeping pills and poured them down the toilet.

Earlier that day she had left her apartment on the Upper East Side and walked down to the Plaza—for some reason, which at first she couldn't explain, she found Central Park to be a kind of magnet in her depression. She had scrupulously avoided watching television or reading a newspaper, for her own bad news about her marriage—Jay would still not agree to a divorce—was enough to cope with. And it was a long time before she realized why she had been going to Central Park, often at night. It was dangerous, a punishment for her failure in her marriage, at college, at living. Then, whether she liked it or not, the world came crashing in on her.

She had been standing by the park wall, across from the Plaza's north entrance, barely noticing the traffic sliding by—a young man showing off, coming out of the hotel, crossing over to the horse-drawn cabriolets, bowing deeply before a bejeweled blonde, twice his years. Soon she would grow old and he'd still be young.

The whip struck the horse's flank and it began the lover's walk through the park. Raucous rock was booming from the band shell, and roller skaters with ghetto blasters weaved by. Why didn't they get Walkmans or earphones or something and just blow off their own ears? she wondered. Then she saw someone nearby reading *The Times*, its banner headline telling the world that an American warship, the USS *Blaine*, had been hit in the East China Sea. She had felt her heart racing with the shock, yet simultaneously she felt a surge of exhilaration for the overwhelming fact was that for the first time in over a year of utter defeat, she knew exactly what to do. A few days in New York to get things set up and then to California.

David Brentwood hadn't had the freedom that his sister, for all her troubles, had, and as one of the six-month reservists, his was a stark choice. Deferment, then go where the army sent you, or volunteer now for the marines. Stacy and Melissa, who had come over to the ROTC office, were watching him.

"Marines," he said. The look on Stacy's face was worth it—or so David thought until David got to Parris Island and quickly came to the conclusion that it had been the dumbest decision he'd ever made. He'd said "marines" to impress Melissa. That bastard Stacy had conned him.

CHAPTER THIRTY-FOUR

"A BAKER'S BAR?" asked Lana.

"No, a bak*er* bar," explained the surgeon at San Diego's War Vets Hospital. His Swiss-German accent was clear, his manner polite, coat as white as the gleaming walls of the intensive burn wing. "It is named after its inventor." After all night on the red-eye flight from New York, Lana still wasn't getting an answer to her question of why she was unable to see Ray. God, if she'd known this—hadn't been so impulsive in the first place—she could have stayed in New York. But then, she didn't want to be in New York. That was the whole idea—to get out, away from the pills, the corrosive self-pity. To be with her brother. The plan was to comfort *him*, to think of someone else for a change. Their parents.

Beth and the two children had come down again from Bremerton. It was supposed to be what a family did in crisis, but here was Dr. Franz Lehman, maxillofacial specialist, seemingly still unmoved by their pleas to see him.

In fact, he understood their concerns very well. It was for that reason he was so insistent on them *not* seeing Ray so soon after the first of the long series of operations that would be necessary to reconstruct something that resembled a face. Skin grafts from the arm and hip seemed promising, but in any event, it would be a long time.

The "baker bar," Dr. Lehman explained, would be only a small part of it, a curved four-inch gold rod cemented to the top jaw, replacing bone that had been smashed by shrapnel. After building the bar, anchoring it either side in the remaining bone, a partial denture of five teeth could be attached to the bar by two small inverted U-shaped clips inside the denture, the gum line being a meld of plastic and gold to reduce expansion and contraction coefficients. A millimeter difference could cause maxillofacial strain on the mandibular joint, and the telltale click, in most people merely the sound of the joint functioning, would in Ray Brentwood's case cause massive malfunction, the precursor of severe headaches that would involve the whole head and radiate deep into the neck, shoulders, and lower back.

"Couldn't we see him for just a second?" Lana pleaded, though she knew well enough from her premed days that it was useless to push. But she felt somehow that if she could see him, it would help—just say a few words to him.

But in the doctor's experience, it didn't help at all. It was doubtful whether, under the pain medication, which was only partially effective in burn cases anyway, the patient was fully cognizant of what was going on, and often the confusion only added to the burden of inner anxiety being suffered.

It was better to wait awhile, not because the patient didn't need support or, like so many Vietnam vets, was refusing to see anyone out of a sense of bitterness, of having been rejected by his country. But the ugliness of the tight, polished skin where the fire aboard the *Blaine* had burned him made Ray Brentwood want to withdraw into a deep cave of cool silence, where the pain from the searing fire that made every breath an agony might finally abate. In that cave, Dr. Lehman knew, the greatest fear of all could be hidden for a while, to prepare him for the shock in the mirror of the visitor's eyes, the eyes that could never lie despite the visitor's determination not to betray shock or any other kind of surprise. In Ray Brentwood's case there was an added torture to the constant searing pain of his disfigurement: As captain of the *Blaine*, he was solely to blame not only for his own condition but for all those, one hundred forty-three men, who had died in the attack on the *Blaine*. Like the parent of a lost child, he constantly replayed the attack. If only he'd given an order for left-hand rudder, perhaps the missile might have missed them altogether. If only he had picked up the skimmer on radar—or was it fired by a ship? In which case he couldn't be blamed. Could he?

What stoked the nightmarish obsession and knotted his stomach with anxiety was the thought of the coming inquiry, when he would have to explain to the board why he, the captain, had abandoned the USS *Blaine* while she was still afloat. Of course, he must tell them the truth—but what was the truth?

He remembered being conscious at the time, but couldn't remember giving the order to abandon ship, and the worst of it—the worst possible outcome for a captain—was that the *Blaine*, though badly holed, listing dangerously to port and spewing flame, had her fires doused by monsoon rains that night only to emerge, her fires out, towed in ignominiously by the *Des Moines* to the shipyards at Nagasaki, where she was now undergoing repairs.

Whatever had happened, Ray Brentwood knew that as captain of one of the United States Navy's most modern warships, he was the man responsible. And along with the dead skin of his body now blistering and peeling, revealing pink and custardlike mucous below, Ray Brentwood saw the death of his honor as well as his career and all the dreams he had brought with him from Annapolis, which during the fifteen years he had worked so hard to attain. Exonerated or not, in that deadly phrase of the navy's, his being named a "party to the inquiry" meant he would never command again.

Early one morning he had managed, against hospital orders, to ease himself out of bed. It had taken him fifteen minutes, his legs in poor shape but not broken, to shuffle with the stoop of an old man—to straighten up was to stretch the skin—toward the washroom. He had sat for twenty minutes trying to defecate, but the painkillers had constipated him badly and it was impossible for him to do anything. When, slowly, he got up, he risked a look at himself in the mirror, now that some of the gauze had been removed. The only discernible features were his eyes, the rest of what had been his face an angry red of loose and taut skin, the two orifices of what used to be his nose obscenely turned up, one hole smaller than the other where the cartilage had collapsed, his breathing hoglike because of the melted flesh obstructions inside, his lips peeled back, suture scars on scabrous skin seeping. His speech, they told him, should be all right after much of the internal swelling from the operations had diminished. In any case, he would have to continue being fed through straws, daily astonished at how different food tasted when pureed to slop. In the vortex of pain and loneliness, only death seemed to offer any solution, but he'd been brought up a Catholic, and though no longer a practicing one, those early years had marked him too, teaching that despair was a sin itself. Even so, had it not been for Beth and the children, he doubted whether he could have held on.

He refused to see any of his family, and Robert, the one member who John and Catherine Brentwood thought might have the best chance of getting through next to Beth, because of security considerations covering submarine whereabouts at sea, could not be informed, his Sea Wolf 2-SSN USS *Roosevelt* being on patrol.

Beth had been allowed to see Ray only once. And he asked her not to bring the children. She didn't argue with him, and after she had been carefully robed in antiseptic gown, cap, and mask, she was only grateful that he could not see anything other than her eyes. She could not sit on the edge of the bed, because the slightest change in surface tension over the sheet would cause Ray's whole body to stiffen in pain, which the doctors called "discomfort."

"Do you good in here," Beth had said bravely. "Lose a bit of weight before you come—" But she couldn't say "home." Home had been another place, had been another time, when the world was another place. Everyone and everything had changed forever.

"I love you," she said, gripping the bed rail, willing herself again against her heart not to touch him. He mumbled something, his voice nasal and rough like someone with bad flu. Even a cold, they said, would kill him in his present weakened condition.

The very worst thing, she told Ray's parents and Lana, was that because of the suturing they'd had to do on his face, he was unable to smile for fear of tearing open the now-closed fissures.

It was a thing you took for granted, she told Lana as they sat together in the visitors' room, waiting for the doctor's latest report. With Ray unable to smile, Beth told Lana, there was no way of knowing he was all right, no comfort for her or for anyone else around her. Lana wasn't really Beth's type, Beth had always thought, too busy . . . always wanting to do something . . . not bossy exactly but a doer who made you mad with your own inaction . . . ate all the right foods, and smart. Not that there was anything bad in being vigorous, healthy, or clever, but sometimes after a day with the kids, she felt such a flop compared to Lana. And a little bit of jealousy—here was Ray, disfigured for life, and there was Lana, wife of Jay La Roche. What could she possibly know about their kind of life, with her big cars, always dressed like a fashion plate? Some said she looked like a brunette Grace Kelly. Not a hair out of place. Never disorganized, and nice on top of it all, especially about Ray. But then, she could afford to be. Next week it would probably be Palm Springs or some charity ball for President Mayne's war bond drive. She could walk away from the pain.

". . . Of course, they might not take me," Lana was telling Beth. "I used to have asthma as a child. At least, that's what they told me. Personally I don't think they were ever sure whether it was more an allergic reaction. Anyway, I had to choose, Daddy said. Asthma or Prince, our dog. I suppose Ray's told you all about it. I kept the dog. My parents had a dingdong row about that one."

"*Who* mightn't take you?" asked Beth, trying to disentangle the conversation.

"The Waves," said Lana. "I have a year of premed, and they told me before I left New York that—"

"But what about your husband?"

Lana was nonplussed. "Beth, haven't you heard a word of what I've been—" She paused. "Well, they say I'll probably be posted to Halifax. They're so short of nurses that it'll be sort of learning on the job."

"Where's Halifax—that in Canada?"

"Yes, somewhere in Nova—"

Lana and Beth could see some kind of commotion—several patients, including one in a wheelchair, were hurrying toward them to look at the big TV screen in the visitors' room. News flashes were coming in that the Nationalist Chinese from Taiwan, under U.S. pressure, were backing off from getting too close to Mainland China. There was some generally poor footage of tanks in a rainy field, the scream of an antitank weapon and the tank upside-down for a minute, the reporter's rapid breathing caught on the sound track as the camera was righted, the reporter explaining that the tank had just been hit by an antitank missile.

"Oooh!" said a petty officer stopping by the door, looking up at the TV. "Really? An antitank missile?" He looked across at Lana and Beth. "Some of these TV guys, I tell ya—" He looked back again at the screen and saw two wide red arrows spreading north and south over a green map of West Germany.

"Where are the French?" the petty officer asked.

"I don't think," said Lana, "France is in NATO."

"They are and they aren't," said the petty officer. It was only as the man turned round toward her that Lana noticed he had one arm, the other's shoulder stump covered by a pajama sleeve rolled and pinned up.

"They're still in it," continued the petty officer, "but they wanted their own command structure. Withdrew from the joint command structure in sixty-six. They want the NATO umbrella, so they pay membership, but want to use their forces how *they* want." The PO looked at Lana. "Autonomous command," he said derisively. "You know—like I'm a member of the ball team, but when I don't want to play, I won't."

"So long as they help," said Lana, "I don't care how autonomous they are."

"The French help the French, lady," said the petty officer. "Always looking for a backdoor deal."

"Such as?"

"You don't come into France and we won't fight you. We'll keep out of it."

"Who's *you*?"

"Anyone who might muss 'em up. Remember they wouldn't let us fly through French airspace to hit Libya? Only Maggie Thatcher stood with us. To hell with 'em," said the petty officer, walking on. "Only thing I like about the French is French toast. They'll sit on their butt till Ivan's got Germany, then he'll want Alsace-Lorraine and then he'll want France, and have forty divisions all down the line. Then the Commie C in C can ride down their Champs Elysées on his white horse."

"No," said Lana, "France'll come in."

"Well," said the petty officer, turning back, "I won't be there." It was said with relief but also with regret. Beth envied him in a way. He'd lost an arm, but compared to Ray . . . Lehman had told them that in the end they might have to consider a porous skin mask that would allow him to go out in public, but he would have to take it off now and then, like dentures at night, to help keep the skin clean. Be best to remove it at night, let the skin breathe.

What would they do, wondered Beth, if he wanted to make

love? Would he still be able to do it after all this trauma? They often said that after combat, high stress, many men just couldn't do it anymore. If that happened, Ray would get mad. Then there'd be more stress. Maybe she could just do it for him some way that wouldn't—and she thought of what he must feel: a young man, captain at thirty-seven, clearly marked for promotion, and then—so suddenly, so terribly fallen from grace.

Thank God the navy would pay the medical bills. Jeannie had said, "If Daddy loved us, he'd let us see him."

Beth had torn into her. "Don't be so goddamned selfish, Jeannie! He's hurt. Very badly. You'll understand when you get—"

"I know," Jeannie said, sobbing. "But we miss him, too."

Beth had folded and taken the children to see him. The nurse informed her that Captain Brentwood did not wish to see anyone yet. Exhausted, Beth looked down at the two children. "If it was one of us all burned up, what do you think Daddy would do?"

"Come see us," said Johnny.

Alerted by the nurse's station, Lehman intercepted them in front of the burn unit. "I don't—" he began.

"They *want* to see him, Doctor. Once the IVs are taken out, it will be all fine."

He didn't know what she was talking about—she was obviously quite beside herself.

"Your father-in-law was in this morning. . . ."

"And?"

"I'm afraid your husband didn't want to see him either. He wants to make his own recovery, Mrs. Brentwood. In his own good time."

Beth's hands clenched as she held Jeannie and John, straining for control.

Dr. Lehman, flashing a smile, had knelt down next to the small boy. "Your Daddy needs a lot of sleep right now. When you're feeling, how do you say—'yucky,' well sometimes you just want to go to bed and not see anyone till you're better. Isn't that so?"

"Yes," put in Jeannie. She liked the doctor; he was the kind of father figure she always expected of doctors. "I think the doctor's right, Mom," said Jeannie, tugging at her worriedly.

"Yeah, Mom," chimed in Johnny. "We shouldn'ta come."

Beth had about-turned, her high heels striking the hard, highly polished floor, echoing the full length of the ward. She spoke only once, at the entrance of the hospital. "Okay, that's it, you two. We are *not* going to the hospital again until your daddy asks for us. Understood?"

John nodded. Jeannie became "little miss proper when you're out." "Yes, Mother."

Beth jerked Jeannie's arm. "I brought you all the way down here because you pleaded, *begged* me to do it. Then you embarrass me like this."

"We're sorry, Mom," said Jeannie.

Johnny thought about it for a minute. "Mom?"

"Yes?"

"If Daddy dies, do we get all his money?"

Once again realizing she hadn't been listening to Lana's plans for a new, hopefully more useful, life, Beth was forced into noncommittal murmurs, trying to cover her inattention.

CHAPTER THIRTY-FIVE

PARRIS ISLAND. SOUTH Carolina.

David had heard all about it, laughed about it, and determined it wouldn't get him down.

As he'd stepped off the bus in the darkness, David could smell the last of the purple oleander blossoms from the trees that had flashed past the Greyhound as it made its way over the long causeway to the island. In the dim glow of the bus's cabin lights he could see a DI, peaked scout hat, strap at the back, khaki shirt and pants pressed with knife-edge precision, and could hear the sound of insects from the tidal flats—then a voice. Demented.

"Shut your fucking mouths! All your shit off the bus into the barracks. *Now!*" David got such a fright that, rushing back into the bus to get his kit bag, he stumbled on the bottom step.

"What's your fucking name?" screamed the DI.

"Me, sir?"

"Yes, you. You're the only goddamned stumble-ass around here. What's your fucking name?"

"Brentwood, sir."

The DI leaned forward. "I can't hear you."

"Brentwood, sir."

"I can't hear you."

"*Brentwood*. Sir."

"Fucking who?"

"*Brentwood. Sir!*"

"Wrong. Your name's Stumble-ass. What's your name?"

"Stumble-ass, sir."

"What?"

"Stumble-ass, sir."

"*What?*"

"*Stumble-ass*, sir!"

"Right. Now get your kit and run!"

David wished he'd joined the navy.

Inside the white building there was a sparse barrack room, an antiseptic smell, a line of double bunks down both sides, the same DI standing, hands on hips, waiting for David, the last recruit in after the delay at the bus. It meant he got the only bunk left—right by the door. There was a whisper.

"Who spoke?" shouted the DI. It was the first thing about a DI that David noticed. They didn't "roar" like lions, they shouted, a tad below hysteria.

A black recruit stepped forward, putting his hand up.

"Put your goddamned fucking hand down until I tell it to move."

"Yes, sir."

"Yes what?"

"*Yes, sir!*"

"You talk again, string bean, and I will personally cut your balls off. Do you understand me?"

"Yes, sir."

"Yes what?"

"Yes, sir!"

There was a long silence, at least five minutes. No one moved. Someone broke wind. The silence continued, then the DI, his voice rolling over the already tired and demoralized recruits, recited the litany of reception in a deliberately unemotional monotone, which made it even more foreboding for the recruits, though it consisted of telling them the obvious: where they were, Parris Island—though some thought it was a nightmare they'd woken up in. Most wanted to get back on the bus, but the bus had gone. The DI informed them that because of the small numbers of the peacetime standing army, training would have to be completed in a much shorter time than usual. If he had his way, he would work the miserable maggots eighteen hours a day, but insofar as Congress in its wisdom had decreed that a maggot had to receive seven and one-half hours of "uninterrupted sleep," the maggots would have to work much harder than was usual for maggots in order that they *might* qualify as members of the "world's finest fighting organization," the "United States Marine Corps," and that they were to do exactly as they were told to do, would not speak unless spoken to, and always end a sentence with *"sir."*

Strange insects buzzed and flapped against the gauze, reincarnated recruits, though David, trying to remember from his zoology course whether it was the female mosquito that made the zinging noise or the male. Anyway, it was like the DI: the one you didn't hear was the one that got you, the one you'd like to kill. One of the flourescent lights had started flickering. They were still at attention.

"Stumble-ass!"

"Yes, sir!"

"What's in that plastic bag?"

David was so rattled, he couldn't think. The sweat was pouring down his back. "Ah—candy, I—" It was some candy Melissa had given him when he'd left.

"Candy!" bellowed the DI. "Candy, shit! You're not going to have fucking time to chew fucking candy, are you, Stumble-ass?"

"No, sir."

"Now what?"

"No, sir." David couldn't believe the obscenity. The only person he'd ever heard talk like this was an Australian.

"Take off your fucking pants!" the DI bellowed at the recruits, and began walking down the rows. "Oh—lookit this!" A recruit quickly risked a glance to see what the DI was looking at.

"Keep your fucking head up, turd." The recruit's head shot back up and stared ahead, eyes glazed by fear.

"Oh, look," said the DI, his bellow daring anyone to look. "What have we got here, limp dink? Valentine shorts." David could see far enough up the line without moving his head to see the recruit, a Puerto Rican, his boxer shorts covered with valentine hearts. The DI walked around the man once, reversed direction, and walked around the other way. David had never seen a more miserable-looking soul on God's earth than the hapless Puerto Rican. The DI stood up to his full height, his nose almost touching the recruit, whose head was straining back while at the same time trying to remain at attention without tipping.

"Who gave you these, Thelma?"

"Name is Thelman—"

"Shut your fucking mouth. Your name's *Thelma*. Who gave them to you, Thelma?"

"My mother, sir."

"Mommy. Long as you're here, you'll wear standard issue, Thelma. If you're good enough to be a marine, *which* I fucking doubt, you will continue to wear standard issue. Do you read me?"

"Yes, sir."

The DI turned. "Get your fucking plastic bag up here, Stumble-ass. On the double! Dump it on the floor." David did so. There was a card from Melissa that fell out with the candy. David was beet-red from embarrassment and anger. He knew what the DI was trying to do. Everyone knew. It didn't make it easier that you knew. The DI handed him the card.

"You're the worst fucking lot I've seen yet. War brings out the best and worst in men. And you are the fucking worst. Empty your pockets and kits of all shit. Now. Understand?"

"Yes, sir!" reverberated the barrack.

"Shit for you stupid assholes is any substance contrary to regulation. Aspirin is shit, narcotics are shit, vitamins are shit,

prescription drugs are shit, candy is shit. Booze is shit. Anything
I don't like is shit. Understand?"

"Yes sir!"

"Only wedding rings are permitted. What's *that*?"

"A radio, sir. Transistor."

"That's shit."

The pile of drugs, combs, neck chains, condoms, gum, cigarettes, filled the plastic bag. But this was the easy part. Next
day, beginning with the usual breakfast by numbers at 0430,
David saw his hair, the last vestige of his individuality, falling
from him in great gobs unceremoniously pushed away by an
enormous broom into a garbage bag. Then there were the ob-
sessive "forming" rituals of induction week, the humiliating
"asshole inspection," the endless grueling day of DI abuse, the
numbingly repetitive use of every weapon from M-16s to the
laser-guided TOW, the "Alfa Bravo Charlie Delta . . ." alpha-
bet so that messages might never be misunderstood, the stifling,
nauseating forced run through the gas hut, the issue of standard
condoms along with other standard equipment, the bare dining
room with the words "TACT, LOYALTY, GUNG-HO, COUR-
AGE, TEAMWORK, HONESTY, KNOWLEDGE, MO-
RALE" stenciled on the support posts above the painted
footprints where you had to stand.

What David remembered most was the kind of small incident
that for some reason stays with you for life. It was one night just
after lights out, the moon cold comfort over the Carolina low
country, when the Puerto Rican called Thelma turned to David
and said, "I came here to be a marine—to fight for my country—
not this kind of crap."

From near the "mouse house," the DI's small room with
basin at the end of the hut, a silhouette appeared, its peaked hat
sharp against the halo of the moon. The DI's voice, for the first
time that David could remember, was not shouting. "We dish
out shit, son, because you're going to get shit. The Russians and
their helpers aren't going to throw flowers at you."

The recruit said nothing. The DI turned to walk away to his
billet, stopped, and looked over his shoulder. "One more thing,
you scumbags. If this thing doesn't get sorted out over there,
the Red bastards will end up in your backyard. I want a marine
next to me. I love this country. I love the Corps."

* * *

By the time David Brentwood, Thelma, and the others had gone through basic training, their platoon of seventy-five had only purple streamers attached to their platoon's standard. This meant that, unlike the "superiors," they were merely run-of-the-mill marines and had not distinguished themselves above the other marines in platoon marksmanship and other battlefield skills. But they were marines and were told they would be among the first used to plug "the gaps in the dike."

"Questions?" asked the DI, his voice now approaching normal, not friendly but not as sarcastic as usual.

"What dike's that, sir?"

"The dikes are everywhere, marine. You just remember everything you've been taught and you might just stand a chance."

Another recruit put up his hand. "Sir, can you tell us how it's going in Korea?"

"Chongju fell last night."

After chow, with graduation next morning, they began the difficult informal good-byes, laced with false bravado lest they be thought too sentimental.

It made no sense to David. He'd hated the place, but now that he'd qualified—in fact, he thought his shooting had been pretty good, though the final marks were not in on that yet—there was a feeling of belonging. In the relatively short time they'd known one another, the men who had made it formed a bond that they instinctively knew would last a lifetime and which marine tradition told them would. And each of them as they packed his kit, knowing he would soon be called to war, to face the fear and all the dangers of the unknown, felt ready, toughened, and in each man's thoughts the honor rolls from Montezuma to Iwo Jima to the fighting retreat from Chosin Reservoir all rang with glory, for as yet none of them had known the smell of it, the feel of it, or the horror. They knew it only in the abstract, and for all they'd been told of what to expect, each of them held the young man's eternal secret: it would not be him, and in that he found his bravery, his willingness to go forth.

On the last day the DI read out the numbers of assignation to the marine platoon of seventy-five. "Devane . . . zero three zero zero. Least you can fire a fucking rifle." There was subdued laughter—training over, the future before them. "Brentwood . . . zero three zero zero—looks like you and Devane are for MAGTAF."

"What's that, sir?"

"Sounds like a disease," said someone else.

"Marine air-ground task force, idiot. Out of Camp Lejeune. Means they think the fuckers can shoot, scratch ass, and jump from a plane at the same time."

"Whose ass?" someone put in.

For the first time the DI did not bawl out at the goofing off. It was as close as a DI would come to showing affection. "Thelman, zero three zero zero—MAGTAF." The DI ticked him off the list. "Do I train 'em or what?"

"*Yes, sir!*"

"Webster, one eight zero zero. Engineers—means you might be fixing toilets."

There was silence, the DI's clipboard in front now that he'd finished reading off the whole platoon. "All right now, here's the score. You made it through boot camp. You're marines. Don't ever forget it. Wherever you go, you'll be marines. Marines don't give ground—they take it. I'm not gonna shit you— right now the gooks and the Russians have got us by the balls, our backs to the sea. I don't know whether you'll be going in, but you're all going to end up in it somewhere. Korea or Europe—'less some dumb bastard pushes the button, which so far hasn't been the case and which I personally don't think they will—anyway, not so long as they haven't blown their wad in conventional arms and got nothing left to throw at us. But wherever you go, you've got one big thing in your favor. You believe in what you're fighting for. When they get tough, and they will, your belief in your country, in the Corps, is an extra shot up the spout. Good luck, God bless."

On the final night, after the graduation, packing his kit for Camp Lejeune, David Brentwood was feeling down. Thelman's stepmother had attended the graduation parade, but neither of David's parents could make it. There'd been multiple bomb scares at JFK and La Guardia, and by the time the passing review rolled around, John and Catherine Brentwood were still on the ground in New York. It was one of the great disappointments of David's life. He was quite surprised at how disappointing. No mail from Melissa either, but then, mail right across the country seemed to be in a shambles these days—more bomb threats and one big explosion in Chicago's central post office, killing three inside workers and scores of others. They'd said it was sabotage. As David took Melissa's photo down from the "hog" board in the barracks room and Thelman took down his girl, he asked Brentwood whether he believed all that crap the DI had said " 'bout us having an extra shot in the spout. I figure the Russians tell their guys the same thing."

"Probably," said David.

"Yeah, well," said Thelman, "any Russian gets on my two hundred line, I'll blow his head off."

David said nothing. He was still thinking about his parents not being there to see him graduate. He knew it wasn't their fault and they'd had a bad time of it just trying to cope with Ray, his Mom wanting to go over to California every time he had an operation. It became too expensive, and anyway, she'd promised Ray she wouldn't come see him until he felt ready. Truth was, Ray was getting on David's nerves a bit.

David didn't want to admit it, but damn it, the only praise he'd received from anyone in the past ten weeks was during the days when the platoon had gone out for qualification to try for the requisite 190 out of 250 on the firing range. It was the one time when the DI stood back and a coach was assigned to every two recruits. Some were more sarcastic than the DIs, and a few, like the one that David and Devane got, Sergeant Osborne, really cared about the young recruits under them.

Osborne spoke quietly, and rather than push them, he led them into doing a good job. He even spent time with them on the Smith & Wesson .45 side arm practice, usually dismissed as "fun" or "fuckin' useless for marines" except, the DI said, "if you let a fucking Gorby get that close to you—then you don't deserve a side arm. Use your hands. Stiff arm with the left and grab the private's privates." But in an age besotted by automatic wonders, Osborne's approach imbued them with a respect for the old-fashioned .45 side arm. He himself approached the weapon with the respect of a novice taking his first communion.

"Can be used for good or evil, boys," he said. "No matter what the damned liberals say, Commies can come and take you away in the middle of the night. In this country you have to have a warrant. That's a big difference. That's a difference worth dying for." What made him give them that little speech, neither Brentwood nor Devane had any idea—maybe he'd overheard Thelman's cynicism on the night of the full moon. He *was* a little odd, they said, but whatever, he did teach them how to fire the .45: breathe in, hold, "No, no, son . . . squeeze, don't jerk it. I've seen guys qualified with the M-16 couldn't hit a DI's butt with a .45 jammed up him. Way I figure it, there's gonna be a lot of use for these little babies. You take good care of 'em, hear?"

David had worked hard for Osborne because the man not only genuinely cared about what he was doing but reveled in their success. "Way to go, Private," he called out to Brentwood when he made qualification with the .45. "Out fucking standing!"

As the 747 banked in preparation for landing at Camp Lejeune, both marines proudly wearing their marksman's badges on their tunics, Thelman was thinking about Osborne and his love affair with the .45. "Why we going to need pistols?"

"I don't know," said Brentwood, looking down at the brownish-green scrub that was the training area for the marine air-ground task force. "But I've got a feeling we're going to find out real soon. Lejeune's pretty tough. Probably give us a can of beans, one mag each for the .45, and see how we do."

"One can of beans?" laughed Thelman. "Man, I'm way ahead of you on this one. They don't give you *nothin'* in this outfit, is what I hear. They send you out and you have to live off the land for five days."

"What the hell's the use of that?" asked Brentwood. "We aren't supposed to be runnin' from anybody. Marines take ground. Right?"

"It's called tactical retreat. They drop you in a zone. Before your supplies are unloaded from the chopper, it's all blown to hell. Then you gotta make do—live off the land till they can send a whirlybird in for you."

As the 747 descended, they could see marines taking off in harness from the parachute towers, small khaki figures beneath them shaking their fists and seeming to be in apoplexy. DIs.

"Jesus," said Brentwood, "I should have gone for deferment. Latrines."

"Nah," said Thelman. "You're a mover, Stumble-ass. First time I saw you crash that bus step, I said to myself, 'Now, there's a mover.' "

"Shut your face."

CHAPTER THIRTY-SIX

NKA'S FOURTH DIVISIONAL HQ was now situated in the shell-pocked Catholic church in what had been the bustling city of Uijongbu. The smell of the American and South Korean rout was heavy in the rain-slashed air, hulks of three-ton trucks and tanks still burning, but barely, as if even the fires were exhausted. Fresh motorized columns of NKA troops, stony-eyed but flushed with victory, wove through the wreckage of the U.S. Eleventh Corps's fighting retreat. The bodies of American and South Korean soldiers lay strewn about beneath the gunmetal monsoon sky. The heavy cloud cover was continuing to make it difficult, despite infrared look-down/shoot-down scopes, for the few American pilots flying out of Pusan's sabotaged and potholed airstrips to distinguish friend from foe beneath them—especially given the NKA's cannibalization of U.S. trucks and jeeps.

From his crowded cell in the partially destroyed school opposite the church, Major Tae could see over fifty M-60 Pattons that had been knocked out by wire-guided tank missiles, the wires now strewn across the roadways like so many abandoned fishing lines.

General Kim was overwhelmed with logistical problems, the advance in a crucial stage. Because of the monsoons bogging down the American M-1s, he had made enormous gains in territory in the three weeks since his invasion of the South, but the supplies he needed were slow in coming. What he wanted now was to crush as many American and South Korean troops as possible by smashing through the protective triangle stretching from Yosu sixty miles away on the south coast through Taegu ninety-seven miles from Ulsan on the east coast. In all it was an arc of about 160 miles behind which the Americans, after losing Kyongju, were hastily trying to regroup their stunned and exhausted army, forming a defensive perimeter with Pusan, sixty miles away from the arc's outermost limit, as the linchpin.

Kim was haunted by the American counterattack from the southeast corner in the war of the 1950s, and knew Pusan must be captured as quickly as possible in order to deny American reinforcements en route from Japan a beachhead. Of all the NKA commanders, their confidence further boosted by their infiltrators' sabotage of the giant Hyundai shipyards in Ulsan, Kim had the clearest understanding of the Americans' incontestable ability to organize a huge logistical effort on short notice. Like all democracies, the Americans were, of course, "degenerate" and "flaccid" in peacetime, Kim told his staff officers, but galvanized by war, their industrial capacity was *kŏch'anghan*—"awesome."

The only certain way was to deny the reinforcements a foothold, and for this he called for the Kim Il Sung *chasal putae*—"suicide squads"—to spearhead the infantry and armored wedge that he intended to drive toward Pusan. Feints would be made first on the southern flank toward Yosu and on the northern toward Ulsan to dilute the American defense, the main wedge with air cover moving two hours later against Pusan.

Both Pyongyang and Moscow, which had been furious with Pyongyang for initiating the action, believed that, as in the case of Vietnam, once the Americans were driven into the sea, the American peace groups, heavily infiltrated by now-activated "sleepers," together with public opinion, simply would not support President Mayne in any venture that would risk so many more American lives to retake the peninsula. Mayne would fight where he had to, in Europe, but not in Asia.

Despite his preoccupation with what he was sure would be the coming victory over the Americans, Kim took a minute to light a fresh American cigarette and watch Major Tae being marched out of the school to join the line of other captured ROK officers outside the interrogation tent.

"What's his interrogation status?" Kim asked his chief of intelligence. There were only two categories of prisoners in the NKA: "cooperative" and "reactionary."

"We do not know yet, General. He should have been questioned at Kaesong, but apparently he was incorrectly diverted. . . ."

"Have we any leverage?" cut in Kim. "From my experience at Panmunjom, Tae was very stubborn."

"Yes. We're attending to that now. We've been able to keep close surveillance on him. His daughter's boyfriend was an active member of the Reunification Party."

"He's a *migook* lover."

"Yes," said the intelligence chief. Hadn't the general heard him? "We're prepared."

"Never mind about any information on Seoul," said Kim, tapping the bone cigarette holder impatiently on an ashtray made from an American howitzer shell. "We'll ferret out their underground units now that we've taken the city. What I need to know, and quickly, Colonel, is who are the KCIA's counterespionage chiefs in Taegu and Pusan—those who could provide local leadership, sabotage, railway demolition, anything to slow our final attack on Pusan. We cannot attack yet, but as soon as we get enough supplies to Taegu, we will move. But I don't want even a day's delay because of sabotage. Even a day's delay for us could be critical. You must understand the Americans are not only in disarray but thoroughly demoralized by our success in taking them by surprise. They are now at their most vulnerable. It is essential that the civilian population realize we have totally infiltrated their defenses. We must give them no hope that helping the Americans will change anything. We must discover who the chief undercover KCIA counterespionage chiefs are and execute them immediately."

"Of course, General. I understand."

A hundred yards away, standing in the mud outside the interrogation tent with the other prisoners, Major Tae saw Kim watching him from the church's narthex, its canopy scabrous from the shelling, long twists of reinforcing steel rods protruding. From one of the rods the bodies of two men, one American and one South Korean, dangled, turning slowly, tongues grotesquely black, eyes bulging obscenely, signs around their necks marking them as *pihyŏpnyŏchŏkin pantongpuncha*—"uncooperative reactionaries." The thing that struck Tae was that the signs were so neatly made, it was highly unlikely they'd been painted on the spur of the moment or in the heat of the battle but rather prepared long beforehand—as part of a carefully thought-out

NKA policy of terror. Tae felt weak, unsure as to how much of it was due to sheer fear, how much to hunger, trying to remember, as NKA guards moved down the line of prisoners, giving them mugs of weak tea, how long it had been since he had eaten. Two civilian women, eyes carefully avoiding any contact with the guards, were brewing the tea in a large copper washing tub outside one of the tents.

The NKA soldier handing out the tea was smiling at each man as he gave them the steaming liquid. "You help us," he told them, "and there will be no trouble." The carrot or the hangman's noose.

Most of the officers in the line, Tae noticed, had removed their intelligence insignia. One of them, a captain, he recognized as one of those who had been brought up from Seoul for interrogation.

Kim had disappeared from the opening in the church. After tea, Tae thought of the American murdered before his eyes, the man's tiny cross, and after getting the guard's permission, he took off his boot, on the pretext of shaking out a pebble; then, putting the boot back on, he used his forefinger as he retied the lace to feel for the gumlike sliver of potassium cyanide hidden in the tongue. Christians, he thought, would feel for the cross as their talisman; he would feel for the cyanide strip. The knowledge that in this battle against the NKA he would have the final say continued to fortify him.

"Major Tae!" It was an NKA lieutenant.

Tae stood up. "Yes?"

"What are you doing?"

"I—I was fixing my boot—something—"

The NKA lieutenant slapped him across the face. "You are lying. Take off your boots."

Tae did so. The lieutenant handed them to the guard, but he was still watching Tae. "You will follow me."

CHAPTER THIRTY-SEVEN

IT WAS A bright, clear day, a cobalt-blue sea and sky—not a day for war.

Twenty-four hundred miles northeast of Newfoundland in the thirty-four million square miles of Atlantic Ocean, the first British convoy of World War III was under way. Dispatched by SACEUR—Supreme Allied Commander Europe—the convoy, consisting of twenty fifteen-thousand-ton container-type ships escorted by twenty-five NATO warships, primarily British, was negotiating its way past an iceberg floe, for though it was early autumn, the ice sheets still extended from Greenland.

No difficult task for each merchant ship, it was a major headache for SACLANT, the Supreme Allied Commander Atlantic, Admiral Horton in Norfolk, Virginia, as the convoy, designated R-1—Resupply One—was just the first of dozens that would have to be made in the first four-week period. After that, the NATO reserves, particularly of fuel, would be dangerously low, and the Russian monolith, with no such problems of cross-sea reinforcements, would clearly win.

The deadly Cold War game of ASW, antisubmarine warfare, of hide-and-seek between the Soviet sub fleet of four hundred and NATO's 270, had been waged with deadly seriousness ever since 1947, NATO's aim having been to demonstrate to the Russians that the cost of a Soviet sub offensive would be disastrous for the Soviet Union.

But all that was before the sudden surge in Soviet submarine technology in the late eighties, due largely to the American Walker spy ring, who, with other highly sensitive material it had gained access to, had sold the Soviets top information, including the location sites of NATO's SOSUS—sound surveillance system—an underwater network of microphones, or hydrophones, which picked up the movements of Soviet submarines throughout the world.

The Russians also knew that, quite apart from trying to sink a whole convoy, they would win the war if their navy could sink allied shipping at a faster rate than lost ships could be replaced. This was especially true for tankers carrying fuel, which, unlike the Russians, Western Europe had to import. Even the British, who didn't import their oil from the Middle East any longer but from the North Sea, were dependent on the transport of that oil by tankers vulnerable to the Russian subs. It was a simple enough equation, but a devastating one for NATO's forward defense in Western Europe. Admiral Horton explained it by quoting Patton: "My men can eat their boots, but my tanks gotta have gas!" And if Europe went, so would America.

The sub packs from Russia's Northern Fleet came out of Murmansk and down from the Kara Sea—153 of them, the Americans' K-12 satellite picking up their thermal discharge patterns. It was an underwater armada of HUK, Hunter/Killer, submarines, the satellite photos suggesting their course was set for the GIUK, Greenland-Iceland-U.K. Gap, in effect two gaps, one group of subs heading for the Denmark Strait, the other for the Iceland-Faeroe Rise, through which they must pass if they hoped to intercept the convoy. The "sound prints" picked up by NATO's SOSUS hydrophones on the ocean bottom of the GIUK Gap, together with magnetometer readings taken by low-flying Norwegian PB-3 ASW planes via their long-trailing wire antennas, confirmed the satellite projection of the subs' course.

The noise of each sub, as peculiar to itself as the noise of each automobile, gave off different sound signatures or "fingerprints." Fed into the computers of NATO's naval commands, matchups were made with the known noise signatures of all Russian subs ever recorded by NATO. Within an hour Norfolk, Virginia, and Convoy R-1 knew via satellite burst message that of the 153 subs from Russia's Northern Fleet heading for the convoy, 43 were modern snorkel-breathing diesel-electric HUKs, 100 were nuclear, and the remainder, 10 old diesel-electrics, used for training purposes. The dispatch of the old diesel-electrics by Arctic TVD military theater naval headquarters at Severomorsk was viewed by Norfolk as an effort to throw everything at the convoy—not simply to maul it but annihilate it, to demonstrate to NATO that the cost to *them* of reinforcing Europe wasn't worth the candle.

Norwegian air patrols were taken over by American air trackers out of Iceland, where the U.S. Navy escorts would take over from the Royal Navy to see the convoy safely to Halifax on Canada's east coast, the largest northernmost port of North America best able to handle the huge shuttle and storage of materials and ordnance for NATO resupply. Most of the cargo, now being assembled on the Halifax docks, was in containers, one modern container ship carrying as much cargo as twenty-seven World War II merchantmen. This meant that Convoy R-1 would, if it arrived safely in Halifax, equal the cargo-carrying capacity of twenty World War II convoys.

A lieutenant commander at Norfolk, Virginia, asked, why the "Brits?"

"NATO," he was corrected sharply by Admiral Horton. "We're in this together, Commander. And it's not the play-offs. It's the World Series. Right here. Right now," he said, tapping the last-reported position of the convoy.

"Sorry, sir, but why's NATO sending over twenty cargo ships? Wouldn't it make more sense to simply load or expropriate container ships here in the U.S. and Canada—load 'em up and move out?"

"We're doing both," replied the admiral, waving the lieutenant commander's suggestion aside. "We've no time to lose either end. And remember—extraordinary security procedures are in effect. We find a leak from this side of the Atlantic and I'll deep-six the son of a bitch." He was edgy; the Russians knew the GIUK Gaps were the choke points, transiting lanes, deep in places but relatively narrow, that had to be negotiated before the Soviet Northern Fleet could hope to break out into the deeper vastness of the North Atlantic.

Twenty-three hundred miles north of Newfoundland's Cape Bauld, Convoy R-1 was proceeding southwest, the square of twenty empty cargo vessels, five a side, a mile between each ship, surrounded by a larger square-shaped U of twenty-five ASW Sea King helicopters. In front of this double square there was a fan-shaped deployment of Gruman EA-6B electronic-countermeasures Prowlers. Each plane, its telltale proboscis sticking out in front of the cockpit for midair refueling, was so jam-packed with detection and jamming electronics, it was simply too heavy to be armed; the main business of its crew of four—pilot, copilot, and two electronic warfare officers—was to be constantly on the lookout for visual as well as "dipstick" sonar evidence of submarines.

Between the advance screen of Grumans, ASW helicopters, and patrol aircraft and the protective outer square of frigates and destroyers that surrounded the core square of twenty container ships, there was a wide arrowhead formation of ten British Trafalgar-class nuclear-powered attack submarines. Behind these were four older Oberon-class diesel electric subs and a lone Dutch glass-reinforced, plastic-hulled minesweeper of the HMS *Wilson* design, with its corrugated hull of fiberglass on plastic formers looking distinctly ungainly as its twenty-seven hundred tons plowed through the medium chop, and it became the butt of many jokes. The minesweeper had been sent along by SACEUR simply to be "on call," though the NATO commander of the convoy, British Admiral Woodall, suspected that as the ship had been designed for coastal defenses, it had really been dumped on the convoy as a tryout, for it wasn't the Russian Northern Fleet that would now be fretting about mines. In the GIUK Gap, NATO had lain both magnetic and "signature primed" mines, sometimes only a hundred meters apart.

Woodall's SUDO—submarine distribution officer—was matching noise signatures recorded by the Norwegian PB-3s and the SOSUS hydrophone networks that had not been affected by the electromagnetic pulse that had blacked out NATO's computers along Europe's central front. The SUDO tapped the computer's keys with the unhurried competence born in the nexus of long training, self-confidence, and expertise in state-of-the-art submarine disposition programs. There they were on the monitor: the hitherto white sub signals indicating "vessel country unknown" changed immediately to red, subdivided into "DHs"—diesel hostiles—and "NHs"—nuclear hostiles.

Forty miles ahead of the convoy, the blue wrinkle of sea beneath him glinting with sunlight, one of the Prowlers picked up six blips. Instantly alerted, two ASW Sea King helos peeled out from the fan, racing ahead of the convoy, rotors catching the sun, toward the six blips. Four seconds later Admiral Woodall

received coded burst messages from SACLANT, confirmed by
Buda, the underground listening bunker in Norway, and by
CINCEASTLANT, Commander in Chief Eastern Atlantic,
Northwood, England, that the first group of the Russian subs
was approaching the GIUK Gap. The six blips on the Prowler's
radar, however, were not coming from the west but from the
south, forty miles ahead of the convoy. Admiral Woodall instructed his SUDO to "Code Top Secret to C in C East Atlantic.
Where are Soviet minesweepers?"

An integral part of the whole NATO forward fleet flexible
response policy at sea was to cause a major "traffic jam" at the
GIUK Gaps, literally cutting the Russians off at the pass, buying
time for the NATO convoys and forcing the Russians to send in
time-consuming minesweepers. Woodall reminded his OOD that
when in 1984 one mine layer dropped its load in the Red Sea, it
had taken over seventeen minesweepers, eight large helicopters,
plus dozens of support vessels from six countries more than
three weeks to clear the area. NATO had laid the equivalent of
twenty mine layer loads in the GIUK Gaps. Of course, Woodall
pointed out, "the Bolshies might send in remote-control metallic barges, four abreast, to detonate the magnetic mines," but
that was easier said than done, and even so, it would still leave
hundreds of sound-activated mines already coded to home in
and detonate on the noise signature of each Russian ship that
would have to pass through the narrow channels. Woodall was
anxious to hear just how many minesweepers had been sighted.

As the two Sea King helicopters approached the six blips well
ahead of the convoy, they recognized them, with great relief, as
Norwegian purse seiners, their high poop decks and nets that
are pulled in into the cone-shaped purse quite visible from five
hundred feet for a distance of three miles, the Norwegian flags
flying stiffly in the breeze as the Sea Kings now descended to
make sure, in the words of the Sea King leader, that there was
"no bloody hanky-panky." The fishermen, however, looked
positively relieved to see the Sea Kings, their voices on the radio
band clearly conveying their alarm at not knowing where they
could go. In one sense it seemed a ridiculous question to the
crew of the Sea King, the blue expanse of the Atlantic all around,
but the fishermen told the pilots that on their way back from
weeks at sea since hostilities had begun, they simply did not
know which coastal region was safe. The natural answer, of

course, would have been to head for Iceland, but despite the NATO Alliance, the Sea King pilots were aware that there'd been some nasty "cod" wars within NATO over fishing rights around Iceland and Greenland, and the fishermen reported that they'd rather head back to Norway if it was possible or attach themselves to the convoy for safe passage. The lead Sea King pilot said he'd have to verify with the convoy commander but told them that meantime they should move over to the eastern flank of the convoy—a polite way of telling the fishermen that the convoy would not alter course for the time being and that Admiral Woodall would not take kindly to any trawlers getting in his way.

"Please," explained one of the trawlers, "we have yet to bring in our net."

"All right," reported the Sea King leader, the helo rising, the fishermen looking fat, their yellow wet gear ballooning in the rotors' wash. "But you'd better hurry it up."

Upon returning to the convoy, the pilot was ordered to land his craft on the helicopter carrier ship nearest Admiral Woodall's destroyer.

Stunned, the pilot was severely reprimanded for breaking radio silence without express authorization from Woodall. The point that messages had been sent to C in C East Atlantic by the admiral himself did not absolve the pilot. The messages sent to C in C East Atlantic were "burst" coded—a matter of milliseconds and of import to the convoy's safety—while the chatter with the Norwegian trawlers had been long enough for the enemy via satellites and/or listening posts from Murmansk to the North Cape to get a vector fix.

It didn't do any good for the pilot to explain that, given the low transmission power he'd used, it would have been all but impossible to pick up the exchange with the Norwegians.

"Hell," complained the pilot to his comrades, "if they don't know where we are now, their satellites aren't worth a tinker's damn." He looked at his copilot, asking bitterly, "How the hell else should I have communicated with them? By bloody semaphore, for Christ's sake?"

"Should have lowered down one of our crew," said the copilot gamely.

The Sea King pilot snorted, went to his cabin furious—at Woodall but mostly at himself.

* * *

Of the fourteen submarines in the advance fan screen, four of the nuclear-powered Trafalgars had now sprinted ahead at thirty knots, submerged to a point thirty miles ahead of the convoy. Reducing speed to ten, then five knots, thus eliminating their own noise, they deployed their hydrophone arrays extruded astern like a long tube worm. In neutral buoyancy the subs sat and listened, their computers automatically subtracting any noise they emitted against incoming noise received by the passive radar. Active radar would not be used, as while this would bounce off any other sub and give its precise position to the listening Trafalgars, the Trafalgars themselves would also have been identified as a noise source. In passive mode, however, no noise was sent out by the Trafalgars, their operators listening intently to the noisiest place on earth. The sea's high density of life gave off a cacophony of sound, everything from the snapping noise of swarms of shrimp to the muted hornlike calls of seals and whales and other mammals. In addition, there was the din of currents in concert, currents in opposition, turbulence of small and enormous mud slides, jets of superheated mineral-rich water steaming out of thousands of vents after traveling through the hot volcanic aquifers far beneath the seafloor interface. One of the loudest noises was that of the plankton which rose with the coming of night and fell with the coming of dawn, their sizzle confounding the sonar operators ever since World War II, when the noise was first heard in the sonar war against the Nazi Wolfpacks. The plankton layer still interfered at times with even the strongest and most sophisticated electronic filters as the billions of microscopic creatures created a massive blanket of sound, distorting all other. In the same way, different density layers that never mixed created, through temperature differentials, warm oases teeming with life in depths once thought uninhabitable.

All this meant that to detect any particular noise, sound moving much more quickly in water from liquid molecule to molecule than in the air, was as much an art as a science for a trained operator. To detect another submarine was as much art as science. One had to develop the *feel* for the sounds, the ability to eliminate all similar sounds, to sort out one propeller's cavitation from another, listening carefully on the narrow-band receivers for the SFP—sharp frequency peaks—and on the two-hundred- to two-thousand-hertz band for the "singing" sound of a propeller shaft's vibrations, these varying within the same ship in proportion to rate of speed, blade shape, and hull curvature—all affecting the overall noise signature. In addition to all these concerns of the sonar operators, the NATO navies were not really sure of the full extent of the damage done by the American Walker spy ring or by Toshiba's sale of the Toshiba-Konigsberg quiet-propeller-making machine to the Soviet Union. In any event, to find their way through all of this, the sonar operators, of course, needed excellent hearing. It was something that Capt. Robert Brentwood, skipper of the USS *Roosevelt*, now back on its Norfolk-to-Scotland patrol, made a point of double-checking whenever a new sonar operator was assigned to the sub.

It was usual for skippers to acquaint themselves as quickly as possible with each new man aboard. But as well as finding out something about their home town, family, hobbies, and such, Robert Brentwood always made a point of asking sonar operators what kind of music they liked, pointing out that there were all kinds of tapes aboard the subs for off-hours earphone listening. Robert Brentwood was an honest man; "straight to the face," his officers described him, or "no horseshit," as the crew put it. But he did not want to prejudice the operator's answer by making it seem like a very serious question, and so he did practice, on these occasions only, a willful deception on his men. If a sonar operator said he liked rock and roll, he would smile accommodatingly and, as if he were an aficionado himself, inquire, "Hard rock?" If the operator said, "Yes, sir," he would never be first choice on sonar in any crisis situation. No aspect of submarine warfare had escaped Brentwood's attention, and he knew that, though in all other respects a person's hearing might test normal in boot camp and training school, sustained hard rock—especially as experienced through headphones—inevitably damaged hearing and as a result, unknown to the operator, "high-tone differentiation" would be lost. The failure of a sonar operator to hear such a tone, as sometimes emitted by the high electric whine or "cue tones" of homing torpedoes and SUBROCs, surface-to-sub missiles, could cause the death of 161 men.

At a thousand feet, the *Roosevelt* sat listening, her tow array hydrophones weighted for three thousand feet, an optimum sound channel depth according to computer readout of Gulf Stream salinity, temperature, and current strength. In ideal conditions, sound could travel through this layer for over four thousand miles. *Roosevelt* was also quietly leaking cold water from its ASREC, antisatellite recognition emission control, the cold water neutralizing the radiant heat from the sub's machinery, which would otherwise produce a "hot" spot on the sea surface recognized by satellites' infrared cameras. The problem for the *Roosevelt* now was for its operators to separate the noise of the convoy, which, according to Brentwood's calculations, it should pass near the halfway mark between Newfoundland and Scotland.

The sonar operator informed Brentwood that he could hear no "hostiles."

"Very well," said Brentwood. "Phones in."

"Phones in," confirmed the OOD, Peter Zeldman, and Brentwood could hear the faint, soft sound of the big spool hauling in the two-inch-diameter oil-filled hose, a long, pale yellow snake containing the series of tiny wristwatch-size black microphones.

Two thousand miles northeast of *Roosevelt*, the sonar operator aboard the Trafalgar suddenly threw his headphones down, hands clutching his head in pain. The six Norwegian-flagged trawlers were on the port beam of the convoy when a mine exploded beneath a merchantman, the rapid expansion of carbon dioxide and methane gases combining with vaporized water to buckle a starboard plate of the MV *Clyde*, creating a jagged four-meter-long hole below the ship's waterline. Next to go was the MV *Baltrain*, the explosion directly beneath her bow. As the cold waters of the Atlantic rushed in over red-hot steel and stiffener beams, they produced plumes of high, hissing steam mistaken by some among the twenty-five escort ships for smoke of the kind expected after a skimmer or air-to-ship missile had hit. It was an assumption that sent the escorts' crews to their antiaircraft missile consoles.

However, Admiral Woodall, who, as a very young midshipman, saw action in the Falklands War, immediately noticed that neither of the two merchantmen that were hit and sinking had seemed buffeted sharply to one side in the telltale manner of a missile, whose blast wave punched its target with such high speed that it usually crumpled much of the upper deck or superstructure.

One of the British frigate's tracking radar operators picked up a blip coming in abaft, on the starboard beam. In a millisecond the signal flashed from the tracker's office to command center in the ship's middle, then through the computer to the swing six-rocket Sea Wolf launcher on the weather deck forward of the bridge. One of the console's hinged jaw flaps opened, and out streaked one of the Sea Wolf's antimissile missiles, which in 1.2 seconds blew a Sea King helicopter out of the air. The helo's fiery debris further cluttered the radar screens of the NATO escorts, whose firing of chaff, or aluminum foil, deception rockets caused further disaster as some foil, due to moisture absorption in one of the rockets, stuck together in a ball, its size causing overanxious radar operators to report, "Incoming missile."

In seconds the confusion of antimissile missiles and radar jumble, including a spray of high-speed depleted uranium coming from the Dutch minesweeper's in-close weapons system, added to the chaos. Two heat-homing Sea King rockets wiped out a destroyer's launcher, stripping the ship's missile consoles' fuses in the process so that soon more missiles on the automatic feed stack below began exploding. From the Sea Kings miles ahead on forward screen high above the white-flecked blue of the sea, it looked like a daytime fireworks display gone wrong. But no one was laughing as men from the two merchantmen were calling for help, desperately trying to swim out of the wash of oil and flotsam bubbling up from their sunken ships. Nearby, a Dutch destroyer's broadband filters and circuits were reported so severely damaged by "friendly fire" that a HERO warning—hazard of electromagnetic radiation to ordnance—was flashed through the ship for technicians to take appropriate action before unprotected circuits could prematurely detonate all depth charges aboard.

One of the trawlers was three miles to port, already burning fiercely from a Sea King air-to-ship missile. As its crew and the men from the merchantmen struggled for their lives in the burning slick, the merchant sailors screaming and waving for help, Admiral Woodall, aboard HMS *Newcastle*, his helo/VSTOL—vertical takeoff and landing aircraft—cruiser, issued orders for the entire convoy to turn about and to withdraw, as near as conditions would allow, along the same course as that on which they had entered the minefield. Strict orders were given that no ship was to stop to pick up survivors, for if Russian subs were in the area, the covering noise generated by the convoy to mask each ship's exact position would be imperiled by any ship slacking off from the convoy. And if enough ships stopped, they would be picked off one by one. All Sea King helos and the advance Grumans on screen were ordered to return and form a closer-in protective perimeter about the convoy as soon as possible.

On the bridge of HMS *Newcastle*, the officers and men didn't have time to realize the full extent of the calamity that had befallen not only their convoy but the entire NATO convoy strategy, for now that radio silence had of necessity been broken, the air was filled with coded message bursts from R-1 to SACLANT in Norfolk, Virginia, and ACCHAN—Allied Commander in Chief Channel—forces in Northwood, U.K., only further confounding the post–World War II years of argument between proconvoy and anticonvoy tacticians. Those against convoys were now pointing at R-1 as stark evidence against convoy strategy and for IMS—independent merchant shipping—strategy, with smaller high-tech, high-speed boats assigned escort duty. This, they argued, would reduce risk both in terms of cost and men, and more important, would free subs and surface vessels from escort duty, giving them the freedom to spread out in search-and-destroy missions rather than being inhibited by an overconcentrated and slower convoy.

Even as the convoy was turning, there were two more thunderous explosions, mushroom plumes of oil and boiling water rising high into the sky, then collapsing in on themselves. Three more merchantmen were going down, and when Woodall saw one of them had been at least a mile to his port side, the other a mile or so starboard, he assumed for a moment that at least two Russians had joined the attack.

There was another explosion and the calm voice of a British captain aboard a Sheffield-class destroyer reporting to Woodall that he was "taking water abaft" the starboard beam. The trawlers' mines, set for individual merchantmen's signatures, now became obsolete, but there was still a question of whether "magnetic/pressure" mines reacting to water displacement and magnetic fields passing over them had been set for the heavier merchantmen, thus allowing the lighter escorts, including the Dutch minesweeper, to pass over before being triggered as, unlike most of the escorts, the merchantmen did not have "self-degaussing" or "magnetic wiping" systems that could give them anti-magnetic protection against such mines.

Woodall gave orders for the escorts to form a single line as the best hope of getting out of the minefield and to fire at will at the trawlers.

"Beg pardon, sir," said the captain of the command cruiser. "One of them's on fire. I wouldn't imagine—"

"Sink them, Mr. Rees!"

"Very good, sir."

The Sheffield destroyer, holed abaft and sinking quickly, listing dangerously over, her pumps working overtime, led the attack, with her 115-milimeter forward gun pumping away at the first big ocean-going trawler, flecks of paint and rigging spitting up into the air above her. The fish boat turned tail, its stern now to the destroyer; the high-piled netting seemed to shrivel up and fall away. There was an orange wink. Two other trawlers were doing the same.

"Skimmer midships!" shouted the destroyer's starboard lookout. A split second later the destroyer's radar, which activated the close-in Phalanx system, began firing. The destroyer's radar mast collapsed on the bridge as the Sea Dart roared off from its weather deck mount. But the list of the Sheffield was so acute that only the 115-milimeter gun could depress far enough to do any damage, the Gatling gun effectively raking only the trawler's wheelhouse. The trawler suddenly bucked, its stern lifted clean out of the water by the force of a British destroyer's Exocet—but not before the trawler and two of its sister craft had fired four fifty-five-hundred-pound Styx surface-to-surface missiles. Two of them missed, or rather were exploded by in-close Gatlings. The other two hit. The entire superstructure and bridge of the next ship in the line, a sleek Leander-class frigate, were engulfed in fire, her radar and radio masts collapsing into the hot maze of twisted steel like a long-legged insect, the crescent-shaped radar antenna aglow as it struck the water, and temperatures generated so high that the port side lifeboat was incinerated amid the reek of cordite, gasoline, and burning bodies, other men spilling into the sea, many of them afire. And methodically, above the sound of the screaming men, the steady *pump-pump-pump* of cannon fire pulverizing the remaining trawlers.

Several of the officers aboard the long line of British, Dutch, and German escorts had difficulty stopping their gunners even after it was obvious that the trawlers were well and truly done for. Among some men it had been an unwritten contract in a war that they knew would be waged with the speed and force of missiles. There would be no time for such old-fashioned notions as rescue. Better a bullet than to be left drowning in oil.

By now it was dusk, and as the convoy re-formed in squares, the Dutch minesweeper leading, Woodall ordered all ships to turn south again in a wide arc, avoiding the area where the trawlers had sown their deadly harvest. But only he, among the entire complement of the cruiser and all the other ships, knew that R-1 had been an experiment—with the empty container ships as decoys. As another admiral before him, Mountbatten, had sent the Canadian Corps to invade a beach in Normandy to test the theory for D-Day, to see if it could be done, Woodall was now seeing if "rollover" was feasible. And as Mountbatten had hoped to draw out the Luftwaffe during the Dieppe raid, now R-1 was to draw out the Russians for the killing. The Dieppe raid had been a terrible failure—more than two-thirds killed, the rest taken prisoner—but from it came the invaluable lessons of D-Day.

The men like Horton in charge of "rollover" had to know as quickly as possible whether the square base, fan-shaped screen convoy was workable in real combat. But the awful thing for Woodall was that now that it had been tried in actual battle conditions, the first time in modern missile warfare, he knew he could not give SACLANT or anyone else a definitive answer, other than to say that the Soviets had very effectively attacked the convoy by deception, using mines to devastating effect. Without question, it was a terrible loss for the convoy, eight of the container ships sunk, two escorts, a total of over four hundred men dead. But the Russian subs, the more telling test for the long-run strategy, had not appeared at all. So far.

Again Woodall wondered what was happening at the GIUK Gap, where NATO had laid its noise-signature-primed mines.

ACCHAN in Northwood, U.K., had replied that as yet no

explosions had been picked up by the GUIK SOSUS network or by any towed sonar arrays.

The only good piece of news Woodall received was that Greenland's ice sheet was farther out than usual for September, making the Greenland-Iceland Gap even narrower, so that if the Soviet subs were going to break out, it would most likely be through the Iceland-Faeroe gap as they rounded Norway's north cape.

"That would narrow the field," the cruiser's captain commented to Woodall.

"Possibly," answered the admiral, "unless they used the ice sheet as cover."

"Then they'd have to bust through, sir. Make a hell of a din. Could hear it in Picadilly."

"No," said Woodall, "they could use the sheet as cover until they're well south of the main channel, then break out on their left flank at speed—into open water."

"There's still our mines," said the cruiser's captain.

"Then why haven't our listening posts in Iceland heard them popping off?" pressed Woodall.

"Yes, unless—" The captain could see that the admiral had already thought of it, too—the possibility of it—the trawlers being the tip-off. "Bloody hell, sir. Special forces?"

"Yes," replied Woodall. "Bastards might have wiped out the listening posts. Either that or been digging up the damned SOSUS lines."

"Dragging them up would be tricky," commented the captain. "Take an age, too."

"I agree," said Woodall, worriedly, his eyes roaming the leaden horizon as night began its descent. "I'd say we're not hearing anything because the—"

"Lines have been closed," put in the captain, as eager as Woodall for an explanation.

Woodall was pacing back and forth across the cruiser's bridge, oblivious to the winking lights of the steering console and the phosphorescent sweep of the radar's arm. "I'd say the posts are still operational but are being run by the SPETS. They've de-activated the mines from shore-control relay. We're being *fed* silence." He stopped walking, looking across at the cruiser's captain. "Everything's seeming normal to us—even a little static on the line."

"How about the call-in code checks they'd have to answer?"

"Broken the code, old boy. Or more likely they've been sold the bloody things. Some pretty East German secretary in Bonn, no doubt—slipping her boss a bit more than rollover."

"I don't like that much," said the captain.

"Neither do I. Those subs could be breaking out right now."

They were both quite wrong. No listening stations had been overrun—NATO had given top priority to defending them—nor was the sub fleet that had come out of the Kara Sea and around the Kola Peninsula in the process of breaking out. They had already done so. Hours before.

CHAPTER THIRTY-EIGHT

Uijongbu

IN THE MORNING Tae had told the NKA interrogation officer, Major Rhee, albeit politely, that he would not cooperate. The NKA pulled two teeth with pliers. Tae blacked out for a second or two, but once conscious again, shook his head, refusing to tell them anything beyond his rank, name and ROK identification number. Rhee told him *kŭsaramtŭl orarŭl chikae kŭchinkuhante sikanŭl nangpihae*—they weren't going to fuck around with him—and sent him back to the schoolhouse cell, his hands still bound behind his back. When they came for him later, it was late in the afternoon, the sun out briefly, making steam for the rice paddies. He was pushed roughly inside the bleak, musty-smelling interrogation tent, its hanging light bulb swinging slightly in the wind.

The moment he entered, seeing Major Rhee sitting behind a folding camp table, writing pad and pencil before him as he sat staring at Tae, Tae noticed another smell, something instantly familiar—the fragrance of plumeria.

It was Mi-Ja, standing quietly in the far corner of the tent to his left. The light was poor but not so dark that he could not see the tears streaming down her face.

"Appa"—"Daddy," she began.

"Be quiet!" ordered Rhee without looking at her, then addressing Tae, a beefy NKA guard behind. "You will tell us," instructed Rhee, "all the names of all underground counterespionage leaders in Taegu, Yosu, and Pusan. Otherwise we will give your daughter to our soldiers—to do with as they please."

Tae shook his head, unable to speak, his head now bowed in shame. The NKA major walked over to Mi-Ja and took her long, dark hair, wrapping it around his wrist, jerking her head back sharply and tearing her bodice open with his other hand, her breasts naked in the dim yellow light as they rose and fell sharply in her panic.

Tae knew that if he gave them the names they wanted, not only would the chief underground counterinsurgency agents be rounded up and shot, but also their families and everyone who knew them. Hundreds. Rhee's left hand flashed up and grabbed Mi-Ja's breast, his finger and thumb squeezing at the nipple. She cried out and Tae turned to help her only to be knocked down with the guard's rifle butt onto the earthen floor, the taste of mud now mixing with the metallic taste of blood.

"The names!" yelled Rhee. "Now!"

Tae had intended giving them false names to buy time, the name "Kim," for example, as common as "Smith" in English. The NKA major, the guard's eyes popping out with surprised delight, moved his left hand down and began rubbing it hard between Mi-Ja's legs.

"Leave her," screamed Tae, frightening the guard, who stepped back from him before he retaliated, smashing Tae in the face with his rifle butt. Mi-Ja heard the bone crack and saw her father fall, unable to get up with his hands still bound behind his back, his foot slipping on the muddy floor even as he tried. Rhee let Mi-Ja go and she ran to Tae crying, pleading hysterically to tell them.

Tae, on his knees, was shaking his head in a way she had never seen before, like a stricken dog trying to rid himself of some internal noise which he couldn't locate and for which all prior experience had not prepared him, stunned, and as full consciousness returned, wondering how they had found Mi-Ja. Perhaps through her boyfriend in the Reunification Party, which explained to Tae why they'd taken several days before deciding to interrogate him.

"Up!" ordered Rhee, and Tae, using one of the table's legs for support, struggled unsteadily to his feet.

"Untie him," ordered Rhee, and the guard used his bayonet to cut the knot. The major held out the pencil to Tae. "Print!" he ordered. "And you!" he shouted at Mi-Ja, who was now cowering back in the far corner of the tent like a whipped dog, trying to cover her shame. "Be quiet or you will be sorry!"

Tae heard his daughter's terrified sobbing and wrote down the names.

Rhee was under no illusion that Tae had given all the names to him, but it was good enough. Once you had one or two names, you could start breaking down the various organizational "cells" in each city, and then, usually more quickly than you anticipated, everything would start to unravel, some of them surrendering even, lured on in part by terror of what would happen if they didn't and in part by public promises of forgiveness and "reeducation." Rhee looked down at the list—nine names for three cities. About right. He pulled out his revolver to shoot Tae. Mi-Ja screamed and ran to her father. She was checked by the guard and fell back as if bouncing from a solid wall, the bottom of her dress now covered in mud.

"Tie him to the chair!" ordered Rhee, letting his revolver slip back into the holster. Walking over to the table, he kicked two of the folding legs from under it so that the table was now inclined like a child's slide. He ordered the guard to hurry up and tie Tae securely to the chair, and when he finished to tie the girl to the table. "If you resist," he said, flicking a hair away from her face, "I will shoot him as we did your boyfriend. You are all lackeys of the Americans."

Mi-Ja closed her eyes and began reciting a Christian prayer. This infuriated the major, who slapped her hard. Unbuttoning his fly, Rhee smiled hatefully at her and, behind him, the defeated ROK major lashed to the chair. "You are all whores of the Americans." She began to fight, trying vainly to get away from the table, her muscles tightening, but was unable to loosen any of the bonds.

"I will kill him," Rhee warned her, and she stopped—still and staring at the mold-spotted roof of the tent that seemed to be breathing as he mounted her, grunting his pleasure, her father screaming at him, the blood from Tae's mouth running over his mud-spattered chin.

"Gag him! Tape him!" ordered Rhee. The guard stuffed an oily rag into Tae's mouth and taped his eyelids back so that he was forced to watch. Mi-Ja kept looking up at the tent breathing over her.

After—staggering back like a drunkard from the table, almost falling against the side of the wind-bulged tent—Rhee, out of breath, told Tae hoarsely, "Now the men can have her." Tae said nothing but looked over at his daughter with all the love he could muster, Mi-Ja's wail of despair no louder than the squeak of a small animal in pain.

"I was mistaken," Major Rhee said, smiling at the guard. "She wasn't a whore." The guard's eyes were bright again with expectation. He asked the major what he should do with her. The major put on his cloth camouflage cap. "Whatever you wish," he said. "Give her to the men."

CHAPTER THIRTY-NINE

ON THEIR WAY down to the White House situation room, Mayne asked his press secretary, Trainor, whether Senator Leyland had accepted the president's offer to join the White House war advisory committee.

"No answer as of an hour ago."

"But you did make the offer public?"

"Yes, sir. This morning's press conference. Some of the

southern papers were a bit smart-ass about it, but *The Times* took the high road.'' He showed Mayne the leader about NATO units falling back in southern Germany. ''Quite frankly, Mr. President, I don't see how he can say no. If he declines, he looks like he doesn't want to help the country when it needs him. If he accepts, he's on our side.''

''There aren't more than two sides in this country, Bill,'' said Mayne. ''Not in this war. You see the people in the streets? Everywhere I go you can feel the momentum. They don't want war—they know I didn't want war. But this isn't any hazy Gulf of Tonkin a million miles away. This is out-and-out aggression. This is the Communists stepping over the line, and the American people know it.'' He paused for a moment. ''How many listeners did you say Voice of America has? Seventeen million?''

''Thereabouts, Mr. President. With all due respect, Mr. President, I don't think any fireside chats are going to sway the Russians.''

''I'm not an idiot,'' said Mayne sharply. ''Point is, our intelligence reports from different sources can seldom agree on anything, but they're all saying the same thing on this one. There's a lot of restlessness out there among the republics. We too often think of it as just one big bloc. I guess it was always a weaker federation of states than we thought, but when Gorbachev got in, started promising them *perestroika, glasnost*, and all the rest of it, he gave them hope. That's a pretty potent force, Bill. This country was built on that in our Revolution.''

''Maybe, sir, but hope can go either way. Doesn't necessarily mean the Ukrainians, the Azerbaijanians, and all the rest of the republics will side with us to overthrow the Russians. Ukrainians hated Stalin's guts and he slaughtered eighty percent of his officers in the purges, but when the Nazis hit him, what did he fall back on? No call for the revolution to be defended—no, sir. It was all Mother Russia.''

''He was lucky,'' countered Mayne. When the *Wermacht* drove into the Ukraine, the peasants hailed them as conquering heros.''

''That didn't last for long, though, Mr. President.''

''No, and you know why? Sent in the SS after the army, and they started their usual horseshit—worse than the Russians—so the Ukrainians went back to Uncle Joe. Missed opportunity,

Bill. Could have changed everything—which is why I want Voice of America telling those people the truth—that with all our warts, our side's the right side and we'll help them get their independence if they throw in their lot with us.''

"Problem there, Mr. President," said Trainor as they passed the situation room's marine guards, "is anyone with a radio or TV set on the wrong channel is going to end up in a salt mine in Siberia.''

"They don't have salt mines in Siberia, Bill. Old wives' tale." Mayne paused, glanced at a White House secretary, and nodded at the chiefs and aides. But his mind was still on the Voice of America possibility and the whole range of propaganda that might help him. What they needed in Europe—hell, what they needed everywhere—was time. If they could get the Hungarians and Poles to do something—not actually fight the Russians, which would cause massive reprisals, but perhaps get them to use go-slow tactics—that would help. Solidarity should know how to arrange "accidental" breakdowns in the war industries.

Problem was, the puppet states had tried it a few times already—Hungary in '56, Czechoslovakia in '68—and only got their faces kicked in. And with the Soviet-WP pushing NATO back . . . which brought him to the first question of the meeting with the Joint Chiefs of Staff. NATO's communications as well as VOA and Radio Free Europe were in a shambles from Austria to Belgium. Why was it, Mayne asked, that the only hard intelligence the Pentagon was getting the last few hours was coming from the French?

"That's because," said Army General Gray, "Suzlov, or whoever's in charge over there, is playing footsie with France. The Russians want France left out of it—that way they can concentrate all their forces—" Gray turned toward the map stand.

"I can read a map, General," interjected Mayne. "What I'm asking is, haven't *we* got underground units in Eastern Europe that should be doing the same thing with Soviet communications?''

There was a pronounced silence.

"Well, don't all speak at once," said Mayne, looking about, his gaze shifting from General Gray to Admiral Horton, U.S. Chief of Naval Operations and NATO's SACLANT. "Well?" pressed Mayne. "Why aren't the bridges in eastern Germany and up there on the Northern Plain cut? Why haven't they been taken out?''

"Most of them have been, Mr. President," said General Gray—after all, *he* was the army authority—"but the one thing Ruskies are very good at is quick pontooning. They can get four tank regiments across in less than an hour. They open the valves, sinking the pontoons a few inches, and we can't see them from the air.''

"Can anyone tell me if we're holding ground *anywhere*?''

"West Germans and some of our forces are holding the Thurian Alps, sir.''

"Well, why *wouldn't* they be? Why in hell would the Russians try climbing over mountains when they've already got us on the run? Are we holding our own in the air?''

"Yes, sir,'' answered Air Force General Allet. "For now.''

Next the president turned around to the Marine Corps commandant, General Barry. "I hear your boys are giving a hundred and ten percent at Fulda, General.''

"Yes, sir.''

"Are we holding them?''

"I don't honestly know, Mr. President. It's an hour-by-hour situation—their tanks are breaking down faster than ours, but we're getting hurt, too. At Fulda we need a six-to-one kill ratio just to stay even.''

The use of the phrase "kill ratio'' got the attention of the secretaries present. Up until then, things had been so frantic, the war meetings were much like any other major White House crisis—hostage taking, attacks on embassies—but "kill ratios'' was not the usual pre-press-conference banter.

In all, the meeting was short and gloomy, each service giving its report. The central front was sagging but holding here and there. The question was, for how long. But Korea was all bad, ROK and U.S. units still in what reports were pleased to call "retreats in force.''

"Which means,'' the president responded, "we're getting our ass kicked.'' He turned to Admiral Horton. "Can you give us what we need, Admiral?''

"We've been practicing this one for a long time, as you know, Mr. President.''

"And what were your casualties during practice, Admiral?''

"High, sir. Considerably higher than we anticipated.''

"Well, fill me in. Is anything moving out there?"

"There's a convoy under way this minute, sir. Out of South-ampton. NATO escorts, which we will take over from halfway, northeast of Newfoundland."

"Our sub fleet?"

"Already at sea."

"Very well." The president stood up and left the room. One thing he had learned in politics was that it was essential you husband your time. Right now, this night, there were vast armies of men locked in mortal combat, and until the situation changed sufficiently to warrant his intervention, there was only one sensible thing to do.

"I'm going to bed, Bill. Wake me if it's Code One."

"Will do, sir."

CHAPTER FORTY

WILLIAM SPENCE WAS a cook's helper aboard HMS *Peregrine*—not yet a chef, but determined to become one. Cooking was the thing he loved to do best because he'd seldom seen people happier than when they were enjoying a good meal. His parents, Richard Spence, an industrial chemist for a large heavy industrial adhesive company in London, and Anne Spence, a retired grammar school teacher, lived in one of the upper-middle-class green belt housing estates near Oxshott in the south of England—forty minutes by train from Waterloo. Young William had never intended to join the navy—certainly it had not been his father's intention for him. But with the middle class increasingly distrustful of the secular state schools, demand and fees for private schools had gone up dramatically. Richard and Anne Spence had scraped and saved early in their marriage so that their two eldest children could go to private school. For Rosemary, now thirty, it had been the school best equipped to get her into teacher's training college, and for Georgina, now twenty-five, the school best suited to win her entry, via scholarship, to the markedly secular but reasonably prestigious LSE, the London School of Economics and Political Science.

William, on the other hand, had not been "planned," and when Anne in her early forties had found she was pregnant, there had been a frightful row between her and Richard, but one conducted in the absence of the two girls. Anne finally decided not to abort, but now Richard, on the verge of his sixties, when both he and Anne had anticipated early retirement, was faced with paying the bills for William to be at a private school. It meant delayed retirement for Richard for at least another five years. Resentment of his predicament, however, had long ago given way to a love for his son that he had not thought possible, and certainly the kind he had achingly missed with his father.

Then one day shortly after his eighteenth birthday, William announced he didn't want to go to university—he wanted to be a cook.

"A chef!" corrected Richard in astonishment. Even then William could see his father was at once disappointed and relieved. Relieved because, erroneously, Richard Spence expected it would cost less money to train his son in the culinary arts, and disappointed because the Spences had always been of professional stock—solicitors, doctors, even the odd barrister. No criminal briefs, of course, mainly mercantile law. It was one of these relatives who, before the war broke out, had advised Richard of a "solicitous compromise" which he believed would satisfy both Richard's desire to see his son in a respectable profession, rather than merely a trade, and William's choice. Richard demurred, however, on the subject of a child of his being in, well—manual work.

"Being a chef's not like being in a trade these days, Richard," William's great-uncle had advised in High Church tone. "More of a guild, I should think. Point is, if you want both, he'll have to don uniform. Have to pass the entry exam, of course, but he'll get his O levels."

"Shortly," Richard assured him. "What do you mean by uniform?"

"Not a bad arrangement at all," the uncle had continued.

"And they're desperate these days. No offense, Richard. But they do want volunteers if they can get them. William seems bright enough. I see no reason why after a while he couldn't apply for officer training school. Rather rushing them through these days, I should think, with all this talk of trouble brewing in Europe." The uncle had looked satisfiedly into his dry sherry. "Yes, I should think it would suit him admirably. End up with a commission and—" he sipped the sherry "—I shouldn't be surprised if he was running a large hotel in years to come. Could do worse."

"I suppose," began Richard, "if he wanted—"

"Richard, old boy, once he gets his one stripe, he's way ahead of other applicants for any hostelry business. Officer, *cordon bleu*, and all that. Doesn't do any harm, Richard. I do think that given his rather limited aspirations, it would be best for him."

Richard was coming around, slowly. "Any of the services will do, I expect?" asked Richard.

The uncle came as near as he ever had to swallowing sherry without savoring it first. "Certainly not. I strongly suggest the senior service."

"The air force," said Richard.

"Don't be fey, Richard. The navy, of course."

"I wasn't trying to be fey."

"Then your ignorance on these matters is lamentable."

"But I've never thought of William as a sailor. Anne won't go for it," Richard had said. "I can tell you that now. All this business about the possibility of war breaking out . . ."

"*War?* Richard, old man, you've been watching too many of those dreadful 'Insight' programs. Either that or reading the *Mirror*." The uncle took his brolly and hat from the front stand, using the unfurled umbrella as a pointer. "You send him to the navy, mark my words."

Richard was right—Anne didn't like it—but he told her it was most likely that, unless the unthinkable happened, William would be posted to a shore establishment. In any case, if there was a flare-up, with modern weapons it would be like the Falklands so many years ago—over very quickly.

Eleven months to the day, William Spence was Leading Seaman Spence, cook's helper, aboard the destroyer escort HMS *Peregrine*. After a very rushed, rather peremptory training drill ashore, he now found himself aboard one of the latest DD escorts, his job one of the least glamorous, most important jobs in war: to prepare food for convoy under attack, to guarantee, no matter what the conditions, that everyone in the ship's company got his NATO-required three thousand calories a day.

For all the destroyer's modern technology, hot meals were, as the cook quickly explained, ill-advised at most and "sheer bloody impossible" in the maelstrom of an engagement: hot stoves, soup tubs spilling despite their gimbals mountings, steaming coffee and tea that would burn, and ovens that unattended could cause a fire—as lethal as any missile. Yet if morale was to be kept up, food was fundamental, providing the high-sugar, high-adrenaline level necessary for any kind of sustained battle.

William Spence had heard but seen little of R-1's action against the trawlers. Apart from the bridge-wing lookouts, no one was permitted on ship's decks, the 115-millimeter gun and the Australian IKARA SUBROCS going off along with the Limbo depth-charge mortar unloading its deadly ordnance off the stern of the seven-thousand-ton ship. The Bristol-class destroyer, her twin funnels astern behind the rotary bar radar her telltale markings, had fired her 115-millimeter at two of the trawlers, but her angle in the close pack of the convoy prevented her from launching torpedoes. After the sinking of the Russian-manned trawlers, the men who did not have their stations overlooking the well deck and so had not see the carnage of broken bodies adrift in the icy waters of the Atlantic were the only ones who were hungry. But the cook, a chief petty officer, assured William Spence that later that night, when the others' shock of seeing their first "dead men" wore off, they would be ravenous—especially if the big fish came.

William Spence didn't get the connection.

"Torpedo attack," explained the cook. "Night's still the worst time—fancy radar or no. And if that happens, it'll be bloody mayhem, laddie. Ship darkened. CIC dimmed—so you make sure you've piles of boxed sandwiches, and keep those thermos cups bunched, ready to go in the elastic basket. We go into search or evasive pattern, this tub'll be swinging from starboard to port, port to starboard so fast, it'll make your head spin. And it'll last hours. And no onions or garlic. Old man'll go spare—can't abide 'em.''

"Hardly haute cuisine, Chief," said Spence. The cook had seen many a recruit come and go, but there was something more likable about Spence than most—perhaps it was his unabashed naïveté, an eagerness that assumed the best in everyone he met, and the cherubic face that was in stark contrast with the salt-leathered scowls he got at times in the mess. Not all of them, like Johnson, who was peeling spuds for the freezer, were volunteers like Spence.

"And that Yank bloke we have aboard," said the cook. "NATO liaison fella. No Marmite for him. They don't understand it."

"Can't say I'm mad about it myself," smiled Spence.

"Ah," said Johnson, "puts hair on your chest. Right, Chiefie? Iron in the old pecker," said Johnson. "Cock stiffener."

Spence blushed. The cook said nothing—they were sending him choirboys, they were, all keen and woefully inexperienced in the ways of the world—but unlike some of his ilk, the chief cook aboard HMS *Peregrine* took no delight in watching the transition from recruit to leading seaman.

"Never mind him," said the cook, pushing the big thirty-two-once jar of black beef extract spread toward Spence. "Just don't put it on till you've made all the other sandwiches. Most crew don't like it when it's been sitting around too—"

"Action stations!"

The cook's voice was drowned out as the sound of the alarm and men running, grabbing life jackets, asbestos balaclavas, and gloves, thumped quickly through the guided missile destroyer. In an instant the high whine of abrupt start-stop electric motors could be heard bringing weapons into line with radar guidance. *Peregrine* heeled sharply to starboard at thirty knots, the flare of her bows lost in a gossamer of spray, phosphorescent with plankton. William Spence could hear the sudden *dump! dump! dump!*—the 155-millimeter—and then the hard-running-faucet sound of the IKARA torpedo-missile, *Peregrine* turning so violently to port that coffee spat out of the hot twin Silex pots that had been shoved hard against their metal guards.

"A sub," said Johnson, either very brave or feigning indifference.

"Yes," said the cook, "a sub, and you'd better get on with it. Soon as you've finished with that lot, you can put them in the freezer, give Spence here a hand with the sandwiches." Johnson was getting mad as he was forced to hold hard on to the sink as the ship rose, bucked hard astarboard, and fell through a belly-wrenching slide into a deep trough. "Only the British bloody navy would have you peeling potatoes. On the Yank boats . . ."

"Ships," corrected Spence good-naturedly, more in the way one might help a friend rather than criticize.

"Quite right, lad," said the chef. "Ship."

"Ship, shit, what's the difference? We aren't sailors. I didn't join up to peel—"

The *Peregrine* now bashed its way through a wave, the heavy spray like fine rain above them, the second escort a lump against moon-tinted sea a quarter mile to port.

" 'S'-pattern," said Spence.

But the chef was looking at Johnson, handing him back the scraper he'd dropped in the heavy, sharp roll. "That's where you're wrong, Johnson. We *are* sailors. Without food, lad, this ship can't function." He handed Johnson another potato. "All right then?"

Johnson grunted.

"Besides," continued the cook, "if you can't take a joke, you shouldn't have joined."

"I didn't," said Johnson, his tone turning surly. "It was either this or a year in the nick."

"What for?" asked the cook. Spence was amazed; he'd never actually seen a real live criminal before, let alone worked next to one.

"I found some silver," said Johnson defiantly.

"Where?" asked the cook.

"In a house. Where else?"

"What's done is done," said the cook, unscrewing a peanut butter jar, face going red. "Just so long as we don't have any silver missing around here. Because—" continued the cook, handing the jar to Spence, "if we find anything missing, we'll cut your bloody twinkie off. Like one of them ayatollahs. Right, Spence?"

Spence didn't know what to say.

"Well it doesn't matter anyway, does it?" Johnson continued, unrepentant, swinging the french fry cutter toward him. "I mean we're all for Davy Jones." He saw Spence's alarm and smiled. "Yeah, that's right, mate. Food for the fucking fishes, we are. What flamin' chance 'ave we got next to one of them

Russian subs? You answer me that." The ship was slowing down, the bell signaling end-of-action and standby stations.

"See?" said Johnson, waving his peeler in the general direction of the combat information center in the heart of the ship. "They don't know what's fucking going on."

"Probably just a drill," said the cook.

Johnson tossed another potato into the bucket. "You know how many miles we've got to go yet?" he asked them ominously.

"Next couple of days," said William, "the Americans will take over. Midway point."

"Oh," said Johnson. "I see. Once the Yanks take over, we'll be all right. Don't you know we'll be taking their convoy back?"

Spence didn't reply—Johnson seemed so jaded about everything that no matter what you said, he'd pick fault with it.

"You married, Spence?" asked Johnson.

"No, I'm not actually."

"Well, *actually*," said Johnson, "it's just as well. No widow." The cook shifted off the safety sleeve on the automatic meat slicer, then swung it around, Johnson's grooved face distorted in its shining surface.

"Stow it!" said the cook. He was the boss of the galley and preferred informal rules, despite the British navy's long tradition of tar and feathers, but when yobbos like Johnson started upsetting people unnecessarily, then he was prepared to pull rank. For a second Johnson said nothing, and in the uneasy silence the cook thought of *his* wife and two children, teenagers, in Portsmouth—and ruminated on the fact of how things had changed. Oh, there'd always been the shipboard whiners like Johnson as long as he'd been in the navy, but he couldn't have imagined a rating daring to speak with such a defeatist streak in him since the first day out. Fortunately, for every Johnson out there, he hoped—believed—there were two or three Spences, otherwise it was going to be a long, grumpy business in *Peregrine*'s crew's mess.

It wasn't only Johnson that he wondered about. With NATO there were foreigners you had to cater to—a Yank or two at the table—usually one would like his meat rare—and a sprinkling of Scandinavians, all blond and looking as if they had just been skiing. And there were Dutch hippies who smoked a lot—"not always tobacco, mate"—and had everybody wondering whether, when push came to shove, they'd be up to it. "Democratic disease," the chef had explained to young Spence. And the Krauts, of course, always liked the British ships best. More beer rations. Spence was too friendly, too young really, to be on a ship with all these other blokes—and always asking questions—what wine was best with this and that, and the cook telling him no wine was any bloody good on ship because everything ended up getting sloshed and corked anyway.

"Wait till the war's over, laddic," the cook had finally told him. "Get this lot down pat and next thing you'll find yourself on some shore establishment doing the hors d'oeuvres for the admiral's party." But William Spence had a theory—that if he could learn to make dishes for everyone, for "all the sixteen nationalities in NATO" coming from all kinds of different backgrounds, then he'd have a head start when he was demobbed. He had told the cook that he'd started a list of what wines *did* travel best—now, that surely had to be of use if you were going into the cruise trade after the war.

Sometimes Spence's zeal just plain wore the cook down, but he tried not to dampen the kid's enthusiasm. He'd seen too many go the other way. Maybe the kid had a point about the wines as nowadays they were trying all new bottling techniques anyway. In any case, the cook knew the boy had the "gift" of all great chefs. Organization. Being the cook of HMS *Peregrine*, one of Britain's star hi-tech destroyers, the chief petty officer had seen hundreds come and go through his charge, and he'd known many of them who could cook meals that you'd never forget. But he hadn't met *many* who could do that and who also possessed the ability to pace themselves, never to have one dish rushing in the wake of another, or too far apart, but just to appear naturally, and always, but *always*, at the right temperature. That's where art came in.

"Now, when you've finished with those spuds, Johnson," said the cook, "I want you to put this vitamin C on them before you start the next lot."

"Stops them going brown," said William. "The vitamin C."

"I fucking know that," said Johnson. He sprinkled the vitamin C around and tied the heavy plastic bag with double twist. "Good as dead!" he said. "Subs. That's what we need. This surface shit is a crock—"

"We've got subs," said Spence before the chef could tell Johnson to shut up.

"Nine," said Johnson. "Jesus Christ, the Russians have hundreds."

"So have the Americans," answered William.

"Right!" joined in the cook.

"You know—" said Johnson, his hand grabbing the cold stove rail as *Peregrine* climbed up out of a trough.

"Know what?" asked William Spence, feeling a little seasick in the closed-off and overheated air that was being recycled through the galley.

"Moscow's only got to move all their crap down the road. Yanks have to move their shit across the whole friggin' Atlantic."

"You should be in comedy, Johnson," said the cook nonchalantly. "You've missed your vocation, laddie. We ought to send you round to the hospitals, we ought. They could do with a cheery bastard like you."

"Haven't you heard of rollover?" asked Spence challengingly.

"Oh 'cor. Spare me, will you? *Rollover*."

"Yes," said William Spence. "We roll over them. Just push through."

Johnson finished peeling the potato, stared at it for a moment, and let it crash to the bottom of the bucket. "*Rollover* Beethoven. You sound like one of those fucking admirals. They do the rolling, we do the over."

"What would you do then?"

"I'd leave it up to the Yanks and the Russians. Their war, not ours."

"But we're part of NATO," said Spence.

"Listen, mate—in this world it's everyone for himself. NATO, TATO, who gives a shit?" He was using the peeler as a pointer. "You don't look after Number One, sweetheart, nobody will."

"Perhaps you should have gone to jail instead," said Spence—the first time the cook had heard Spence angry.

"Now, *that*," rejoined Johnson, "is the first bright idea you've had, Sunshine."

"Then why didn't you?"

"Because, ducky—" Johnson savagely extracted a rotten spot from the potato "—I didn't know some silly bastard'd start shoving—did I?'' He moved the bucket of potatoes over toward the sink. "Well, now I know and I'm telling you, mate—any friggin' thing hits this ship and I'm first off—Beaufort raft and all.''

The cook heard the buzz on the bridge-to-mess intercom, and as he picked it up, wondered whether he should put Johnson on report. "Right, yes, sir. Right away, sir.'' He clipped the phone back on its cradle. "Sandwiches and coffee to the bridge. Corned beef and lettuce for the old man. No pickles.'' He paused before giving Spence the plastic tray, checking that all the indents for cups, plates, and so forth were spotlessly clean. "Sub pack ahead of us.''

"How far?'' asked Spence, trying not to sound frightened.

"A ways off yet,'' said the cook, "but it'll be a long night ahead of us, boys.''

"What did I bloody tell you?'' said Johnson. "I thought our mob were supposed to knock 'em off up near fuckin' Greenland with all the super-duper mines we planted down there. Christ—now we're in for it.''

During the trawlers' mine attack on R-1 and R-1's defense, ocean noise was such that it shook the fine instrumentation of sonar buoys and towed arrays for thousands of miles. By sheer chance it provided a noise cover that seemed heaven-sent by the Soviet sub pack approaching the GIUK choke points in two groups. The first group was heading for the Greenland-Iceland Gap, close in to the extended ice sheet, using it as added protection against which ASROC and other antisubmarine warfare missiles could not penetrate other than by blowing themselves up. The second group of seventy subs, using the static and a heavy sea for cover, was going for the Faeroe-Iceland Gap. NATO's mines in both the narrow Greenland-Iceland Gap and the Iceland-Faeroe Gap had been beaten by the Soviet subs, who, with the help of the Walker spy ring secrets, had found out how to best "baffle''—or alter—their noise signatures— similar to altering sounds from a car.

But once through the ice-free Iceland-Faeroe Gap, the submarines were detected through "thermal patching,'' the Soviets' COMONES—computer-controlled emission systems—not being nearly as sophisticated as the Americans'. Here the Russians' luck ran out and there was a terrible slaughter.

"Prigotovitsya vsplyt!''—"Prepare for surfacing!'' was one

of the oft-repeated phrases that morning of the NATO attack on the Russian Northern Fleet.

Protected by F-111As—Ravens—from Upper Haywood and the Norwegian bases, NATO's Nimrods came out of Scotland's Kinross Air Station, with searchwater radar and aerial-release depth bombs. The American Lockheed search-and-attack Vikings, with infrared sensors, magnetic anomaly detectors, and homing torpedoes, closed with thirty F-15 Sea Eagles out of Keflavik. The attack spread out from the shallow 190-mile-wide gap to Wyville Thomas Ridge two hundred miles south, where the water depth increased to two thousand meters. It was the "high," as one of the Viking pilots put it, of all their years in NATO, as one Soviet captain after another ordered, "Prepare to surface." Nineteen subs, eleven nuclear and four diesel-electrics, were outright kills, and four forced to the surface, white smoke pouring out of them high into the pristine air, their crews having no alternative but to ditch into the ice-cold Arctic waters. Some managed to get rafts inflated in time to drag themselves, half-frozen, aboard, but it was the first time since World War II that the Russian navy, at least its submarine branch, had come under such attack and proved so wanting. The American Vikings' under-wing ECM—electronic countermeasures—worked superbly well, not only in jamming the Russians' "snoop tray" radars but also in feeding the submarine force and its Russian battle group false over-the-horizon echoes. This caused two of the Soviet ASW helo carriers to fire Cruise missiles at empty air, and one of the Tango-class subs to send back an advisory "burst" message that was picked up by two of the Vikings. Thus identified, the submarine never stood a chance as a Mark 46 torpedo streaked through the water at twenty-five meters a second, its explosion rupturing two of the Tango's forward watertight compartments, torpedo room, and crew's mess, the sub driving nose-first to the bottom, its implosion registered by a Sea King helo from HMAS *Invincible*.

"Like bloody Clapham Junction," said one Nimrod bombadier after the fourth sub had gone down. Neither the Turkish nor Greek NATO radar operators, two nationalities that normally would never share the same console, had heard of the British Rail choke point in London that was notorious for terrible train crashes. But both Turk and Greek operators knew what he meant. Carnage. It was on a scale predicted by only a few of the sonar experts who, in the not-so-cold war, had gone hunting, "pinging" the Russians until they withdrew in confusion because of inferior sonar.

The public relations assistant to CINCHAN—Commander in Chief Channel Forces in Northwood, U.K.—handed Admiral Newsome the information in a jubilant mood. "It's all coming home to roost, sir."

"What is?"

"Soviets' deficiencies. We knew their sonar was bad, but— well, no matter how many SSNs they have, subs aren't much use to them if they don't know precisely where we are."

The admiral knew this and he also knew the other Soviet deficiencies: inferior repair facilities and not nearly the same number of overseas bases or coastal listening stations as the Americans had. The admiral also knew that the NATO forces had been lucky; the absence of large numbers of Russian fighters due to Soviet "surge" tactics now being used on Germany's central front had allowed the British Nimrods and American Vikings to go about their sub killing unmolested, the Russians having concluded, correctly, that if they won Western Europe quickly enough, the NATO sea lines would be rendered useless. Consequently, the admiral did not share his aide's mood of exhilaration.

"We can both see and detect one another's battle groups from four hundred miles away," the assistant was explaining to a member of the press, admitted only on the understanding that all details of the battle would then be quashed. The admiral's assistant was explaining to the newspaperman that while both the Soviets and Americans at times *did* have comparable early-warning radar on their fixed-wing planes and helicopters, the American carriers were so potent in terms of air cover that they could search four times the area as their Soviet counterparts "in any given time frame."

It *was* an enormous advantage. And it now became clear to CINCHAN, in Northwood, and ACNE—Allied Commander Northern Europe—in Kolsas, Norway, why in the prewar years the Soviets developed an obsession with shadowing any Allied ship they could, often using their fishing trawlers. It had been an attempt to make up for their lack of bases on the continental shelves, which the Americans possessed. The Soviets had been shadowing the NATO ships in those years not just for information about size and armament but really operating as seaborne early-warning stations in the event of war.

"Yes," conceded the admiral, looking over at the chart of the North Atlantic. "They've taken a drubbing, all right, but that doesn't preclude a trap. They know they have inferior search capabilities. Question is, gentlemen, what do they have in mind to compensate for it? What are they up to?" He reached over the broad map of the North Atlantic, his hand brushing the 170-mile-wide Iceland-Faeroe Gap. Had the Russians feinted here on the western flank while using the unusually extended summer ice sheet as a roof to slip their best subs through the Greenland-Iceland Gap to the west? NATO's bombs and torpedoes couldn't penetrate the ice, other than by blowing holes in it, which pack ice quickly refilled. "Using one choke point to take punishment, one to slip their right flank past us, dividing our force."

"Rather a bad mauling for a trap, I should think," suggested Newsome's PR assistant, a commander who, rumor had it, had risen very quickly because he'd married another admiral's daughter. The commander glanced at the tally sheet. Of the eleven nuclear subs sunk, nine were Alfa II–class nuclear attack boats out of Leningrad's Sudomekh yard. With titanium alloy hulls for deep water, a submerged speed of over forty knots, and fifteen thirty-mile homing torpedoes, the Alfa II was the "Rolls-Royce" of the Russian attack boats. The commander was telling the reporter that the Alfa could dive below the crush depth of most other subs, including many of the Americans'.

"Does the depth make that much difference to a torpedo?" asked the reporter. "I thought those Mark-48s could get anything."

"They can, old boy. Problem is, if you get deep enough, you're much safer. At three thousand feet they can even beat our Caesar network." He meant the North Atlantic section of the SOSUS network, and Admiral Newsome was getting tired of him. Perhaps he *was* promoted because he was married to an admiral's daughter. God help us, thought Newsome, if he gets to flag rank.

The officer of the day walked in, and the PR commander gave him the tally sheet with the same bonhomie with which he'd been nattering away to the reporter. "Bloodied their nose a bit!"

The admiral was frowning, still looking worriedly at the

GIUK Gap, eyes flitting back between the shallow shelf about Iceland and down toward the deeper Labrador and Newfoundland basins. A lot of water to hide in there.

The OOD looked down the score sheet, letting out a low whistle, joining the commander's spirit of celebration. It was as if they'd both sunk the lot themselves. "Seventeen!" he said. "I say, Freddie. Well done!"

"Yes," said the admiral, without looking up. "That only leaves a hundred and thirty-six." The two commanders looked at each other abashedly as the admiral continued. "You mustn't get caught up too much in the numbers, Freddie. Don't want to damper your enthusiasm. I understand—it's a good start. Eleven nuclear subs would mean crippling the U.K. fleet, or any other European power, for that matter. But remember the Russians lost twenty million in World War Two, a colossal number of tanks and ships—mainly given them by the Americans, of course. Point I'm making is, it's a big country. An enormous country. It can absorb big losses. What we have to worry about is those blighters who got through here under the ice." The admiral told the reporter he'd have to excuse them. When the reporter had gone, Newsome asked the OOD to tap in the intercept vectors, given the Soviet subs' average rate of speed, forty-two knots for the SSNs, seventeen for the diesels. "Let's have a vector first for the nuclear subs alone."

"Nuclear subs . . ." The OOD entered the information into the computer. "East of the Labrador Sea—approaching the edge of the basin. Five hundred miles south of Greenland's Kap Farvel."

"English designations?" said Admiral Newsome. He was a stickler for the use of English in NATO—some horrible mistakes had been made because of similar-sounding names.

"Ah yes, sorry, sir. That's five hundred miles southeast of Cape Farewell."

"Yes . . . well, I just hope it won't be farewell for our first convoy," said the admiral. "Let's have the vector for the diesels, will you?"

Behind him he could hear the array of computers and telexes as the NATO commands were feeding in not only results of the naval battle at the Iceland-Faeroe Gap but the SITREPs in Western Europe. Fourteen Soviet–Warsaw Pact divisions had broken through on the North German Plain and were now attacking the low countries into Belgium toward the channel ports. If they weren't pushed back, where the hell would the NATO convoys from the United States dock? France still hadn't come in.

"Diesel-electrics, fifty-five hours at least—unless they run on the surface. Unlike the nuclear boats, of course, they'd pick up a few knots on surface. Anywhere from two to five knots. That could put them ahead some."

"I hardly think they'll risk open running," proffered the commander.

"I agree," said Admiral Newsome. "They'll keep the diesels for a return convoy, I suspect—if we have any bloody ports left in Europe by then."

"Well, if they do, sir," said Freddie, "I'd say they're asking for trouble. Being diesels, they'll need service boats, and we could pick them off like ducks."

"They have snorkels," the admiral reminded him.

"Of course, sir. I wasn't thinking of the snorkels giving them away, but refueling's another business. Then they really are sitting ducks. Sir."

"Twenty thousand miles on one tank, Freddy. Those chaps can go a long way before they need to come up for refueling. What's this Three Lib?" asked Newsome, pointing to the printout.

"Well, they're too slow to hit Convoy R-1, sir—if that's what you're worried about?"

"Can't do much about the R-1 now, Freddie. Too far south. Afraid it's on its own."

CHAPTER FORTY-ONE

IN THE PENTAGON, Chief of Staff General Gray was on the telephone with the president. The chief executive's question was, "Can we hold the Taegu perimeter?"

"I honestly don't know, Mr. President," replied Gray. "The Seventh Fleet is in a much better position now to strike the peninsula, but the weather's not so good." The next question was, how did bad weather affect Smart bombs, infrared guided missiles, and so forth?

"They're superb, Mr. President, but as yet we can't be sure we're hitting enemy targets. The NKA are continuing to move rapidly, and they have civilians on the munitions trains as well. If we could restrain some of the television networks and press photographers from showing refugees holding up signs and—"

"I want up-to-date contingency plans for withdrawal, General, as well as for reinforcement. National Security Council meeting is at four-thirty."

"Very good, sir."

General Gray had a plan for both situations, but each one, as everything else, depended on securing safety of movement between Japan and Korea, and the Seventh Fleet was busy fending off incoming attacks from NKA MiG-29s from fields so close to the Yalu—the Chinese border—that they might as well have been in China itself. But the fighter pilots of the Seventh Fleet understood that to cross the Yalu was to invade China.

Gray's aide, a major from Logistics and Supply, came in with more messages from the Taegu perimeter. Till now they'd been decoded automatically and piped in onto the TV map screen overlay on his wall.

Gray took the sheaf of paper. "President wants to know, Major, if we should cut our losses and run. Or reinforce. That's not the way he put it, but that's what he means."

Apart from the military position, both Gray and the major knew a lot of careers now hung in the balance. Gray was looking at the worldwide distribution of forces, from fighter bases in Japan to AWACS with the U.S. Third Fleet headquarters in Hawaii, from which he could move at least one carrier to Subic Bay in the Philippines, and Guam, where Communist mortar attacks had wreaked havoc among the B-52s. But there was no way he could tap any of the resources tagged for Europe with the East German and Russian divisions still pouring through the Fulda Gap and engaged in broad, sweeping armored thrusts south and north. He simply did not have the forces available from either Third Marine Division in Japan or from the Third Fleet to do a MacArthur, to buy time for the twenty thousand Americans and forty-six thousand ROK forces and many thousands more of refugees from Yosu to Taegu. To buy time and to try another day; that was his plan.

"Maybe Doug Freeman has some ideas," suggested the major. "That letter that he sent you about predicting the Soviet–Warsaw Pact breakout was right on the button."

Gray grunted, fighting his tendency to withhold praise from subordinates that might dim his own halo. "Well, they weren't simultaneous breakouts. Yes—well, he had the general plan right, I suppose. No use to us now, however." The general was exhausted after having slept only two or three hours in the last twenty-four; his petty reluctance to give credit to Freeman for the spot-on prediction about Europe told Gray he was more fatigued than usual. "Doug Freeman's a good infantry and tank man. Airborne-qualified to boot, but he's still a colonel because he talks too much. Always telling people what *he* would do."

"Isn't Freeman's tank corps in New York," asked the major, "waiting for the convoy to Europe?"

There was a pause. "You send for him?" asked Gray, not so tired he couldn't smell a setup.

"He's at the Washington Hotel."

"You think I should see him?"

"It wouldn't hurt, General. Doug's record at the war college was outstanding."

"Yes, I know. Thinks he's Patton resurrected. You know what he's got in his tank?"

The major was tempted to answer, "A tiger," but didn't risk it. "A one-twenty millimeter I hope," he answered.

"Got a goddamned index card in the commander's cupola. Has 'You have three minutes to surrender!' in ten different languages."

"Yes, I heard something about that," conceded the major. "Only, I thought it was in every tank."

"Oh, it is. He does spot checks. Every loader and gunner in the battalion has to know them—otherwise it's a fifty-dollar fine."

"Well, he's confident, all right."

"Funny thing is," the general ruminated for a second, "I've seen him on social occasions—with his wife. When we were in California. Perfect gentleman—wouldn't think he had an ego big as one of his M-1s."

When he came in the door at 3:07, Colonel Freeman was carrying his map case of the European central front. He had a plan for a counterattack from the Jutland Peninsula following an amphibious landing northwest of Kiel, supported by B-1 bomber strikes out of Southeast Anglia. He was shocked by General Gray's appearance, the chief of staff's eyes so dark from fatigue, it looked as if his nose had been broken. As they shook hands and Gray gave him a peremptory smile, Freeman thought the general needed a damned good tonic, and he had it in his map case.

"You know Major Wexler," Gray introduced him to his aide.

"Of course—" But there was more a professional than personal tone to Freeman's greeting, and Gray recalled Freeman's terse response to the circular sent out by Wexler notifying officers of the possibility of the Supreme Court ruling in the near future that, other than submarine duty, women might be permitted a wider range of combat roles. "Goddamn it!" Freeman had written back. "No room to piss inside a tank except in your helmet—let alone having a woman in there!" Major Wexler had responded that as women had been dealing with such problems for years, he had no doubt that if, as Colonel Freeman had put it, his men weren't "allowed to stop to have a pee"—they just did it in their helmets, threw it out, and kept on fighting—the female members of a tank crew would find this a great motivator "to win battles quickly."

"He's a smart-ass, that Wexler," Freeman had told his wife. "A Washington smart-ass."

"Douglas," continued General Gray, "Major Wexler here was struck by your prescient abilities regarding the Soviet–Warsaw Pact invasion of Europe. He pulled your file and suggested we call you in."

The change in Freeman's manner was dramatic, the smile now genuine for a man who, even if he didn't' agree with Freeman about not having women in tanks, hadn't let it cloud his ability to see a brilliant tactical mind at work. "Yes, I remember the major well. Good to see you."

"Douglas," Gray informed him, "I read your letter and I must say I was a little surprised at it not coming through regular channels."

Freeman grinned. "I didn't think that would surprise you at all, sir. After all, you were the one who taught me about initiative. I figured if I sent it by regular post, last thing a Commie agent would think of is trying to penetrate our mail service—it being such a balls-up."

Gray motioned him to a chair; with men dying as he spoke, he was in no mood for another one of Freeman's harangues about the mail service. Freeman had once suggested that the postal service be run along military lines—any letter not delivered anywhere in the United States within four days would render all employees in the post office liable to a fifty-dollar fine.

"Douglas, I have to tell you up front your manner is considered extremely abrasive by many of your colleagues. And especially by the State Department. By all accounts, your record, militarily speaking, shows you should have had your first star two years ago."

Freeman was wearing a scowl but nodding; he could feel possibility in the air. If he could only keep his cool. "Yes, sir, I understand that, but I've—"

"Goddamn it, Douglas, let me finish!" The figures on the TV screen of wounded and missing were changing. Getting worse, especially on Germany's central front.

"In one hour," continued Gray, "I have to present a contingency plan to the president. At that meeting there might well be several more members of the cabinet than usual. Transport and Communications secretary included. Military needs their help if we're to have these NATO convoys loaded and shipped out on time, so I don't want you getting anyone's dander up unnecessarily. You're a first-class tank and infantryman, Douglas, and God knows we need more like you. I'm giving you a chance to show your stuff, but if any questions are directed at you, remember you're not Randolph C. Scott—''

"I think you mean George C. Scott, General."

"What? Oh, yes. Christ!—Douglas, that's precisely what I mean. It's not—'' Freeman affected a lot of people this way, bringing out the fight in them at the drop of a hat. It was precisely what was needed in battle, Wexler knew, but deadly to smooth sailing in Washington.

"Randolph Scott's fine with me, General," said Freeman, flashing a smile. The general was shaking his head, surprised at his own reaction to Freeman's personality. The tank commander seemed to carry a charge in the air about him that stirred up everything it passed.

"What I need, Douglas—in simple, straightforward terms— is a plan for a tactical withdrawal from—'' At the word "withdrawal'' Freeman stiffened, all sense of humor, his earlier air of accommodation, gone.

"From Europe? General, this would be catastrophic—''

"What?—No, Goddamn it! Korea.''

Gray's aide looked quickly at Freeman. Was he as good on his feet with an entirely new situation thrown at him? Freeman stared at General Gray.

"May I smoke?''

"No. Well—?''

Freeman swung his hard gaze up to the green fluorescent map of Korea, as if it were an assassin towering over him, daring him to risk a career. Gray was telling him the situation was much worse than the newspapers or anybody else outside the Pentagon had presented it. But whether Freeman had heard him or not, the general didn't know, Freeman taking out his bifocals, leaning forward, looking past the casualty figures at troop dispositions in the Yosu-Taegu perimeter. "How up-to-date is this intelligence, sir?''

"Satellite," answered Gray, turning to Wexler. "Real-time or delayed?"

"Real-time, General."

"Air superiority?" asked Freeman.

"Not as yet. Holding our own, but that's about all. Hope to get better as the Seventh Fleet moves further north, but Europe gets first call on everything." Freeman was tapping the series of half dozen or so red lights flashing on the big screen on Japan's west coast from Shikoku to Hokkaido. "What's this? Air strikes?"

"Yes. Japanese fighters are doing well, but their main function is defense and they haven't the carriers. Combat time off the North Korean coast is very short." Now Freeman pointed to the position of the Seventh Fleet steaming north midway between South Korea and Japan's main island of Honshu. He zeroed in on the cluster of blips behind the Seventh Fleet. "Reinforcements?"

"Yes. Nine Corps. It's based in Japan."

"Hmm—" responded Freeman. "Soft in the belly. Too much sushi and pussy. They won't last."

Wexler looked across at the general, who calmly responded, "Well, the Third Marine Division is part of the reinforcements, too."

"Well—that's good news. Problem is, we might not have any perimeter left by the time they get there. Those newspaper reports right about the NKA using some of our captured M-60 tanks?"

"Afraid so."

"Goddamn it! That's sacrilege." Freeman shook his head like a medieval bishop might upon hearing his church had been sacked by vandals. "Course, the trouble is, we're not up against Hitler here. This toad won't hold back the tanks. He'll drive us right into the sea if he can—which I suspect he's close to doing right now."

"Well, Kim's no fool," said Gray, "whatever else you might think of him. George Cahill found that out when—"

Freeman stood up, his stature growing in the reflection of the big screen. "Hell, no, General. I didn't mean General Kim. He's run-of-the-mill Commie trash. Learned all he knows in Beijing, where it's all numbers—just keep pushing the bastards at you. And in Moscow—echelon attack with the tanks. No, I mean Kim Il Sung's progeny. He's blood-crazy. Killing all those people like that in Rangoon. Civilians. Blowing up women and children in airliners. We should have shot that bastard long time ago.''

"Then you think withdrawal's the best bet, Douglas. Don't be afraid to say so. Everyone else here's come to the same conclusion.''

Freeman stood back from the screen, eyes moving quickly up to Korea to Japan to Manchuria, down to Korea again. "No, sir. I do not concur.''

"Then, Colonel, you're a minority of one.''

Freeman took off his bifocals, grinning broadly as he slipped them back into his top pocket. "I know, General, I know.''

"You seem pleased.''

Freeman's smile was gone. "I don't like being beaten, General. Not by anyone. And 'specially not by that goddamned psycho. Little runt needs a good kick in the ass.'' Freeman held up his hand as if halting oncoming traffic. "General, I don't mean to be disrespectful. I'm sure you understand that. But would it be too out of line to say that the policies of the majority of the general staff got us into this situation?'' Gray said nothing. The screen flickered again and the perimeter had grown smaller, the NKA spearheads, however, reportedly stopping to draw breath before the final assault. Or was it, Freeman wondered, that the hard-pressed U.S.-ROK headquarters in Pusan had decided to give up a little territory in return for a smaller, tighter perimeter? Either way, it was shrinking dangerously.

Freeman had his bifocals out again, using them as a pointer on the screen. "Kim's supply line,'' he announced, pointing at Taegu on the western side of the Sobaek Mountains, which ran north/south between Taegu and Seoul. "Airborne attacks, General. Here at Taegu, where they have to haul freight through high country, and further up—at Taejon, halfway down from Seoul. This toad has got too big a mouth and not enough belly, General. Grabs more than he can hold. He's overrun so much territory—damn near two hundred miles in a little over ten days— that's why he's cannibalizing everything he can. That's why he's using the M-60s. They can't maintain sophisticated equipment like that. Haven't got our ground support, technical backup. They're using oxen carts to move half—''

"How do you know that?'' interjected General Gray. "Oxen carts?''

"New York Times.''

Wexler looked out the window at the Potomac.

"And," continued Freeman, "that's why he's raping the goddamned countryside. Feed 'em as you go."

"Well, he's getting a lot of civilian support, I'd say," put in Gray.

"Don't buy it, General."

"Well, I do. We're not seeing any scorched-earth policy from the satellite photos. All the fires are the result of military action. He's getting help from South Koreans, Douglas. I know that mightn't be palatable, but you of all people surely aren't blind to the—"

"Disagree, General." Freeman's bifocals were sweeping the air. "That fink is getting support because the son of a bitch has had over fifty years to plan underground networks right across the country. All his goddamned spies doing the spade work. He's getting food and water from those civilians same way as Napoleon did in Dubrovnik in the Balkan campaign. That's why, other than Uijongbu and Seoul, you're not seeing too many cities on fire in the satellite photos. Like Napoleon. Sent his boys ahead, infiltrated the city. City fathers did a deal. We'll feed you—leave our city and us alone. Quid pro quo. That's why you can still walk around the walls of Dubrovnik. Message gets out fast. But we hit him with those airborne attacks just when he least expects it, and *we'll* get civilian support as well. Everyone knows—the Americans come back."

"We didn't in Vietnam," put in Wexler.

"By Christ—" began Freeman, "that's because we put up with that Fonda woman and all her cronies. When she sat on that NVA gun and told our boys they were war criminals for bombing those sons of bitches, we shoulda dropped her from a B-52 right in the middle of the goddamned—"

"As soon as the B-52s are patched up in Guam," interjected Gray, "then we won't have to use any troops at all. We can go in and bomb his supply lines and—"

"No time, General. That's what the board tells me. That's what you're telling me, isn't it? No time. Nothing more we can scrounge from NATO-designated supplies. Besides, Kim's no dummy. He might overextend his supply line—almost everyone does when he gets the bit in his mouth, sees the other side hightailing it. But it's my guesstimate, general, that he's just about shot his wad. Biggest mistake he made was destroying that oil pipeline outside of Taegu. Caused us a lot of damage, but now he's got no oil for a while at least. Oh, he'll hold, but he won't be advancing for a week or two. His supply lines are well over two hundred miles south of Pyongyang now. I say chop it in two at the places I've indicated and we'll stop him for a week or two instead of a few days. Give us time to rush troops into that perimeter. Is the strip at Pusan operational?'' He looked at Wexler.

"It's rough, but it's operational."

"Hell, even if it isn't. We could ferry a lot of Hercules across from southern Japan under Seventh Fleet umbrella in twenty-four hours—use pallet if they can't land. Around the clock. Christ, that's what we're best at. But—'' He held his finger up. "The coup de grace, gentlemen. Clear an air corridor for me up here—'' his hand shot north of the Seventh Fleet's battle group, beyond the brown spine of the Taebaek range ''—and I'll turn this thing around. Christ—I'll take prisoners!'' He was pointing deep into North Korea. At Pyongyang.

General Gray sat still for several seconds, leaning forward in his chair. "You have any idea of the casualties, Douglas? I mean—what would you expect?''

"Seventy—eighty percent."

Gray glanced quickly across at Wexler, then back at Freeman. "Douglas, I think the Seventh Fleet could give you that corridor—for five or six hours anyway—enough time for your air jumps. But to lose men at that rate is simply unacceptable—''

"General,'' said Freeman, his voice even, unhurried, "we'll lose *sixty times* that number if that perimeter's punctured.''

"*We,*'' General Gray said, "we won't be losing our lives, Douglas. It'll be the men in those choppers and Hercules that will—''

Freeman was stunned. "I assumed I'd be in command, General.''

Freeman's audacity left General Gray speechless. "You're a colonel, Douglas. This would be brigade.''

"Sir,'' said Freeman. "I think we can solve that problem right here and now.''

"How?''

"Promote me.''

Gray looked across at Wexler, who was biting his lip.

"The president,'' said Wexler, "would have to authorize it.''

Freeman wasn't sure whether Gray meant the promotion or the plan but quickly cut in, "I'm sure he will, General."

Gray shook his head and looked down at his watch. "Douglas, you wouldn't by any chance know the Korean phrase for 'You have three minutes to surrender,' would you?''

"Sampun inaeĕ hangpokhae."

CHAPTER FORTY-TWO

IN THE ROUGH ballet of the *Salt Lake City*'s flight deck, danger was everywhere.

The Seventh Fleet's battle group's heart was the carrier itself, and the heart would need protection from aerial and sub attack. To provide early warning, prop-driven Hawkeye AWACS, their rotodomes giving 360-degree, sixty-target-at-once capability, were already in the air together with the relatively slow but long-range and effective Grumman A-6 Intruders, each of these armed with twenty-eight five-hundred-pound bombs and sophisticated antisubmarine detection and attack systems. The Intruders' periscopic booms for in-flight refueling glinted in the late afternoon sun as they passed over the advance screen of destroyers and frigates that surrounded the Seventh Fleet on its mission to "secure the integrity of the sea lanes" from Japan to Korea's east coast—in other words, to tell the Soviet Eastern Fleet it came south at its peril.

Aboard the carrier, as one Hawkeye AWAC was pushed off the elevator amid the scream of jets and hundreds of other pieces of equipment, crewmen in padded brown vests were already unfolding the plane's wings, its pilot engaging the hydraulic line that lifted the two-thousand-pound "pancake" dome from flat storage to raised position. In the cramped rear of the plane, its three "moles," electronic warfare operators, were already going through their preflight checks amid banks of consoles.

It all seemed chaotic to any new men on the ship, but out of the six thousand sailors aboard the carrier, those who worked the flight deck had of necessity to develop the ability to work calmly yet quickly in the sustained roar of sound, yet stay attuned to alarms of their own equipment in conditions where one missed step or the slightest reduction in concentration could cost a man his life. It was a world of screaming engines, of planes taking off and coming in, flashing lights, rising steam from catapults, hot, stinking engine exhausts, and a maze of hand signals from different colored jackets, a world of hookup chains and "mule" tractors.

Inside the carrier's island, to starboard, it was less noisy but every bit as stressful as anticollision teams in primary flight control, or "prifly," had to know where any plane on their computer screen was at any moment while staying in contact with the pilots as they were guided in by flight deck control.

As the pilot of one of the returning Hawkeyes brought his aircraft down in the controlled crash the navy calls a landing, its hook seeking the two wire, or arrester cable, the Hawkeye's twin Allison turboprops were roaring at full power, the plane's flaps down, ready to lift off if his alignment, or any one of a hundred other things, was not right. The pilot's concentration was on centering his plane in the "meatball," the big orange-lit mirror on the carrier. If it came in sight, he was halfway there; if he saw the meatball arrangement of lights was too low, he would have to ease the nose up to center and maybe go for the three wire. He saw the meatball was askew. A green jersey, "A" on its back, turned and waved a "no go." In a split second the LSO—landing signal officer—pushed the button for vertical red, cutting through the meatball, sending the Hawkeye screaming past the island as the pilot kicked in maximum power, pulling the plane off the deck with only inches to spare. The three wire was showing a stress split visible to only one of the catapult and arresting crew, whose thick ear protectors and jersey disappeared momentarily in the cloud of kerosene exhaust and salt particles that flew up from the deck, stinging his face, the Hawkeye climbing, a blast deflector now going up on the starboard catapult in preparation to launch a jet fighter to begin its patrol even as the Hawkeye was turning for the rerun.

As the Hawkeye banked, its rotodome a golden disc in the fading sunlight, another AWAC, its green-jerseyed catapult crew sliding under and attaching the restraining and launch bridle forward and aft of the fuselage before scrambling away, readied for takeoff as more AWACS bunched up behind it, the control tower unforgiving in its insistence that at least three launches' lead time had to be maintained. The unlettered green jerseys of the specialist technicians or "troubleshooters" could be seen nearby through the quivering heat curtain in the event that any of the plane's electronic components suddenly needed replacing by slide-in, slide-out "black box" units.

The carrier's commander, seeing the fighter was ready and receiving confirmation of no obstacles on deck, signaled, "Clear deck."

"Landing light is red, sir," repeated the executive officer in the tower.

"Very well. Turning takeoff to green," said the captain, pushing the button for the harsh, metallic "tweedle" sound warning.

"Takeoff is green, Captain."

"Very well." Now the captain pushed the backup "horn," whose sound was so powerful, it blasted its way through the line of roaring, waiting AWACs, above the whining elevator bringing up more planes, and could even be heard by the five-man crew of the orange-silver rescue helo.

The rotodome of the Hawkeye about to be launched was a platinum disc under a partially cloudy sky. Its pilot showed two fingers, signaling he was approaching full power; the propellers made of fiberglass to protect the plane's radar from metallic-induced Doppler effect were now two black blurs. The pilot, his cockpit already splattered with sea spray, saw the yellow-clad catapult officer's knee drop, left hand tucked close in behind and against his back, right leg low, right arm thrust forward and seaward. The catapult shooter pressed the button, and one deck below, the controller let her go, the force of the release throwing the plane aloft and leaving a long trail of steam rolling back over the carrier's deck. The Hawkeye's pilot was already flying his zigzag pattern at low level to prevent any enemy AWACs detecting his takeoff and thereby pinpointing the carrier's position.

Down in the pilot's ready room, the TV monitors were giving the pilots of the Tomcats up-to-the-minute weather information for the first of the patrols that would begin to clear the corridor for General Freeman's three-pronged attack to be launched from the *Saipan* and other LPHs—Landing Platform Helicopters—in *Salt Lake City*'s battle group.

"Why the hell don't they just bomb the shit out of the supply lines?" asked one of the pilots. "Get some of those B-52s up from Guam. That'll cut their supply line fast."

Frank Shirer, one of the F-14 Tomcat leaders, was idly flipping over old magazines, glancing now and then at the monitors. His would be one of the last of the patrols, not due to go out until early next morning before dawn, but often off-duty pilots would sit in on another briefing merely to get the feel of the weather and bone up on any added information that might come in handy. The weather was deteriorating, visibility having dropped from thirty-five to ten miles, heavy cumulus in places, freezing level twenty thousand.

"So why don't we bomb the crap out of them with the BFUs?" He meant the big fat uglies—the B-52s.

Shirer, twenty-seven but looking older, dropped an old *Newsweek*, its cover bearing the promise of "New Peace Initiatives in the Middle East," back into the magazine rack. It struck him as one of the supreme ironies of this war that the Middle East, most volatile area of concern before the war, was not yet involved, at least not directly, as it was generally believed that Israel was doing what it could to help the West. Meanwhile it was surrounded by ever stronger Arab states. Wait until we run out of North Slope oil from the Arctic, thought Shirer, and have to tap the Gulf.

"Hey, Major?" A lieutenant, his Tomcat's RIO—radar intercept officer—asked him again. "What do you think?"

"Ever heard of the Ho Chi Minh trail?" said Shirer. "We dropped more ordnance on those gooks than we did in all of World War Two."

The other pilots were now listening attentively. Shirer was held in high respect, for despite his relatively young age, he had been the pilot of one of the three big 430-ton "Doomsday" Boeing 747s on constant alert at Andrews Air Force Base outside Washington. It was a Doomsday plane that the president would issue his orders from in the event of a nuclear war. Shirer was called "One-Eyed Jack" aboard the carrier because of the requirement of the Doomsday pilot to wear a patch on his left eye so that in the event of a nuclear flash blinding him, he would still have one good eye to fly by. But at the outbreak of war in Korea, Shirer had immediately requested transfer to a combat wing.

The truth was that Shirer, after the initial excitement and prestige of being "the president's pilot," had soon become tired of the routine and disillusioned with what he saw as a role of little more than highly paid chauffeur. At twenty-seven he craved some action, and after a while Washington had just gotten to be too small a town. Everybody there had SEXINT, sex intelligence, on everyone else. Not good for the president's pilot, and why he was careful to "have it off" away from Gossip City. Trouble was that after more than three nights, the women always started talking about serious "relationships," especially with the new AIDS strain on the march. Some of them were so businesslike about it. In New York two beauties had asked him to have a blood test—in their presence—to see if he tested positive. One of them even had an over-the-counter test kit ready. Put him right off, especially when he was prepared to take precautions anyhow.

"So what about Ho Chi Minh, Major?"

"Oh—supplies kept coming. Boat, oxen, you name it. Disassembled whole artillery pieces and transported them on bamboo poles. Four guys would carry a wheel for a howitzer. Air force always thinks you can bomb everything into submission. It's not just the gooks either. More bombs Hitler dropped on London, more the Brits dug in."

"You saying they don't make a difference?"

"Not saying that, but bombers are only part of the triad—sea, land, and air. We bombed Ho Chi Minh's city flat till it was nothing but rubble. They lived underground."

"You think this Freeman guy's plan'll work any better?"

"Don't know. But on the ground you can see more sometimes. High tech's good, but hell, you bomb out a bridge, next day they float a pontoon link across right next to the old busted-up one, sink the pontoon a foot or so, and from the air, looks like there's nothing there. Then they move stuff over at night."

"Our guys'll be wiped out," said the RIO.

"Quite possible," shrugged Shirer, "but it'll probably buy time. That's what it's all about. Anyway, we can give 'em support. We can make a difference."

"Jesus," said another pilot, opening a well-thumbed issue of *People Today*, flown in with fleet mail the evening before. "Those poor bastards on the *Blaine*." The pilot showed the color shots of the frigate in Nagasaki and the wounded being unloaded from hospital planes in San Diego after the flight from Tokyo.

"Old man alive?" asked another pilot.

"Stateside," said the pilot reading the magazine. "Burned up pretty badly, according to this. Interviews some of his family. Hey—says here his old man was in the navy. Admiral. Brother's on an SSN."

An RIO was looking over the pilot's shoulder. "Who's the broad?"

"His sister."

"Man, look at this. Would I like to get into her pants."

Shirer glimpsed the photo and held his hand out for the magazine. It looked like her, the girl he'd met at a Washington ball—some military outfit had put it on. He couldn't remember her last name, though, or whether she'd said she had a brother on a U.S. frigate. Taking the magazine, he looked at the photo more closely. It *was* her. Hair all different—a more sophisticated look than he remembered. The caption said, "Mrs. La Roche."

For a moment he was back with her. It had been one of the gentler nights. She was beautiful and shy and not sure whether she wanted to do it or not, but along with the shyness there was a grabbing hunger, as if she couldn't wait, wanting love but afraid, holding back. Then she got all serious and he had her. She'd closed her eyes, sighing deeply when he kissed her, shivering with excitement and fear at the same time—and need. He'd been as tender as he could, but it wasn't very good. It had soon become evident to Shirer that it was her first time and it had turned into a production, her grimacing, trying not to show the pain but clearly hurting like hell. He'd eased off and she'd been sorry, apologetic—how she hadn't been fair to him—how she felt like a slut. He tried fooling around a bit to lighten her up for a repeat run, but she'd almost freaked out when he'd put on the patch. They'd gone out a few more times, but it didn't seem to work. She was too highly strung anyway—beautiful and innocent, but her sensitivity was too fragile for him to handle. Now, from the photos at least, it looked like she'd had a bit more experience, knew who she was, what she wanted. She'd be great.

"Who's this La Roche joker?" he asked, glancing through the article.

"Her husband. Some cosmetic poof."

"Well—" Shirer answered. "That's that."

"What d'you mean?" asked his RIO.

Shirer handed back the magazine. "I mean that's it for the skipper of that frigate."

"Oh, yeah," answered the RIO. "Yes, sir, he's down the toilet."

Shirer was sitting back in the high-backed pilot's seat, trying to remember the last time he'd had a woman. Felt a hard-on coming. Tried to put her out of his mind, but something about it bothered him. Completely irrational, he told himself, but somehow he felt as if she should have told him when she'd decided to get married. But why should she? A short, brief fling.

"Man," said another navigator. "She could sit on my face anytime. Anywhere."

"Look at the board," said Shirer, nodding toward the TV monitors. Visibility had been cut to near zero about the carrier, one of the things that really spooked the pilots, though they would never admit it. The carrier was no bigger than a postage stamp when you were coming in to hook the wire at several hundred miles an hour. It was nerve-racking enough when you could see.

"You'd better get your minds out of your shorts," advised Shirer. "And think about Charlie. Now he's got *two* things going for him—distance and heavy cloud cover." He turned to another pilot. "Fisher, that drop tank of yours. Got the release fixed?"

"Yes, sir."

"Good."

The briefing officer came in. They stood up and he immediately waved for them to sit down.

"It's still on for tomorrow morning. We'll be riding shotgun for the choppers and Prowlers and the Hercules. Drop tanks to give us extra time for strafing and rocket attacks."

The RIO called Fisher leaned over to his Tomcat's navigator. "How the hell did One-Eyed Jack know we'd be using drop tanks?"

The navigator shrugged. It meant that they were going in deep. A long way inland.

"Target, sir?"

"You'll be told later, Fisher. Meanwhile I suggest you get some rest."

"With this noise?" someone asked.

As the ready room emptied, Fisher turned to Major Shirer. "Sir? How'd you know about the drop tanks?"

Shirer looked around so as none of the others could hear. "I have the knowledge, Fisher," he said, tapping his head. "Know what I mean?"

CHAPTER FORTY-THREE

IN WESTERN FRANCE, autumn cast a russet spell over the countryside, and poplars were turning half-golden in the breeze, the only sign of war being increased traffic on the road to Coquelles as apprehensive Frenchmen began lining up for hours, waiting their turn in the creeping traffic line heading for England. They were not going by roll-on, roll-off ferries, as these had been stopped two days ago when an East German fighter, out of control over Holland, had plummeted into the channel midway between Dover and Calais. There were no injuries, the pilot picked up by the Calais-to-Dover hovercraft. He was not popular, however, and was roundly booed in several tongues as, dripping wet, he was fished out of the frigid water and taken to the bridge for safety's sake. Sitting wrapped in British Sea Link blankets, he was a forlorn figure, torn between gratitude for the British having picked him up and anxiety about what would happen to him later on.

The London tabloids gave prominence to the fighter "attack." Overnight the ferry traffic from France dropped away to a trickle. Now the twin twenty-five-foot-diameter undersea tunnels of the "Chunnel" through which rail-borne cars, passengers, and freight trucks moved under the channel from Coquelles outside Calais to Cheriton outside Folkestone, a distance of thirty miles, became the preferred way of crossing.

It was shortly after 10:00 A.M. the following day at Cheriton when a lorry driver, having to leave his truck on the rail wagon because of a false fire alarm, arrived in a foul mood at the Cheriton terminal. Agitated and mumbling to himself after having to walk three hundred meters from inside the Chunnel, he complained bitterly to the British Eurotunnel public relations officer on duty. This was the third time, the lorry driver told the official, that there'd been a false fire alarm. In addition, he protested that when he tried to call London on one of the emergency phones inside the Chunnel, to tell his employer that he'd be late, "the bloody thing wouldn't work." He'd been jinxed, he told them, by inefficiency. At the beginning of his journey in France his truck had broken down and he'd been cursed "to Kingdom Come" by damned Frogs who were backed up behind him. And when he'd tried to get help, there was no one available at the French terminal. If Eurotunnel couldn't keep the phones working, he charged, and provide assistance when needed, then they shouldn't have built the bloody Chunnel in the first place.

The British public relations officer did not handle the criticism well, insinuating that perhaps the truck shouldn't have been on the road in the first place if it was "mechanically unsound." This infuriated the driver, and the official didn't improve matters by grudgingly telling the driver he could use the office phone but would have to pay for any "trunk"—long distance—call to London. The driver stormed off, saying that he wouldn't use Eurotunnel's damn phone, and was last seen hailing a taxi.

Five and a half minutes later, one of the cross-service tunnels connecting the two main traffic tunnels under the channel began to shake as in an earthquake. Light fixtures popped, cement debris began falling, and there was an enormous rumbling. Seven seconds later a huge, vomiting stream of rolling fire like a napalm bomb roared out of the Chunnel at the Cheriton end. Cars and trucks came spewing out like so many toys as Centrex explosive, together with the NATO-placed wartime contingency explosives, collapsed not only the cross-service and ventilation tunnel but the two main rail tunnels as well, millions of tons of rock and water cascading in.

Over eight hundred people were killed. The newspapers reported that they had been drowned, but Department of Defense coroners ascertained later that most victims had in fact died from the concussion of the explosion even before the tunnel had collapsed.

For days after, bodies were still washing up on the beaches between Folkestone and Dover, many children and pets among them. The minister of transport resigned, and had it not been for the war conditions in Europe, the whole government might have fallen following the informal yet traditional rule of ministerial responsibility. But with the country at war, it was considered essential for national security that the war cabinet stay intact.

In one blow England's and America's strategic land link with Europe had been severed.

In Moscow, in STAVKA—the Soviet Supreme High Command—former Colonel, now Brigadier, Kiril Marchenko, at fifty-five, one of the youngest high-ranking officers, was again receiving congratulations, for it had been his plan for SPETS units to sabotage the Chunnel using the "lorry" attack. After Marchenko's successful suggestion of putting the Far East Fleet to sea in order to stabilize the Sino-Soviet situation around Vladivostok and to dissuade the Taiwanese navy from "adventurism," Marchenko had risen even higher in the Premier Suzlov's estimation.

The destruction of the Chunnel would prevent U.S. troops and supplies from disembarking in England and being shuttled to Europe now that Northern Europe's ports were being closed by the advancing Soviet–Warsaw Pact shock troops, which included Marchenko's son Sergei at Fulda Gap. But much more important than cutting off the undersea link between Britain and the Continent, the destruction of the Chunnel meant that the vital British oil supplies, particularly Avgas, following refinement of crude from the North Sea, would be cut. The idea had first occurred to Kiril Marchenko when his son Sergei, who was still pressing for a chance to enter the air force's fighter academy, had pointed out that the American M-1s *"proglotal"*— "guzzled," as he put it—two gallons of gasoline per mile; the F-16s, eight gallons a second; and the B-1 bombers burned off twenty-nine gallons every sixty seconds.

The U.S. tanks and those of its NATO allies would soon deplete their one-month reserve stockpiles throughout western Europe. Kiril Marchenko passed Sergei's figures, though he knew his son was quite wrong about the B-1—it used more than 34.7 gallons a minute—to a colleague in the Soviet Air Force Academy. There was no overt pressure, Marchenko simply mentioning in a conversational tone to the Air Force Academy's general that his son Sergei was particularly keen on the air force as a career. The general said he'd make a note of it and had Sergei's initial application pulled from central data bank. Sergei Marchenko, it seemed, had passed all entrance requirements with flying colors except one—the vision in his left eye was slightly deficient and not quite up to the standard for fighter pilots.

"But," said Kiril Marchenko upon hearing the news, "why didn't the interviewing board mention this before?" He had a colleague who knew one of the top men in the Moscow Hospital, where they'd pioneered "laser spot surgery" on a mass assembly line—why, the outpatients could even listen to tapes, from Beethoven to heavy rock, as they were operated on. "I did not tell you, Comrade Marchenko," the general said without a trace of embarrassment, "because then you were only a major in the STAVKA. If there is a possibility, I will let you know."

In Washington it was 5:30 A.M. and President Mayne was in his smaller West Wing office, which he was using more than the Oval Office for the day-to-day war conferences. He refused to go down into the situation room anymore unless it was absolutely crucial, for no matter how leak-proof he thought his White House staff was, the mere suggestion that the president was retiring to the situation room sent tremors through the country. Neither General Gray nor Trainor liked the room very much, but the president noticed Harry Schuman looked rather comfortable in it. But if so, Harry Schuman's contented air was about to be ruffled as the President told Gray that, although he had given presidential approval for the *Salt Lake City* battle group to provide air cover for the airborne attacks on the Korean supply line, under no circumstances was the battle group to support the Taiwanese navy to the south of them.

General Gray did not show it, but he was shocked. "Mr. President, the Taiwanese are superbly equipped. If they start shelling the mainland, they'll have to protect themselves, and that will bottle up the Chinese nicely. Keep 'em off our Seventh Fleet's back while we try to take the pressure off the Yosu-Pusan perimeter.''

Then Mayne astonished not only Gray and Harry Schuman but Trainor as well when he announced, ''I've put through a call to Beijing this morning, to Premier Lin Zhou, and I told him that despite any Chinese logistical support for North Korea, I would not authorize an attack on Chinese soil if he ceased such logistical support henceforth, nor would I endorse or support any such attack on Chinese soil by the Taiwanese navy, and we would actually oppose, militarily if need by, any such attack by the Taiwanese.''

Gray could no longer maintain the pretense of calm. ''Mr. President, I must protest in the strongest possible terms. In my view . . .'' He paused. ''Sir, this is, militarily speaking, extremely unsound. Believe me, Mr. President—''

''General, the Soviet Union has over a million crack first-line troops—I repeat, *over a million* crack troops—all along the Chinese-Soviet border. And ten squadrons of MiG-29s and MiG-25s. Right now that's just where I want them. Those jets could be in Europe in four hours with or without air refueling, and our boys' only hope over there is to hang on until we can resupply.'' He paused, looking over at Trainor. ''How far back now, Bill?''

''Ah, a withdrawal bulge forty miles west at Fulda, Mr. President, thirty miles into the Ruhr and—''

''Deep into Western Germany,'' said Mayne. ''Now, if we can't stall them there, General, if we can't hold till our convoys start pouring in men and matériel, then we're up shit creek and we've lost Europe. I want you to tell Admiral Horton categorically that the Seventh Fleet must step in and if necessary *attack* any Taiwanese incursion—air, sea, or amphibious. Lin Zhou has promised me he'll hold back if we do. He will not cease making public statements about imperialist aggression on our part. But so far, as supplies and men go, he will not reinforce North Korea so long as we hold the Taiwanese in reign. Quid pro quo.''

''Sir,'' said Gray, ''the Seventh Fleet is about to launch the amphibious attacks on North Korea. It will soon be dusk in the South China Sea.''

''What bearing does that have on the Taiwanese?''

''Mr. President, the fleet's going to have its hands full clearing corridors for the airborne attacks without having to worry about—''

''General,'' the president said, leaning forward, an edge to his voice, ''when you people came to me with budgetary requests for updating the AEGIS system, you told me it would be worth it because we could see everything that was going on within a radius of three to four hundred miles.''

''That's correct, Mr. President. All I'm saying is that at the moment our maximum concentration has to be on launching—''

''General, I will not be deterred from this course of action. The Seventh Fleet's battle group was specifically designed to handle multiple targets and, if necessary, cross-referencing missions. Am I correct?''

''Yes, sir.''

''Then let's not start making excuses and get on with it.''

''Very good, Mr. President.''

''Jesus! Jesus!'' railed Gray to Major Wexler on his way back to the Pentagon. ''Always the goddamned same, isn't it? You let a 'peace' president in and they start thinking they're goddamned General MacArthur. My God—''

Wexler did not dare remind the general that the president of the United States was, after all, commander in chief.

''Mr. President,'' Trainor said, ''have you ever read Camus's *The Plague*?''

''What—?'' The president was holding his head, the fingers of his left hand strained as they massaged hard above his left eye. ''I think so—why?''

''There's a character who keeps writing an opening paragraph, and in his mind he keeps envisaging the editor receiving his manuscript and being so overwhelmed by it, the only thing the editor can do is stand up and say, 'Gentlemen, hats off!' Well, it's hats off to you this evening. That bit about China—it's brilliant. How did you ever—''

''I didn't,'' cut in Mayne. ''Senator Leyland's idea.''

''Oh—''

''That make it less impressive?'' said Mayne, looking up.

''No, no, not at all. I mean at least you made the decision. You were for it.''

''Don't worry about who gets the credit, Bill. Bring me a

glass of water, will you? Two Empracet and two two-twenty-twos.''

"Lights out?" asked Trainor.

"Please."

Trainor turned off the light and, walking over to the drapes, shut out the dawn.

Mayne was already seeing the aura: steps, covered in shimmering water, like the water that used to run down fish shop windows, and above it all the most beautiful emerald green he'd ever seen. It was a warning. If he didn't hit it hard now with the codeine, Tylenol, and aspirin, the migraine could get a hold and put him out of action for hours. Sometimes he had nightmares of Trainor, giantlike, looking down at him, holding the pills, threatening: "If you don't give me what I want, I'll leak it." Personally Mayne believed that his determination, the ability he'd developed to work despite the fierce headaches, made him capable of more endurance than most under stress. Marx had had the headaches. So did Ulysses S. Grant—the general so sick with one, he couldn't sleep while waiting for Lee's response to his surrender ultimatum.

Mayne tried to remember the last time he'd made love with his wife, Jean.

CHAPTER FORTY-FOUR

GENERAL KIM, HIS divisional headquarters now moved to Seoul, was pleased when Major Rhee, the interrogation officer from Uijongbu, brought him the names of the KCIA's counter-espionage chiefs in Taegu and Pusan, which he had gotten from Tae. But most important of all, Rhee brought a summons from

Pyongyang for a personal conference between Kim and the NKA's "dear and respected leader."

The problem was that some ROK "bandits" had managed to cut communications between Kim's divisional headquarters and Pyongyang, since the major had arrived, by blowing up one of the microwave relay discs on one of the hills leading down the Uijongbu corridor. This would soon be remedied, of course, but Kim was in a quandary. He had been summoned to Pyongyang, and it was essential, he told the major, that "our great and respected leader" be apprised of the situation—in particular the rapid rate of advance.

"If Pyongyang does not hear from us, they will be concerned that our advance has been halted. It is necessary that they hear firsthand that all is proceeding as planned." Kim did not tell Major Rhee his secret hope—that whenever he returned to Pyongyang, he would not only be greeted as a national hero, even greater perhaps than Admiral Yi, but he would also receive from the leader himself the coveted Kim Il Sung medal for valor.

"I will unleash the final assault on Pusan in forty-eight hours," he informed Rhee. "By then all supplies will be in place. It is essential I be here. However, I would also like to explain to the leader himself how Taegu will be completely overrun, from where we will move quickly to crush Pusan."

Kim put another Sobrainie into the bone cigarette holder. "When Pusan has fallen," he told the major, "then you will have some mopping up to do. As it has developed, the names of the counterinsurgency chiefs may pose no more than an academic question after all, for I foresee a massive American surrender. There will be great honor for the Fourth Division."

Major Rhee said nothing. He was sure of the military victory to come, but under no circumstances would he be drawn into advising the general on whether or not he should go to Pyongyang. Not to go would be to disappoint their leader. It would only take a day there and back, but then again, Kim should be at divisional headquarters when the final drive for Pusan began.

It was a difficult question.

CHAPTER FORTY-FIVE

THE TRAWLERS HAD done more damage than Admiral Woodall or anyone else had realized, as before laying the signature mines in the way of the approaching convoy, the trawlers had sown a line of pressure and magnetic mines behind them.

These mines didn't stop R-1 but caused a great deal of confusion as Woodall, after losing another merchantman, retraced his course, trying to find a gate in what seemed a moat of mines. Four attempts and eighteen hours later, with three more ships, including a destroyer, sunk by the mines, Woodall decided the only way out was to form a long line and "plow ahead" as would a line of soldiers walking through a minefield, following the lead ship, as it were, conscious that if they kept to the same path, the risk of being blown up by a mine would be minimized and that sooner or later the minefield must peter out.

The Dutch glass/plastic minesweeper led the file, but its magnetic anomaly detectors, powerful enough to detect the positions of the magnetic mines, were unable to protect against the pressure type, which reacted to changes in water pressure caused by the heavier merchantmen passing over.

Soon the minefield, designed by the technical experts of the Soviet Northern Fleet merely to delay a convoy long enough for their subs to break out through the Greenland-Iceland Gap, ended up destroying almost a third of R-1, including half of the twenty fifteen-thousand-ton container cargo ships.

By the time R-1 was well out of the mined area, they were still thirteen hundred miles north of Newfoundland's Cape Race, Soviet Hunter/Killers closing two hundred miles west of the convoy. At the same time, American relief ships and subs were

250 miles to the south of the convoy, heading toward it to take over escort duties for the remaining half of the convoy's journey to Halifax. One of the subs that was heading north but not assigned escort duties was the USS *Roosevelt*.

Robert Brentwood was bringing his sub up from a thousand feet to trail his VLF antennae for a "burst message" which, lasting only milliseconds, was designed for a position verification for SACLANT and also for passing on any new instructions to the submarine. Around the raised periscope island in the combat control center, the highly polished brass rail was a ruby sheen, the sub rigged for red.

Brentwood listened carefully to the depth readout from the planesman. "Three hundred feet . . . two fifty . . . two hundred . . . one hundred. Steady at one hundred." The sub's props were now stilled, the *Roosevelt* suspended a hundred feet below the surface.

"Very well," said Brentwood. "Stand by to extend VLF."

"Standing by with VLF, sir."

"Extend three hundred."

"Extending three hundred, sir. Ten—twenty—thirty—forty—"

Brentwood was watching the sonar blips from the *Roosevelt*'s built-in hydrophones; the noise of the oil-smooth VLF aerial extending out from behind the sub was audible to the hydrophones, but the operator was classifying it as "soft," meaning there was only low risk of enemy subs picking it up unless they were closer in than fifty miles, which Robert Brentwood figured he should know about, the *Roosevelt* having dragged its towed array less than an hour before. Even so, the sheathing on the Russian Mike-class was extraordinarily good at dampening prop noise.

"VLF at three hundred, sir."

The VLF would stay out for five minutes exactly, during which the millisecond burst should be received, updating instructions.

Five minutes later the VLF automatic override began winding in the aerial. Six minutes later, the OOD reported, "VLF in."

It was not unknown for a sub not to receive its VLF burst during the prearranged time slots, but it had never happened aboard the *Roosevelt*. Four misses and the sub captains had been ordered, contrary to persistent and vigorous denial by both the administration and the Pentagon, to use their initiative. After the fourth miss or, "Zippo," as the no-burst was known, the captains of American subs entered into the world of TKI or two-key initiative. This meant the captain could, upon "reasonable and repeated confirmations," take it as given that the United States of America was under nuclear attack—HQ unable to transmit.

In such a situation the sub commander was free to fire his nuclear missiles at predesignated targets or at his discretion.

In the case of the ten-thousand-ton Sea Wolf II, it would involve firing nuclear-tipped Cruise missiles from torpedo tubes, and the six Trident ballistic missiles in two rows of three aft of the sail. The killer word was "reasonable." What was reasonable if you couldn't get confirmation? It was a lawyer's delight. The reason for no message being received, of course, could be that there was some malfunction in one of the huge and elaborate VLF signaling and relay aerials on the East and West Coasts of the United States, the other aerial "farm" for the *Roosevelt* and the other subs in the Atlantic Fleet being in Wisconsin. But Washington knew this and allowed for four attempts at ten-hour intervals—ample time, it was thought, to work out any problems on the aerial farms. If these could not be worked out, however, after twenty hours, TACAMOs—"take charge and move out" signal planes—would be ordered to fly over and beam down the bursts at the prearranged times.

"What happens if we're out of TACAMOs?" an off-duty submariner asked his friend. "Then where are we?"

In the control center Brentwood ordered the diving planesman to take her to one thousand, and turning to the OOD, instructed him to resume patrol. There was no point in making any comment about not receiving a message. The thing was to not make a big deal of it among the crew, start them worrying unnecessarily. During the resume patrol mode, the rules of operation were as carefully spelled out as any other mode, including an informal one that said if any member of the crew made a noise above detectable decibel level, the old man would personally stuff him into one of the eight forward torpedo tubes and blow him out to the sharks.

CHAPTER FORTY-SIX

IN THE SEA of Japan, aboard the U.S. carrier *Salt Lake City*, there was a half hour to go, pilots advised there would be a briefing in ten minutes. Fisher, Frank Shirer's radar interference officer, knocked and came into Shirer's cabin. "Scuttlebutt from 'Prifly' says we might have to go north of the parallel for a while. MiG Alley."

"Haven't heard," said Shirer. He tore half a dozen tissues from the Kleenex box, forming them into a cuplike shape, which he put in a small plastic Baggie, reached into his underpants, put the cupped Kleenex over his penis, then reached for his gravity suit.

"I'll bet they want us to hit some of the bridges over the Han," said the navigator.

"Maybe," said Shirer. "You might get an OET bonus."

"Over enemy territory? Hell, the whole of Korea's enemy territory. Just about."

"Only kidding," said Shirer, slipping on the G-suit. "You'll get a thank-you from the old man and a cup of coffee. No bonuses."

"Yeah," said the RIO. "Y'know, this friggin' G-suit of mine is too damned tight."

"Supposed to be. Stop the—"

"The blood from pooling," the RIO cut in. "I know. I think it's worth shit. Just a goddamned girdle."

"Ever been up without one?" asked Shirer.

"Yeah," said the RIO.

Shirer looked across at him, surprised. "The hell you have."

"I have. Pan Am flight out to the coast."

"Stupid bastard." Shirer grinned, the RIO playfully punching the other's shoulder patch: "Salt Lake City's Shooting Stars."

The sun, a red ball that had burned off low cloud, cast long shadows along the flight deck. Three "moles" from the most recent AWAC to land were being led like a column of blind men, their polarized visors down, each man's arm on the man's in front of him as a member of the flight deck crew led them through the hectic, noise-filled activity of prelaunch, the three moles' eyesight not yet adjusted to any kind of daylight after spending four to six hours straight in the windowless twilight in the rear of the AWAC.

In the ready room the monitors were giving all pilots the good news that the Soviet Fleet was not proceeding farther south, intelligence reports indicating all available jets were being thrown into the European theater. The bad news was, a front was moving south into the Sea of Japan and was bringing more low stratus, reducing visibility again and even possibly interfering with some of the infrared systems because of "100 percent" moisture.

"There will be three predawn attacks," said the briefing officer: "Two companies of helo-ferried Second Airborne at Taegu, and two at Taejon preceded by Hawkeyes with fighter cover. I'll get to the third target in a few minutes. First, targets one and two." He called up the computer image of the 180-mile-wide, 200-mile-long peninsula. "Phantoms will be riding shotgun for Taegu, and Taejon troop choppers will be preceded by Apache attack helos and Huey gunships from the helo carrier *Iwo Jima* north of us. They'll be laying eggs," by which the briefing officer meant laying mines around the landing zones.

"Remember the chain guns on the gunship helos are mounted left, so they'll be going in counterclockwise when they start their attack. By then, of course, it will have hit the fan and the MiGs'll come in, trying to chop them up. It's your job to break up the MiG attack and take out as many as you can."

"All right!" said Fisher.

"First wave of airborne will go in via *Iwo Jima*'s Super Stallion helos—thirty-five men apiece. Ten helos to secure the airstrip's perimeter, and then the Hercules out of Japan will land if possible. They're checking satellite photos now—if they can't do that, cargo will be palletized. Just so you know what's going on.'' It meant that if the airstrip was secured by the troops, the 155-millimeter howitzers, strapped tightly to wooden and metal frames, would slide out over the rear ramp of the big aircraft as they thundered in at less than a hundred feet above the ground, the equipment-filled pallets of guns, ammunition, and other supplies braked by the simultaneous deployment behind the pallet of three drag chutes.

"If the *Iwo Jima* can spare them, it will also launch a dozen vertical-takeoff Harriers to act as gun platforms in case Charlie starts bringing up artillery around the airfields.''

"Sir, wouldn't they already have the strips zeroed in?''

"Not from the intelligence photos we have. It would appear they're racing like hell to the south for a final push against Pusan and that they've decided to put all the heavy guns down there to open up a gap through the perimeter. Hopefully we'll be able to take them by surprise. The distances aren't that long, nothing further than about a hundred and forty miles in. If we can secure one of those two airfields for a few days to fly more of our boys in, we can buy time for the guys trapped in that jammed perimeter and hopefully segment their supply line. That's what it's all about.''

The briefing officer took a sip of water. "Also, we've received news that we've got nine B-52s at Guam patched up and ready to go in about a week. Ground crews down there have been doing an outstanding job getting them ready. If we can buy our guys a few extra days in that perimeter, pretty soon the B-52s will be able to pound the shit out of the gooks' supply line. Whatever happens, we can't let them push our guys into the sea. That happens, it may be years—maybe never—before we get it back. After Nam, that'd be two losses in a row.'' He paused. "Third target we've been charged with. Shirer, you'll lead a second wave of Tomcats to fly cover for a combined helo-borne infantry and MAGTAF strike from the helo carrier *Saipan* against Pyongyang.''

There was a low whistle from one of the navigators, and several pilots looked over at one another—a few in silent sympathy for the marines and other infantry who would be going in deep behind enemy lines.

"The psychological significance of this mission, if successful, will be tremendous, gentlemen. Our problem, however, is to get through their radar screen. Now, we can go in low as far as the coast and they won't pick us up, but once we climb over the Taebek Range, we'll be on their radar immediately. From the coast it'll be a hundred and twenty miles in. The helos from the *Iwo* will be following ravine contours as far as possible and discharging flares against heat seekers. But the MiGs are sure to come in before we get halfway there. Shirer, you deal with them, and remember Pyongyang is surrounded with SAMs. We'll proceed with the fly-in no matter what happens, allowing for ten minutes of fuel for low ground-support attacks. The North Koreans had air supremacy when they first crossed the DMZ, but now our fleet's moved up, we hope to even the score.''

"They intend getting those marines back, sir?" asked Shirer's RIO.

"It's been carefully thought out. That's all."

"That's no answer," said Fisher, "that it's all been thought out. So was Carter's attack on Iran."

"Doolittle," said Shirer.

"What?"

"Doolittle's attack on Tokyo not long after Pearl Harbor. Gave the allies one hell of a lift."

"Yeah, but it won't be much of a lift for those marines and other poor slobs."

"Well," answered Shirer, "all I know is that if I was in that Yosu corner and I heard a Commie capital had been hit, it'd boost my morale. Can't underestimate morale, Fish. Sometimes it's almost as good as live ammo."

"Think they can do it?"

"Think *you* can do it?"

"What—piece of cake. I'm worried about the guys who'll be on the ground."

"You worry about the damn radar."

CHAPTER FORTY-SEVEN

Nova Scotia

THE OLD TOWN clock in Halifax began its quarter-hour peals as Lana and three other Waves on their first day off walked along the tree-arched trails in Mount Pleasant Park. Out on the harbor, the replica of the famous *Bluenose II*, her sharp, classic lines undiminished by time, was cutting spritely through the cobalt water, the grace of her design, which had won so many international trophies in the first part of the twentieth century, a striking contrast to the gathering fleet of U.S. and Canadian ships. The smell of the sea was on the east breeze, and had it not been for the dozens of gray shapes dotted about the harbor north of them, the light, invigorating wind coming off the North Atlantic could have convinced Lana all was right with the world. As yet the only casualties they had had to deal with had been a few broken legs and arms and one man badly injured after being struck by a sailboat's boom while assigned to help crew a visiting admiral in one of Halifax's many yachts. His injury had been entered as "DA" by the head nurse, a stout, no-nonsense Englishwoman whom the Canadians called "Matron."

"No, not dead on arrival," Matron had said humorlessly, demonstrating how such things had to be handled if one was to work one's way through the bureaucracy. "DA means 'dockyard accident,' " she explained brusquely.

It was a small enough incident, but it meant that responsibility and costs would be entered against the dockyard rather than the Canadian navy, and it told Lana something about Matron and the bureaucratic system they'd have to contend with even in a harbor that in wartime became one of the busiest and most strategically important in the world, the start of the long convoy runs across the Atlantic and a port that during the Second World War had repaired more than six thousand Allied ships.

Lana had been asked out several times by young doctors at the hospital, but declined, her experience with Jay having been so traumatic that while it didn't sour her against men in general, it made her wary—and the very thought of having to fend off the uninvited and inevitable sexual advances was too much to contemplate. For now it was all she could handle to pass the examinations in a punishingly more concentrated training period than was normal, because both Washington and Ottawa had advised Halifax that a "substantial number of casualties" could be expected.

The scuttlebutt had it that a convoy from England was already en route. And in the pubs around the old cobbled streets of the "Historic Properties," where press gangs had once roamed shanghaiing "volunteers" for Her Majesty's Navy, there was a rumor claiming empty container ships had been sent first so as not to risk any vital cargo. A "guinea pig" run for the British and Americans. The Halifax *Chronicle* printed the story.

Within three hours SACLANT in Virginia and CFB—Canadian Forces Base Halifax—flatly denied the assertion, pointing out that container ships were now in the process of loading some of the millions of tons of matériel that would be needed to reinforce Europe. The *Chronicle*'s publisher was invited to the admiral's house for tea. It was brief, polite, and during the conversation the admiral asked the publisher's advice on whether or not he thought it would be worth, "in the public interest," running a story on the Canadian War Measures Act. He reminded the publisher that the last time it had been used was in 1970—when the then Liberal prime minister, Trudeau, had deployed armed soldiers on the steps of Parliament, and during which time, the admiral noted, newspapers, along with everyone else, were forbidden under the emergency powers to discuss the FLQ—Front Libération du Quebec—the terrorist organization that had kidnapped Labour Minister Pierre LaPorte, shot him dead, and dumped him in the trunk of a car.

The reporter for the *Chronicle* who had broken the "empty cargo" story was reassigned to the obituaries.

* * *

To the other three nurses with her, Lana's personality was something of an enigma, an odd mixture of shyness and assertiveness in her job, and working with the patients in an oddly detached way. She wasn't unpleasant, but rather, distanced, as if somehow nursing for her was less a vocation than a refuge. It was several weeks before one of them discovered from an American newspaper that she had been *the* Mrs. J. T. La Roche and sister of the naval officer who, if he survived, looked like he was going to be court-martialed. This news only confirmed Lana's fellow nurses in their intuitive belief that she was running away.

"But to a hospital?" one of them had put to the others.

"Why not? After a rotten marriage, a hospital's as good a place as any to lose yourself for a while. Get a new perspective."

They were right and wrong—right in that Lana had found a place to retreat, where other people's needs forced her to leave her troubles for a while, but wrong in thinking it gave her a new perspective. The outbreak of war to Lana was but another example of people's inhumanity to others, something she had experienced in her marriage. And she made the depressing discovery that no amount of work, no amount of depressing news from the war in Europe and in Korea, could take herself out of herself long enough to rid her of the feeling that inside she was somehow permanently contaminated, dirty, that in succumbing to Jay's sexual demands, she'd sullied herself more than anyone could ever know. The very thought of it would start her throat constricting as if she were being suffocated, nowhere to hide, no one to help—a heart-thumping terror of suddenly losing control. In those moments she was secretly but deeply depressed. And despite the veneer of self-assurance, she realized that not even the cataclysmic possibility of nuclear war could erase an individual's guilt.

It didn't take Lana long to fall afoul of Matron. On one of their few days off, after Lana and a black girl from Boston had discovered their U.S. dollars were worth at least fifteen percent more in the Canadian port, the four nurses had taken a cab around to see some of the sights. One of the girls wanted to go to Fairview Cemetery on the city's west side, where 125 people who had died on the Titanic were interred and where more than 200, some of them unrecognizable after the blast, had been killed on December 6, 1917, when a French munitions ship, the *Mont Blanc*, on convoy to Europe, had struck a Norwegian merchantman and exploded, razing most of the buildings on the city's northern side, killing over fifteen hundred men, women, and children and permanently injuring thousands of others. To Lana's surprise, she found the cemetery peculiarly comforting—why, she didn't know.

When they returned to the nurses' quarters, they were summoned by Matron.

"I don't want my nurses tearing around town waving dollar bills about like tarts!"

The nurses were taken aback and angry, but none of them stood up to her except Lana.

"Matron, I don't know who could have told you that. But we were on our own time and—"

"Own?" asked Matron malevolently. "No one has their *own* time in a war."

Lana didn't answer that as yet they had seen nothing of the war. Intuitively she sensed Matron hated her for her good looks.

"Well," Matron continued, "you'll soon be too busy for any nonsense." She paused for a moment, was about to turn her back on them, when she cast a steely gaze on Lana. "*Mrs*. La Roche, you may not be aware of it, but a great deal of unnecessary resentment can be caused by people with more money flashing it about in front of others who go without. Canadian nurses earn far less in real terms than you Americans—not nearly enough to hire taxis to roam at leisure—"

"But, Matron," interjected one of the two Canadian nurses, "we didn't mind spending—"

"The point I'm making is that there'll be enough for people to gripe about once these wards start to fill, and believe me, they will. We'll have enough to deal with without petty squabbling breaking out between Canadian and American and British servicemen about whose girl is whose and who has the most money. Don't tell me—I've seen it before in—" She decided not to reveal her age. "I merely want you to act responsibly. That's all."

Elizabeth, the black girl from Boston, shook her head as they watched Matron walk away, heels clicking on the hard linoleum floor. Even the echoes of her sharp footsteps on the immaculately clean floor had the very sound of cold efficiency.

"I've been told about people like that, but I never believed they actually existed. Old bitch!"

"They exist, all right," said one of the Canadian girls. "Can't bear to see anyone else happy. I think she likes the war—gives purpose to her miserable existence." Lana felt herself going red with embarrassment.

"Yes," said the other Canadian. "They should put her on a sub, down and out of sight. Maybe if we get lucky, she'll get torpedoed."

"With what?" asked the girl from Maine. They were all laughing except Lana. Elizabeth suddenly bit her lip, turned, and consolingly put her arm about Lana. "Hey, I'm sorry, Lana. That was some dumb thing to say."

The others had to wait for Elizabeth to explain that one of Lana's brothers was a sub captain in the Atlantic Fleet. But Lana hadn't been bothered by the reference to either subs or torpedoes. She was far more worried about the way she had stiffened when Elizabeth had put her arm around her. Now, even the touch of another human being trying to comfort her only raised her defenses. Had Jay destroyed her that much? she wondered, in which case her days of comforting as a nurse were surely numbered.

CHAPTER FORTY-EIGHT

VIA THE ARMED services radio in Japan, news had reached the LPH *Saipan* that in New York a drug ring had been discovered by military police overseeing the loading of NATO supplies. Among the drugs stolen were substantial quantities of Demerol and morphine destined for the U.S. forces fighting in Europe. The news item had so upset General Douglas Freeman that he had been unable to complete his meal with the usual light banter of the officers' mess.

The general was now looking out into the darkness from the flight deck of the *Saipan*, its destroyers and forward ASW helos invisible about the eighteen-thousand-ton ship.

"Know what they ought to do with them, Al?" he asked his aide. "All those junkies?"

"What's that, sir?"

The general's voice was made sharper by the salty wind whipping it away. "Shoot them. Like the Chinks do." He was looking up through scattered stratus at a spill of stars twinkling over the Sea of Japan. "By God, I don't like that Zhou bastard, but one thing those Commies know how to do is to deal with pushers and all the other scum. I find any in this outfit with glue up their goddamned nose, I'll throw 'em overboard."

"Pity you haven't had time to get to know them, General."

"I'm well aware of that deficiency, Al. They'll get a chance to size me up when I talk with them tonight. And I'll get to know them well enough tomorrow. If we can hold past noon— well, we'll make history. In, bang, and out! That's the ticket."

Freeman was about to go down to the troop deck when Al Banks decided to get something off his chest. "Sir, I haven't had time really to get to know you yet."

"You resent me having command? Don't think I've earned it? Too fast a promotion. That it?"

"No, sir, not at all. It's casualties I'm concerned about. I know soldiers, marines—airborne or otherwise—know they might be called in to do their job anywhere at any time. And they've had sudden changes in command before. But the scuttlebutt aboard this ship is that casualties are going to be exceptionally heavy."

"They've seen the projected figures—or they've seen figures leaked by someone to discredit my plan of attack?"

Al Banks began to object.

"Doesn't matter how they found out," cut in Freeman. "Even if they hadn't seen the figures, they've been studying the satellite photos and mock-ups until they know every square inch. Hell, they'd be dummies if they didn't. What do you expect me to do about it?"

"They're saying the projection is higher than seventy percent, General—way above any casualties that we would normally—"

"Normally?" Freeman seized the word. "Now listen to me, Al." As a stiff northerly was blowing across the darkened ship's flight deck, Al Banks was unable to hear everything the general was saying but could feel Freeman's eyes boring into him. "This isn't a normal time. This isn't a normal war. Besides which, following normal rules is a recipe for disaster for any dumb son of a bitch without imagination. Normal rules won't win the goddamned thing. Commanders are changed all the time in battle because they followed normal rules. My God—" the general paused, like an engine building up more steam, turning to go back inside the ship "—we've got to get this thing sewn up so we can go over and kill those Russians."

"I'm not sure the men share your optimism, General. Taebek Mountains look pretty formidable even on a map, and they're wondering—"

"All right, I'll speak to them about that, too. By God, I tell you, Al—I smell a big Washington rat around here trying to undermine my authority."

"I just think they're scared, General. None of them has been in action before except on one or two missions during the invasion of Grenada."

"Grenada?" The general's face was now visible in the dim light of the ship's interior. They were walking down toward the helo deck, lights ablaze, all solid metal portholes sealed shut as the mechanics performed last-minute checks on the big three-engined Super Stallions and Chinooks, which, carrying forty-five and thirty-three men respectively, would be carrying half the fifteen-hundred-man force into Pyongyang, the other half coming in on the Hercules, right on the airport if possible—if not, circling and making the jump. "Grenada was a shambles," said the general. "Sustained more injuries there from our own foul-ups than from any resistance we encountered."

Either side of them down the cavernous hangar deck, scores of technicians and ordnance men were now "bombing up" the smaller Apache attack helos, each Apache receiving two AA Sidewinder missiles, one on the edge of each stubby wing, two pods each carrying nineteen seventy-millimeter rockets and four Hellfire antitank missiles. Loaders were also laying the belts of high-explosive thirty-millimeter for the unmanned chain gun in the chin turret, which would be slaved to the gunner's integrated helmet and display sight system or IHADSS. The general

stepped aside to let a small forklift truck pass by loaded with
.50 armor-piercing trays for the port-side chain guns of the Black
Hawk Hueys, the prototypes having proved themselves in Vietnam. Here and there the sharp, angular shapes of the fifty-nine-
foot Apaches and several smaller Cobras were broken by raised
engine cowlings, laser sensor boxes half-out, and thermal
imager sensors being tested by army and navy electronic technicians.

Farther down they passed tightly stowed three-in-one off-the-
shoulder antitank Starstreak missiles, which would be used by
the Marine Air Ground Task Force units to stop any tanks that
might be brought to bear if the enemy tanks got through the air-
dropped minefields.

As he entered the huge area of the troop deck, one below the
hangar, where rows upon rows of concertina bunks were drawn
up to the ceiling while not in use, revealing an area half the size
of a football field, Freeman looked out upon a sea of over a
thousand faces. In less than ten hours—if the weather held—
just before dawn, they would leave, and those who survived and
were lifted out would probably number no more than two or
three hundred. The very idea of the desperate gamble, to strike
an unsuspecting blow at the enemy and to rejuvenate America's
confidence in herself, as Doolittle's raid on Tokyo had done so
many years before, filled Freeman with such pride and expectation, he felt his whole body gripped with excitement. However, he knew the men before him didn't share in his happy
anticipation. It was something a commander had to change. He
would, he knew, be relying on that rare ability of his and a few
others like him to convey a sense of immediacy, of intimacy
with the person whom they have just met as if they had known
him all their lives. He carried with him an intensity and aura of
tough trust, and above all, the deep-seated fanaticism that other
Americans, he knew, would not call fanaticism: the rock-solid
belief that you could fix anything—that, God willing, you could
put it right.

Going up on the small rostrum made up of aircraft packing
cases, the Marine Corps and Infantry flags behind him, Old
Glory higher than the others, Freeman knew this was the most
important speech he had ever made. In the front row sat his
company commanders, five in all, and the padre assigned to the
USS *Salpan*. He'd heard Freeman was a cowboy.

Freeman glanced down at the padre, at the company commanders, then up at the waiting faces. He was immediately struck by their youth—most, he guessed, no more than twenty, maybe twenty-five.

CHAPTER FORTY-NINE

IN THE MID North Atlantic the sea was turning angry, the northward flow of the Gulf Stream running up against a cold Arctic front. It was a mixed blessing for Convoy R-1, for on the one hand, it meant that enemy subs' sonar would have a harder time separating ships' engines from the turmoil of the air-sea interface, where hollows had become deeper, troughs more frequent. It was, however, also more difficult for the convoy escorts, or rather those that were left, to detect the noise of submarine props in the turbulence as increasingly heavy seas crashed against the ships, distorting the noise patterns even further. It wasn't impossible for the operators to work, but it made them much more tense than usual, conscious of how false echoes, or blips wrongly interpreted, could cause the launching of a SUBROC missile from the convoy's escorts into its own Sea King helo screen, as happened earlier.

An equal possibility, HMS *Peregrine*'s captain realized, was to mistake one of your screen subs as an enemy sub, or to throw it into the suspicious "unknown" category. Accordingly, at 2200 hours he ordered mirror semaphore to signal all destroyers and frigates to be especially alert to this danger lest they prematurely launch a torpedo or depth charge attack. "The situation's complicated, Number One," the captain added, "because none of our sub screen will, of course, come up and radio their position to us for fear of revealing themselves to Ivan."

For this reason the captain reminded all those on the bridge and in the combat control room that they had to be particularly cautious if one of their own Trafalgar subs from R-1's screen picked up an enemy sub. The Trafalgar's skipper's plan of attack would more likely than not mean he'd have to take the Trafalgar out of the fan-shaped sub screen and, in order not to reveal its position to the enemy, would be unable to notify the convoy. This could result in electronic misidentification of the kind that had caused a USS guided missile cruiser, *Vincennes*, to shoot down a commercial air bus in 1988.

Further down the line, in HMS *Peregrine*'s mess, William Spence was witnessing one of the strangest sights of R-1's voyage so far, indeed one of the strangest sights in the whole British navy. Leading Seaman Carswell, with a full cup of coffee placed carefully atop a small silver salver, proceeded to carry out the first steps in the dance that had won him fame throughout the fleet as the only steward in the fleet who, leaving the galley with a full cup of coffee, could deliver it to the bridge unspilled, no matter how rough the weather. In the worst storms, with the litheness of a prima ballerina, fighting gravity against impossible angles, he was a sight to behold, as he began his trip toward the stairwells, his legs seemingly made of rubber, sometimes walking back, giving a few paces, then recovering the lost distance the next second with a short, fast run, the cup held aloft as if he himself were in a gimbals mounting. More than once, new lieutenants had lost a day's pay by having an unofficial wager with another officer, waiting impatiently upon the news that Carswell had left the galley and was now on his way up, expecting the seaman would trip at least once on the way. But so far it had not happened, leaving Carswell with what the ship's company called an unbroken number of "FCDs"—full cups delivered. Several crew had already written inquiring whether Carswell's feat qualified for mention in the *Guiness Book of Records*. They hoped a reply would be waiting for them when they returned to their home base at Plymouth.

The *Peregrine* leaned hard aport as a starboard wave struck her, the well deck awash athwartships, the bursting white cloud of spray enveloping the bridge with a sound like hail. It would soon be a force eight, the bollards already covered in foam, water rushing down the decks like a spring runoff, spilling out through the stern scuppers and swirling about the aft twin Sea

Dart launcher, the flow broken but not stopped by the Limbo mortar before pouring back into the sea.

Visibility was now down to two miles, more whitecaps evident as R-1 continued to plow ahead.

"Only time I'd like being in a submarine," Johnson said in the galley, he and Spence having left spreading the Marmite sandwiches till last. "All this bloody rockin' and rollin' is for the—"

"Action stations!" came the voice over the "Tannoy," the PA system, and before Johnson could pick up another slice of bread, he, Spence, and the cook could hear the sound of running feet in the passageway outside the crew's mess. Spence had expected a lot of shouting from petty officers and the like, but what struck him was the lack of any harshly shouted orders, the ship's crew reacting more like a well-trained sports team than men at war.

On the bridge, the sonar operator calmly reported, "Contact, bearing two oh five degrees."

"Half speed," ordered the *Peregrine*'s captain, reducing the noise of his own ship's prop.

To help avoid giving away *Peregrine*'s position, her sonar was on passive, and so no range could be registered. But the captain knew that a good operator, knowing his own set and ship, often developed a reasonably good guesstimate. "Can you give me a distance, Sonar?"

"I'd say five thousand yards, sir."

Quickly *Peregrine*'s captain looked up at the computer board, comparing the vector plot of R-1's course and his present position. It put the noise about two miles away, at a wide angle from where the convoy's screen subs should be.

"Contact gone, sir."

"Completely?"

"Yes, sir."

"Odd," the captain said, turning to his number one. "Damned odd. Any thoughts?" He looked about the bridge— it was no time for pride. "Anyone?"

"Wreck, sir—old oil drum or something moving? I mean, sir," he continued, "something hitting it—you know, rock slide or something—"

"Could be, Chief . . ."

"Contact bearing two six one."

"Go active," commanded the captain. "All round sweep."

"Aye, sir. On active. All around sweep." Now the inside of the *Peregrine*'s bow began to twang as its hull-seated transducer, in effect two metal plates buckling under electric charge, sent the distinctive sonar pinging noise into the ocean's depths, the operator turning the echo onto the bridge's Tannoy, the sonar sweeping from zero degrees to 360 every two minutes.

"Contact. Bearing two six one. Range eight hundred yards. Moving fast . . . Contact. Bearing two six oh. Range seven six oh yards. No props noise, sir. I'd say a mine."

"Torpedo motor?"

"No, sir. No beat count."

Without taking his eyes from the sonar screen, the captain waited for the two seconds as the computer digitized all incoming information, telling him that whatever it was, was coming at them at forty miles per hour. Impact in 4.71 minutes—on the starboard side—and countermeasures giving him the IAVS—impact avoidance vector and speed.

"Notify all ships. Steering by IAV." Unhappily he knew that none of the merchantmen had IAV capability. "By voice to transports. Plain language."

"Aye, sir. To transports in plain."

"Follow the ship in front of them."

"Aye, aye, sir."

The computer IAVs were entered into the *Peregrine*'s memory and she swung hard aport, her type eight steam/gas-combination-driven prop driving her at twenty-four knots, her decks now constantly awash as she heeled sharply to avoid impact.

"Same contact. One one seven degrees. One thousand yards."

"Identify?"

"Negative. But different from the other."

By the time the new IAVs had been spat out by the computer, it was too late for the *Peregrine*, the captain realizing it was not an isolated mine coming for them but a series, probably combination pressure/noise, triggered by the active pulse the *Peregrine* had just sent out. Like a man trapped in an ever-shrinking room, the IAVs were now seemingly accessories to the fact, for wherever the ship moved, there was another ping. Carswell had just placed the coffee in the captain's special gimbals-mounting cup holder when *Peregrine* was hit. The second mine she had picked up on sonar was the first to explode as she turned into its path, the mine homing in on the keel at the forward end of the engine room, the second blast cleaving her well below her waterline at the stern and buckling the prop, lifting the destroyer's stern completely out of the water. As her bow rose high, then fell, a breaker came rolling down under her like a leaden gray wall. All main fuses gone, the ship was in total darkness for several seconds before the auxiliary battery lights kicked in. In the galley the plastic crates of sandwiches slid en masse, but none were lost, as the three men were thrown against the cold ovens and one of the huge, shining mixing bowls. As part of a standby fire team, William Spence and, more reluctantly, Johnson moved to their station aft by the hangar door, where the battery charge lights had ruptured in the stern explosion. They heard a noise coming from outside—an enormous gushing sound like so many fire hydrants turned on, crashing in a sustained crescendo against the bulkhead—and felt the ship jerking starboard, then port, and back to starboard, her motion so violent, it seemed nothing less in the dim passageway than the enraged effort of some great leviathan caught in an iron trap thrashing to be free.

"What the bloody hell—?" began Johnson.

"Shut your face!" ordered a CPO sternly.

Young William Spence, already putting on his asbestos suit and helmet with breathing apparatus, looked up at the CPO, a man he'd never seen before, and in that second the unfamiliarity of the man's face and the noise were so disorientating that for Spence it momentarily took on the aspect of a nightmare.

"Get on your 'casper's' pack like Spence here," the CPO said to Johnson. "And follow me." Spence was surprised the man knew his name until he remembered they all had their names on their shirts. "Come on, Johnson—between Spence and me. Move!"

"All right, all right," moaned Johnson, pulling on the fire-retardant suit and tightening his head gear.

As the chief petty officer opened the hatch leading down to the engine room, they were enveloped in clouds of steam that instantly fogged their masks. They could hear men screaming and the rushing, bubbling sound of the water.

"We should abandon ship!" shouted Johnson, his voice nasal in his suit. "Bucket's going to sink!"

"Shine your light over there!" the CPO ordered Spence. "Port side."

One of the big gas turbines was still going, despite the captain ringing the telegraph to stop all engines, the torque on the prop enough to keep it turning, but jerking the ship, pushing and pulling it like an animal still moving though brain-dead. They couldn't hear any more screaming, Johnson urging them to get out of it while they could. Then, just as suddenly as the explosion, the turbine stopped, telling the CPO that someone in the engine room had managed to reach the controls of the manual override after the automatic controls had been severed. Or perhaps the turbine had cut out of its own accord? Spence thought he heard a faint cry above the rushing water and the now creaking sound of the ship—but it was difficult to be certain when one wanted so much to help. Spence saw a subby, or junior lieutenant, walking up the incline of the slanting passageway, blowing high and low on his whistle, telling everyone to report to their boat stations immediately.

"See, I bloody told you," began Johnson.

Spence had never been so petrified in all his life, but he could hear a voice. "Here, Chief," he said to the petty officer, his voice dry, with an almost squeaky quality to it in his fear. "I'll have a gander."

"All right, lad. Here, loop this about you. I'll take up the slack." The CPO started feeding out the yellow nylon rope, taking a turn around one of the ladder's rungs.

"Too short tugs from you," he called out to Spence, "and I'll haul. Make it snappy as you can."

"Righto, Chief."

"Righto, my arse," said Johnson.

Spence was now up to his chest in surprisingly warm water, the ice-cold Atlantic momentarily heated by the dying gas steam turbines.

"Listen, mate," yelled Johnson, "you want to play bloody hero, you go ahead, but I think—"

There was another explosion; this time the ship pushed hard aport fifteen degrees, its whole structure shuddering. Spence was off the last rung, underwater, the CPO and Johnson sucked off the ladder as well, the CPO barely managing to hold on to the rung around which he'd taken a turn with the nylon rope. In the thick fog that now filled the rapidly flooding engine room,

Spence glimpsed Johnson's fire-red air tank going past him, Johnson screaming. Spence made a grab, felt a boot, and hung on with his right arm, his left groping for a hold, any hold, as he felt his air supply cease, his mouth full of salt water and oil. He felt a violent wrenching, his shoulder driven so hard into a stanchion that, putting his left arm out to grab it before he was swept away again, he felt his hold on Johnson weakening, the seaman not helping by panicking and thrashing about. But with all his will Spence held fast to his shipmate. His left boot touched something and he let it take all his weight. It was one of the upper rungs on a stairway thirty feet farther down the engine room from where they'd entered. As he hauled himself up, not yet realizing he had been driven so far down the engine room, thinking he had somehow been hauled back to the first ladder entrance, Spence, straining to hold Johnson's head above, looked about in the fog for the CPO.

He was gone, the yellow nylon rope floating now about him and Johnson like some great water snake in a lake that had only minutes before been *Peregrine*'s engine room. The petty officer's single turn about the rung on the first stairwell had saved Spence, who in turn had saved Johnson from being sucked out like the CPO through the gash in the engine room's side.

Johnson already had his helmet and air tank off as he lumbered up the last dozen rungs to the top, water rising quickly behind them. He swore violently at the inner tie of the asbestos trousers, which his fingers were plucking at frantically, his words a torrent of frenzied invective. Spence was now out of his suit but was still looking back to see if there was any sign of the CPO. A body washed past them, its face puffy, purple, and badly lacerated—an engineering officer, by his arm stripes, the facial wounds remarkably clean.

"My Gawd!" said Johnson, taking the last steps in twos, spinning open the ring lock door, stumbling out as the ship leaned farther to port, the door slamming shut, opening on the rebound, revealing a new hissing surge of water rising in the engine room. Spence, still inside, tripped on the second to top rung, a foot from the door's sill, and instead of catching hold of the ring handle at the door's center, his fall meant that he just managed to grab the sill. "Hang on!" he called to Johnson. Johnson paused for a second, heard the hiss of more water, slammed the hatch shut, spun the wheel, and bolted, knocking down an artificer on the now sharply inclined passageway.

Dazed, lifting himself up, the artificer saw the door of the engine room move, its high, mouselike squeal audible amid the deeper rumble of the ship that was now sinking, as millions of gallons sought to fill every possible space, ironically bringing the ship back to a stable position before it began listing again, this time to starboard. The artificer saw the wheel move again and was about to reach for it when the door flew open on the downward incline and a seaman came tumbling out, slamming against the opposite bulkhead. For a moment as the artificer leaned on the door, pushing it shut and spinning the wheel, water bubbling out about his feet like an overflowing toilet, he thought the seaman was wearing red Day-Glo gloves.

Coming up, splashing behind them, was a CPO from the combat control center. "Come on, you two. Topside. Old girl's had—" He saw Spence collapse onto the deck, and now the artificer saw what he had thought were red gloves.

On the deck they laid William Spence down on a net stretcher, the roaring light above him so bright, it seemed he was entering the sun. A sick bay attendant struggled for several minutes beneath the down-blast of the helicopter's blades and in the spray it was whipping up about them before he managed to give Spence a shot of morphine.

The OOD, his face bleeding, the cuts superficial, looking more serious than they were, cast a glance down at Spence. He thought he'd seen him in the galley once. He saw the seaman's eyes open briefly, then shut. All about him there were men calling for help, some quietly moaning as the fury of the sea continued unabatedly, indifferently, to batter the dying ship.

The OOD was trying to decide the priority cases for the chopper's first run. Amid the noise of the chopper, shouts of men dying around him, some washed overboard and lost already, the other ships unable to stop, the sick bay attendant realized that the officer's glance at Spence was a silent question, but the attendant's grimace was one of agonized indecision as he shouted above the roar of the helo, "Wouldn't put money on it, sir. Then again—"

The officer looked helplessly around, but there was no one to help him decide. He knelt down in the wind that was whistling wildly through sheared metal and over the bodies littered all around, and placed his hand on the boy's forehead, making the sign of the cross, trying to remember the words of the Lord's Prayer, trying to decide whether the boy should be a priority case or not.

"I can't move my legs, sir. I can't feel nothin', sir, nothin' at all." It was Johnson, lying on a stretcher near a starboard davit. "I can't."

"It's all right, old chap," said the gunnery officer. "You just lie there. We'll get you off on the next chopper." Nearby, a bosun, overhearing the conversation, turned to his mate. "Don't see much wrong with *'im.*"

"Shock, I expect," said his mate. "Poor bugger's spine probably crushed, paralyzed from the waist down. That's why he don't feel anything."

"Thought I saw him walking out on deck," said the bosun. A wave smacked the starboard side of the *Peregrine*, a black wave suddenly incandescent, angelic in the cone of the chopper's belly light, water streaming frothily through the scuppers, the ship rolling very slowly now, the water sloshing back and forth, gurgling through buckled decking. "Least I thought it was him," said the bosun, still looking at Johnson.

"Nah," said his mate. "Must have been another bloke. Christ, I 'ope they send more helos. I don't fancy this lot."

Next to him the bosun was zipping up the body bag in which they'd laid the cook.

CHAPTER FIFTY

THE CAVERNOUS TROOP deck aboard the LPH *Saipan* rang with the general's voice.

"My name is Douglas Freeman and I'm here because, like you, I was considered the best for the job. First thing I want to tell you tonight is that I have no intention of dying."

There was a ripple of strained laughter.

"Neither, I trust, do you."

More laughter.

"Secondly—" The general's eyes were taking in the whole hangar with such intensity that every private, section, platoon, and company commander thought the general was staring at him. "I'm not about to lead any dope-heads into battle. I don't give a goddamn what the doctors say, or the surgeon general says—there isn't such a thing as a goddamned *calming* pill that'll let you go into battle like you were going to church—which, looking at you sons of bitches, I seriously doubt you've ever done anyway." In the first row, Al Banks, arms folded, was looking at his shoes. He had personally authorized the issue of .5 mg sublingual Lorazepam before it was known who would be commanding the hastily assembled mobile force.

"Now," continued Freeman, walking, hands on hips, across the small podium, stopping, facing the men, his voice reaching every corner of the hangar deck, "I know you've all been through the drills, the maps, the platoon assignments." He paused. "But there's something else you should know. When you go into battle I want you to know who you are, where you are, and what the hell you're doing. I want you shooting *gooks*, not one another. So—before you disembark, indeed before you dismiss this evening, you will dispense with any pills you were issued with and any other pills you may have in your possession, depositing them with the padre."

The padre, in the front row, looked up, surprised.

"If he runs out of pockets, put them in his helmet. I don't want any space-head shooting up his section because he popped a pill too many and thought you were all gooks—even though you're better-dressed than any gooks I've ever seen, and that includes that runt, Kim Jong Il, who—"

The troops were loosening up, the laughter coming more easily now.

"An evil piece of shit," continued the general, "who, by the way, when you were trying to get your first piece of ass, was trying to figure out how to murder innocent civilians and who is as evil a bastard as Qaddafi, Hitler, that shit Pol Pot, or any other son of a bitch ever hoped to be.'' The anger in Freeman's eyes was so intense, Al thought the general was about to jump right off the stage, his forefinger sweeping across his audience, his lone star glinting in the hangar light. ''It is our duty to go in and give that son of a bitch such a shake-up and hopefully kill the bastard, so that his henchmen will think twice about ever attacking the United States of America again.''

The general paused again and, glancing along the front rows, saw the padre was not at all fazed by the profanity. ''To teach them a lesson,'' Freeman went on, ''namely that they've bitten off more than they can chew because the United States of America will not—I repeat, *not*—give up the ghost in Korea and that if they persist in their butchery, we'll get tougher still and nuke the sons of bitches into oblivion!''

There was a roar of approval. The padre, Freeman noticed with satisfaction, was distinctly uncomfortable. The general's voice dropped.

''Now, I've heard that someone says this is a hopeless mission—a suicide mission, politically motivated. Well, I don't lead suicide attacks, and I won't give you any BS about minimum casualties. We're going into the enemy's belly and I expect casualties to be heavy. But we'll be coming out!'' There was still silence.

''As for it being politically motivated—hell,'' said the general, shaking his head, ''I don't know what that means. All I know is anything we don't like—don't care for—becomes politically motivated. We are instruments of national policy. Our profession is not peace, it's *war*. That's what we're paid for—to go in there—not to walk softly and carry a big stick but to give 'em the stick right up their heathen ass.''

Some whistles and a smattering of laughter. The padre was clearly angry.

''Another thing,'' Freemen continued. ''I know some of you have been wondering why an airborne assault when we could launch bombing runs. Two good reasons. One—Pyongyang, which here on out will be referred to in all directives and communications as 'Crap City,' is *festooned* with surface-to-air missiles and MiGs on alert. Our choppers will be going in low, and the SAMs, which are most effective at high altitude, will have extreme difficulty discriminating among low-flying aircraft with ground clutter thrown in. And we'll be hill-hopping all the way.

Furthermore, we'll have fighter cover from *Salt Lake City* to help confuse the SAMs when they scramble their MiGs to meet our boys. But the most important reason for sending in this force is that we have learned, contrary to *all* the armchair experts, think tanks, and God knows what else, that bombing fails to cut the head from the snake, including *Miss Jane Fonda*."

"Give it to her, General!"

"No comment!"

A roar of laughter. The padre's face had now turned from anger to disgust.

"What we have repeatedly found out, and here I don't wish to malign our comrades in arms in the air force—they do a damned fine job—but we have repeatedly found, most spectacularly with that dung heap Qaddafi, that you can send in the whole damned air force, with 'Smart' bombs to boot, kill everyone, and miss the son of a bitch in his *tent*. That is why the prime target of this foray is to take that runt out of circulation. *Permanently!*" The soldiers were clapping and stomping now.

"So that *your* families, *your* children, no longer have to live in a world where slime-balls rule the roost."

There was thunderous applause. The general held up his hand.

"You're scared. So are the boys who'll go into Taegu and Chongju. I understand that." He waited. "But you keep your powder dry, pass the ammunition as your forebears did, and you'll be all right. Above all, remember this is no mixed-up slugfest inhibited by peace pansies and Fondas running riot in your country. This is a war which all true Americans, together with our allies, know is right beyond the shadow of a doubt. That monkey 'cross the line—he has no honor, no decency, no conscience. This isn't a mission, it's a crusade, and you're the bearers of the flag. God bless you."

As the general walked down from the podium, there was the sound of over a thousand men rising, forming lines according to company and platoon. Before Freeman left the deck, Al Banks introduced him to the padre assigned to the *Saipan*.

"Pleased to meet you, General."

"But you didn't like my speech?"

The padre hesitated, "Ah—no, General. I didn't."

"Quite all right, Padre," said Freeman, pulling on his leather gloves tightly. "Difference of opinion. Was there anything in particular you objected to?"

"I thought your remarks about the North Koreans—though I understand your intention—were—I mean, categorizing a whole race in—"

"Padre. We're not going up against the Sisters of Charity. We're going to do battle with the Philistines—NKA special forces, their best militia, best home guard . . . top-of-the-line killers. I mean NKA *special forces*. Booby-trap their own kin if they thought it would kill Americans. They hate us more than any people on earth. I understand your position, Padre, and it's admirable—in its place. But these jokers my boys'll be fighting have never heard of any commandment, let alone ten of them. And I don't want any of my men thinking about loving their neighbors when they file out of those whirlybirds." Freeman moved so close to the padre that the latter felt distinctly uncomfortable. Freeman was pulling his gloves on tighter. "I've seen the devil, Padre, and the son of a bitch lives in Pyongyang and I'm going after him. I'm putting a bounty on his head." The general looked quickly at the other officers in line. "I want it known, gentlemen, that anyone who brings me the head of Kim Jong Suck will personally get a field citation and ten thousand dollars!"

The padre's eyes widened. "General, none of us have the authority—"

"Padre. I need you in the field. A lot of my boys are going to die. Are you coming along or going to rest your butt in the officers' lounge?"

"The padre's already volunteered," put in Captain Al Banks.

Freeman murmured, nodding approvingly, but the fierceness was still in his eyes. "Good. I need brave men."

"That padre," said Freeman as he left the hangar with his aide. "He's a good man."

"Yes, General, I think he is."

"I say he is. Only one thing wrong with him." They were walking back past the helicopter crews in the forward hangar area.

"What's that, General?"

"Didn't you see it? Stood out a mile."

"I'm not sure—"

"Egg on his vest. I won't tolerate that, Al. You get the word out."

"Yes, sir."

"I ever tell you my father was a keen sportsman?"

"Not that I—"

"Baseball, football, of course. Golf—God, he even liked tennis."

"I like tennis, General."

"You do?"

"Yes, sir."

"Well, can't be helped. Never took to it myself. Son of a bitch sitting on that high chair, yelling out every shot. 'Course, my father wouldn't have it any other way. And dress code—that's what I'm getting at, Al. Uniform instils pride in a man. First thing you learn in basic training. Now, that padre, you see, he thinks he'll get closer to the men more laid-back, more slovenly he appears. Doesn't work."

"It was only a very small piece of egg, General."

Freeman ignored the captain's comment. "You play baseball?"

"Yes, sir."

"I notice you're a sou' paw—left-hander."

"Yes, sir."

"Good. We'll make a good team, Al."

The captain was nonplussed. Freeman stepped out into the cold night air, the blackness all around them like a great velvet blanket; not a light could be seen. "You on my left, me on the right. It's my intention to be in the first chopper. I don't want anyone saying Freeman freeloaded—took the last chopper in."

"I don't think anyone would, General."

"I've got enemies, Al. Say I'm too flamboyant. Too full of myself. Waiting for me to take a fall. You think that?—I'm a windbag?"

"No, sir—I—"

"Yes you do, you son of a bitch. Well, Al, tomorrow at cockcrow we'll find out."

Above them they could hear a faint squeaking as the concave dish on the main radar mast passed through a 180-degree sweep, its signal pulsing out into the darkness toward the enemy's shore.

Inside the hangar, in the long line of soldiers approaching the padre, who had now commandeered a second helmet, a private, first class, a helo side gunner, who, like most of the soldiers in the hangar, had never been in combat before, was veering between sheer fright and the bravado instilled by the general's speech, asking his comrades imploringly whether a packet of nasal decongestant capsules qualified as a pill he should surrender.

"What's it do to you?" asked a corporal from one of the *Saipan*'s Medevac choppers. "Slow you down or jack you up?"

The worried private flipped over the packet. ". . . Says might cause drowsiness, not to operate machinery."

"Aw, shit," said the medic, "take the fucking lot. Any luck, you'll sleep through it all."

"What's your helo number?" asked a marine.

"Twenty-six. Why?"

"Good number to stay away from," the marine said.

CHAPTER FIFTY-ONE

IN GERMANY, THE British Army of the Rhine was fighting for its life in the fifty-five-mile-wide pocket between Bielefeld and Dortmund, where swarms of Sukhoi-24 Fencer ground-attack fighters were wreaking havoc on the British tanks.

The Sukhois, chosen in part for the relatively narrow swept wing width, and painted in green-brown blotch camouflage, had been crated by rail through Poland and the GDR weeks before, reassembled, parked and maintained in the vast Nordhausen tunnel complex in the eastern part of the Harz Mountains, where Hitler had made his V-1s and V-2s out of range of Allied bombers. Rising swiftly, behind a semicircular screen of SAM sites in the Harz foothills between Nordhausen and what had been the West German frontier, the Fencers streamed northeast over Stolberg. Swooping down through cloud over the tranquil bluish green of the Harz, they reached the British army's defensive position a hundred miles west along the Bielefeld-Dortmund line in less than nine seconds, their terrain-avoidance radar and laser target-seekers cutting through the battlefield's smoke cover, setting the Challenger tanks ablaze.

When the Challengers fired, their aim was deadly, and they fought hard, even as they fell back northward behind the Weser and Mittelland Canal. The British were taking terrible punishment as well on their southern flank, where East German motorized shock troops were now crossing in force, so that in the depression between Bielefeld and Dortmund, what had been a fifty-five-mile-wide defense base now shrank to thirty-five miles.

As the Russian Fighter/Interceptors streaked down, pouring a deadly hail from twin-barreled twenty-three-millimeter cannons, a British commander watched his battalion of forty tanks and support infantry being systematically destroyed. Here, unlike Fulda, where American and Russian tanks were too close, the Soviets' gasoline bombs tumbled down—and with such apparent aimlessness, it seemed they could do no harm. Then the big silver "jelly beans" would burst, the saffron fireballs rolling over the velvet green hills. To the men inside the Challenger tanks, whose reactive armor had worked so efficiently in blowing up the Soviets' earlier high-explosive antitank bullets, the napalm was an agony worse than any gunfire, their bodies becoming torches, all but impossible to extinguish. As one man would try to smother a comrade to snuff out the flame, the jellied gasoline would fly onto the would-be rescuer, the droplets of red-hot mercury sticking to flesh and clothing—often as not, killing two men instead of one. On many occasions British infantry caught by the napalm not only died a screaming death themselves, but proved deadly to comrades as the heat from their burning bodies set off their own ammunition.

The Soviet Fencers supporting the GDR advance did not have it all their own way as the NATO air forces rose to meet them. But the overwhelming numbers of Soviet and Warsaw Pact fighters meant that the NATO fighters trying to contain breakthroughs on the Northern Plain, in the center along the Fulda axis and in the south on the alluvial flats of the Danube, were stretched too thinly, the NATO commander not knowing where the twenty Russian divisions held in reserve would be thrown in, therefore unable to concentrate wholly on any one sector.

It was a Soviet–Warsaw Pact strategy designed for the quick win, to push NATO so far, so quickly, that in the case of a cease-fire *ranovato*—"early in the game"—the Soviet–Warsaw Pact would have in its possession at least a third of West Germany and a sizable strip of Holland 50 miles wide and 130 miles long north of NATO's air-control center at Gelsenkirchen.

The NATO pilots were exacting a high price, but against all the other intangibles of battle was the stark mathematical fact, known for years before the war, that NATO could not outlast sheer quantity, even with a kill ratio in the air of two to one in the first seventy-six hours.

On the ground the situation was worse, the tank commanders facing Sergei Marchenko and the other ten thousand Soviet tanks having had to maintain a four-to-one kill ratio merely to hold ground.

In the wake of the Soviet–Warsaw Pact three-pronged attack, the old confident air among NATO commanders about Western quality versus Russian quantity took on a decidedly hollow ring.

Leading three other F-16s, one of the NATO pilots, Colonel Delcorte, his F-16 still climbing though getting low on fuel, was in his fourth sortie within seven hours. Based in Hahn with the U.S. Tenth Tactical Wing, Delcorte now had only fourteen of his twenty-six-plane squadron left. Ten were downed outright; two had managed to return to Hahn but were badly shot up.

Of the ten downed planes, three reported bailouts: one's chute was seen high above Fulda Gap, presumably drifting down over East Germany, the other two pilots picked up southwest of the Bielefeld-Dortmund basin in a heavily wooded section serving as revetment areas for the battered Chieftains. The British-made battle tanks, which, though first-rate and well armed with 120-millimeter cannons and assisted by crack British infantry regiments from the Army of the Rhine, were doomed because of their slower speed compared to the Russian T-90 A's seventy-five kilometers per hour. Had the Chieftains more time to dig in, it might have been a different story, for as Sergei Marchenko and fellow S-WP tank commanders quickly realized, the breakdown rate of the Soviet tanks was much higher than for those of the Allies. But though the British Chieftains were well served by their appliqué Chobham armor packs and first-rate crews, they could not overcome the Russians' three-to-one superiority in tanks.

At Fulda Gap, where the Russians had been unable to use their jellied gas bombs, the latest British Challengers, German Leopard IIs and American M-1s lasted much longer in battles where foam-filled, self-sealing rubber fuel tanks took many direct hits from armor-piercing shots without exploding.

In the Bielefeld-Dortmund area, The Russian T-90s, built to accommodate the shortest Russian tank crews and so presenting the lowest silhouette of any main battle tank, swung out wide through the British Chieftains' heavily laid white smoke screens and regrouped in ambush, isolating several Chieftains at a time. In the process, the smoke- and dust-filled battleground confused the pilots of the American Thunderbolts, who had to make split-second differentiation between Russian and British tanks—this becoming increasingly difficult in the dense smoke as the day wore on. It was, Delcorte thought, his F-16 leveling out high above the smoke, as terrible, or from the Russian point of view, as good, an example of using superior numbers as you could get, quite literally mugging the opposition with sheer brute force and size.

The only hope, Delcorte knew, was to keep going up, keep engaging them until the Juggernauts' attack was blunted, when finally there would be so many tanks off tread, so many automatic shell extractors in the T-90s out of alignment due to overheating and crews' fatigue, that the West's overall superiority in technology, including more tank transports, would start to pay off. Delcorte took some hope from the fact that NATO's Medevac and air crew rescue units were of uniform high quality. This was a crucial factor if NATO pilots were to fly and fight again. Delcorte and the other three F-16s in the finger four formation, resembling the spacing of a right hand's four fingertips, the little finger farther back, entered heavy cloud. At six thousand, Delcorte was still in it, the other three pilots in the clear, the wingman on his left advising him there were "Bogeys—six of them at one o'clock. Seven thousand."

"Keep going for the top," Delcorte instructed, the prearranged ceiling having been ten thousand feet, from which they hoped they could dive down upon the funnel leading to the Fulda Gap.

"Bogeys below—nine o'clock," came the next report.

"Go for the top," Delcorte repeated evenly, though in fact he had a multiplicity of incoming messages and warning signals, all of them conflicting, during his effort to outclimb the Bogeys coming in from the east, ignoring those passing below en route to shoot up the withdrawing Americans of the Black Horse division.

It was tempting to Delcorte to break the four-plane formation and go into two fighting pairs, a leader and wingman, the latter watching the leader's COV—cone of vulnerability. This would leave him, as the flight leader, free from blind side attack if he dove in to break up the enemy formation. He was also tempted to go to a fluid four formation, having two of the F-16s a thousand feet abreast in front, the other two, ten thousand feet apart behind. All these options sped through Delcorte's head in a split second, then, seeing the Soviet fighters still climbing, trying to get an advantage over the Americans, he forgot Fulda Gap and decided to attack.

Once the combat began, the geometry of flight formations was lost, two Soviet Foxbats winking orange, then gone in cloud, the air-to-air Acrid missiles streaking toward the American Falcons in excess of fourteen hundred meters a second, the faster F-16s already breaking, their pilots hitting the superchargers, climbing fast, trying to expose only their cold side to the Foxbats to deny their afterburners' exhaust to the heat-seeking Acrids. One American turned too late, and a dirty orange burst lit up inside a cloud, followed by broken thunder. The other three Falcons were on the northern flank of the Foxbats, who now had separated, three going ahead to Fulda Gap, three remaining to engage.

Watching his HUD—head-up display—afterburners on full, going air to air, Delcorte saw the green impact line on the HUD, where graphics condensed a thousand variables to a single display. The target vector was arcing right, cutting through three parallel lines. There was such a jumble of chatter, radio cuts from the Technicolor spread of the tank battles far below him, fragments of other NATO feed-ins, aerial and ground control, a smattering of Russian, that Delcorte, feeling he was ODing on noise, shut them all down except for that of his own formation. He rolled the Falcon, tried for a scissors, turning hard left, and was on a Foxbat's tail. One hundredth of a second later, the HUD's three green horizontals were cut by a green arc ending in a dot. Delcorte flicked the stick again. The dot moved toward center. The Foxbat banked hard left, Delcorte's lines altered, the dot slipping below the line, Delcorte cursing, flicking the stick again, not worrying about sighting the Russian's afterburner, sliding in with the cannon. He flicked one, twice more, but held his fire—at this angle the Foxbat could outrun his bullets. The green dot shifted, centered. He pressed the red, the .20-millimeter cannon sounding like a rip in the fuselage. He saw parts of the Foxbat breaking off, pulled away before any more came at him.

Three Foxbats had been destroyed by the three Falcons, but the NATO air force's problem, here as elsewhere, was starkly evidenced by the encounter. The American Falcons had killed the three Russian MiGs, but with the Soviet–Warsaw Pact's 3,000-plus fighters against NATO's 2,070, not counting the Soviet fighters in Russia, the deadly equation was still there.

Since the first sortie at dawn, Delcorte had lost sixteen pounds, but now once more he, like the other three Americans, felt exhilarated by the kill as they now headed down to bomb and strafe the long supply columns stretching out across East Germany, the enemy's breakout all along NATO's front from Austria to the Baltic continuing unabated. Delcorte's right wingman fired a Sparrow from twenty miles at a Soviet swing wing. He'd mistaken a Flogger MiG-23, equipped with a "High Lark" fifty-mile range nose radar, for a Flogger D.

"Save the heavy stuff," Delcorte cut in. "Only when we need it." He was reminding the pilot, but not saying it on open channel, of what they'd been told in the ready room that morning—that NATO's supply of all AA and AS missiles was dwindling until the convoys got through.

The MiG-23, at the end of the Sparrow's maximum range, dropped a pod of flares, or "highballs," as the Americans called them, the MiG rolling away at relatively low speed, reducing the temperature of its exhaust. While the MiG pilot hoped to *obmanut'*—to "sucker"—the American missile to one of the flares if it was a heat seeker, he momentarily disengaged his radar as an added precaution, just in case the missile wasn't a heat seeker. It saved him, the Sparrow made to home in on radar signals, not infrared.

"I'm gonna get me a Phoenix," complained the wingman in disappointment. "No evading that baby."

Delcorte said nothing other than to tell the others to follow, peeling off to attack a goods train winding through the foothills of the eastern Harz Mountains. "Note for debriefing," Delcorte told the others. It was the first train they'd seen moving in broad daylight so close to what had been the border between the two Germanys. Running trains in the daytime was dangerous in any war, even though, contrary to public belief, a rail link was extremely hard to put out of action for long, even with state-of-the-art jets. A daylight run might mean, however, that supplies at the front were running dangerously low. If this could be confirmed by Allied intelligence pilots in other sectors, it could be important for NATO to launch counterattacks with fewer NATO troops now rather than waiting later for a buildup. "Who's got the Brownie?" asked Delcorte.

"I have," answered his left wingman.

Delcorte knew he should have known who was on camera, but after three sorties today, five the day before, twelve—no, thirteen—planes down, the detail of who was designated "recon" duty had escaped him. "Get a shot of that?"

"No problem."

For a moment as they went below ten thousand for the attack run, Delcorte saw the humps of mountains where there was no evidence of fighting, even though he knew that on the other side of the range there was the beleaguered British Army of the Rhine, Dutch troops trapped with them. Farther down, more NATO divisions, primarily Americans and West Germans, were trying to hold ground on the central and southern front, so that to anyone west of the Harz, the whole planet must seem afire with war. Only here, high above, could one appreciate the fact that there were areas that the war had not yet touched.

On the first run in, the train, mostly boxcars, rounded fuel wagons interspersed with quads of antiaircraft guns, was going into a long, slow turn over a canal. "Bingo!" called one of the other pilots, signaling his fuel warning light was on and that he was breaking formation, heading back to base.

Delcorte centered his two five-hundred-pounders, released them, felt the plane buffeted by wind sheer as he climbed, the bombs bursting either side of the line but close enough that their craters tore out rail lengths. The train wobbled for a second.

"Beautiful, Colonel!" came another pilot's congratulation as he, too, headed back to base.

The train now looked like a chopped-up worm, the rear section thrown helter-skelter off the rails, only the front cars, about thirty, still upright but off track, one, its side split, spilling a load of sulfur, the yellow in stark contrast to the green fields by the track. But Delcorte knew that derailing the train would delay it only a matter of hours, and so he and the remaining F 16s came in again, Delcorte leading.

He never saw the missile—only felt a thump somewhere on the fuselage and the F-16 shaking violently. He pulled the stick back and gave her full power. He was climbing, but barely, and it felt like a heart-testing machine that, no matter what he did, would not allow him to go faster, sweat pouring from him, instruments jiggling, fuel warning light on, a pins-and-needles sensation in his right leg. He was at three thousand. Quickly he tightened his harness, pulled his legs hard together, reached up overhead, gripped the two ejector rings, and pulled them down over his front.

With the sound of a pistol shot, the cockpit was gone—the ejection the most violent shock he'd ever felt—and all he could think about was whether the pins and needles in his right leg meant he was shot up, the walls of blue sky, spinning cloud, and green fields, the brown of a farmhouse coming up at him. For a second he was convinced the chute hadn't opened, but then, quite suddenly, he seemed still in the air, the drag weight growing heavier under his arms, and he had the sensation of actually moving upward though he was still drifting down, a good seven miles from the train—now a thin, black line in the Harz's purple foothills.

When he landed there was still not much feeling in his leg, but he could see he wasn't shot. He made no attempt to hide the parachute as it was inconceivable to him that though he was in the countryside, anyone would have failed to see him land. But when he spotted the truck, green-uniformed troops standing in the back against the wooden slat side, he began pulling the parachute in, feeling, oddly enough, that he'd be in trouble if he didn't. Littering. He knelt down by the camouflage-patterned chute, took out his standard-issue .45, placed it on the chute, felt for the emergency ration pack, slipped the packet in his flying suit, and stood up, hands held high, walking well away from the sidearm so there's be no possible misunderstanding.

The truck stopped ten feet from him, and a stout GDR "Home

Force'' officer came out, pistol drawn, looking grumpily at him and walking over with the exaggerated stride of a minor official, peering at the chute and the .45 but touching neither, as if there might be some booby trap. Then he started shouting at Delcorte, pointing in the distance to what presumably was the train wreck.

The other door slammed and a woman in khaki came bustling out, yelling at the man in what Delcorte was pretty sure was Russian. The man tried to cut in several times, but she wouldn't let him and kept shouting at him. Delcorte felt ridiculous but didn't figure this was the time to grin, let alone laugh. But for some inexplicable reason, perhaps because of his fatigue, he began hiccuping. He thought of Emily, his fifteen-year-old, who had given him her ''recipe'' for curing the hiccups: ''Swallow a glass of water, hands up high, and hold your breath.''

''That's crazy.''

''No, truly, Daddy. Try it.''

Delcorte heard the Russian woman walk to the parachute, then she was beside him. Cupping her hand over the .45's hammer, she fired once. The hole was very round and she was relieved nothing had spat back on her uniform as the American flyer staggered, legs crumpling like a rag doll. The East German said the American was still looking at them, but the Russian officer told him that was nonsense. He couldn't possibly be alive. In any case, it was quite clear to the East German that she wasn't going to waste another bullet as she walked over and put the .45 in the American's right hand.

She went back to the truck and returned with an old Voigtlander camera.

There was a loud argument about the correct exposure, as it was getting dark and the old Vito B camera didn't have a light meter.

''F5,'' said one of the soldiers.

''No, no. F8—or it will be washed out. You'll see.'' The Russian woman said they'd better make up their minds because soon there wouldn't be enough light left and Berlin wanted just such a photo to show the world how the American terrorist bombers, so shamed by what they'd done, preferred to commit suicide than answer to the people.

''No one will believe that,'' one of the younger soldiers said scornfully.

''That's not the point, Comrade,'' the Russian told him. ''It will tell the people what to do with American pilots." She went to click the camera and discovered the film hadn't been advanced. There was a murmured insult about Moscow know-it-alls. She ignored it.

"He's still moving," said one of the soldiers.

"Suicides don't shoot themselves twice, Comrade," said the Russian.

"What do you mean, Comrade Lieutenant?" the soldier asked her.

"Think it out for yourself," she said. "Put him in the trunk."

He was so heavy, it took four of them to hoist him aboard, his head lolling, completely lifeless now in the lavender light of autumn sky, two of them turning their heads away in disgust, the other two, in their late teens, laughing hysterically.

"What is so amusing?" asked the Russian officer. She was from the Political Corps, whose members were stationed in every Warsaw Pact village and town to help render "fraternal assistance" to Russia's allies.

"He is making a bad smell," said one of the giggling youths.

"It is quite natural," she said.

Earlier that day, in southwestern Germany 450 miles away, a twenty-six-year-old American lieutenant and a Lance missile's crew of three had readied their truck under camouflage netting. They were in a clearing deep in the heavily wooded area on the German side of the French-German border over ninety miles west of the confluence of the Isar and Danube. In the pine-scented air mixed with the acrid odors of the battle raging barely fifty miles east of them, the first of the conventional warhead mobile launch batteries to fire in World War III was in the countdown.

"Five, four, three, two—" The American lieutenant turned the switch, and with a feral roar the twenty-foot Lance streaked up the ramp on its forty-mile journey eastward toward the Czechoslovakian- and Russian-led tank divisions pouring through and over the alluvial flats of the Danube. The moment she had closed the circuit, Lt. Margaret Ford snapped closed the lid of the khaki "shoot and scoot" box. "Okay, let's go!"

The three-man crew needed no encouragement, for all of them knew that within five minutes of the rocket's firing, the Soviet-WP detection units on the Czech-German border would have them vectored in and they'd be under counter battery fire.

CHAPTER FIFTY-TWO

GENERAL FREEMAN WAS dreaming. It was long ago, and in the dream he was gazing down through the shot-dirtied air as French cannonades, steadily walking toward his lines, their noise stunning many a soldier by Freeman's side, continued their deadly fire, and the muskets of his own infantry squares steamed in the rain from countless volleys. And still Napoleon's infantry advanced, smoke from their cannons rolling in front of the big guns like fog. Now and then the sound of a French musket ball striking his sword's scabbard gave off the strange, eerie sound of someone rubbing his fingers about the rim of a champagne glass, and all around him, like an errant ghost, was the sweet smell of wheat trampled under the boots of Napoleon's army. His horse strained at the bridle, and through a gap in the smoke Freeman saw the first line of French cuirassiers, the squadron's second line of cavalry, advancing at the walk some twenty yards behind. Freeman doubted they'd begin the swinging gallop as yet, the slippery ground littered with dead soldiers and mounts, if indeed they began it at all. It was late in the day at Waterloo. Europe held its breath—and still the Prussian reinforcements had not appeared. There was no time left. He drew his sword.

He awoke in pitch darkness, heard voices, unsure of where he was—then felt the yaw of the ship. His eyes sandpaper-dry from the heat of the recycled air, Freeman glanced at his watch. Three A.M. Someone was knocking on his cabin door. He flicked on the reading light, swung his legs off the bunk, and was pulling on his boots as the knocking continued. From long habit and in one fluid motion his left hand reached out, took the nine-millimeter Beretta revolver, and placed it firmly in the left shoulder holster. He stood, buttoning up his tunic. "Come in!"

The moment he saw Al Banks's face, he knew something was wrong. A small, worried-looking ROK army captain was with him.

"General," said Banks, "Washington's cabled us. We're to put the operation on hold."

"*What?* Goddamn it, why?"

Banks handed Freeman the decoded transcript: "INTERNATIONAL SITUATION DEMANDS YOU POSTPONE OPERATION TROJAN STOP . . ." The rest of the message was merely an advisory of the CNO's order to the LPH *Saipan*'s skipper to withdraw fifty miles from her present position, which would place her a hundred miles east of the interdiction of the thirty-eighth parallel and the Korean coast.

"By God!" said Freeman, jaw clenching, crumpling the message in his right hand and turning to the map of Korea taped to the bulkhead above his desk. "They're in bed with the Communists!" He looked quickly from Banks to the hapless ROK officer, who was swallowing hard.

"This is a goddamned plot," said Freeman, the ROK officer looking nervously at Banks, who knew better than to try to assuage the severity of the shock. The general was now pacing between the desk and the blacked-out porthole. "By God, I smell State Department all over this. Those bastards in Foggy Bottom are in bed with the Communists."

The ROK officer was clearly alarmed.

"Don't worry, Captain," Freeman told him. "I'm not mad mad. I'm just plain mad. Those pansies back there in Foggy Gulch are doing a deal with those Chinks. It's tit for tit." Freeman stopped, stared up at the brown-spined green of Korea, his fingers running through his gray hair, then turned slowly to the other two. He appeared slightly dazed, recovered his composure, and spoke slowly, without anger but with the tone of the betrayed, a betrayal so deep, it was only now fully registering. "They're giving up Korea," he said. "Beijing's putting on the pressure, and those yellowbellies at the State Department are telling the president to lay off. Give up Korea for a promise of nonintervention by China elsewhere." He paused, the tendons in his neck tight as cables. "We, gentlemen, are being sold down the river."

"Maybe you're right, General," Al Banks replied. "But we'll have to tell the men right away. Most of the choppers are on the flight deck."

"Yes," Freeman agreed. "Pass the word."

As Banks left the cabin, Freeman put his arm round the ROK officer's shoulder, looking down at the man's name tag. Kim Dae. There were more Kims in Korea than Smiths in America.

"You're no relation to that NKA bastard, are you?"

"No, sir," Kim replied hastily.

"Well, Dae, let me tell you, one of the worst moments in a commander's life is having all your men pumped up, ready to go, and then—nothing. Orders canceled. Like they're all ready to sit their final examinations and you have to tell them it's off."

"Only *postponed*," said Dae encouragingly.

The general sighed, his hand dropping from the Korean's shoulder, exhaling audibly in the small cabin that he knew was but a pinpoint in the darkness.

"Temporarily postponed," said Kim, trying again.

Freeman, hands on hips, looked up again at the map and shook his head in disbelief. "By God, to think that all those brave men in that perimeter might be sacrificial lambs—it's unforgivable." He paused for a few seconds. "And my speech . . . ."

Kim was nonplussed about the speech.

"One of the best I ever made," said Freeman, his gaze so fixed, it seemed as if he could see straight through the bulkhead. "By God, those men were ready for it."

"I cannot believe the American president would do this," proffered Kim.

Freeman said nothing.

"I am sure he will not abandon us, General."

"We'll see, Captain. We'll see."

The Canberra bombers were now over West Germany in high cloud that extended all the way back from the Harz. Below, sixty-three miles south of Hanover and outflanked by the Soviets' Forty-seventh Tank and 207th Motorized Infantry divisions pressing home the attack to secure the port of Rotterdam, one of the West German AA battery commanders requested an IFF— "friend or foe"—identification code from the planes. The Canberras responded, but their radio beams were scrambled by MiGs patrolling out of Rostock 160 miles northwest of Bremen.

The MiGs were jamming NATO ground control as they came round over Lübecker Bay to attack the besieged U.S. Third Corps, who, in an unexpected counterattack, had destroyed three Soviet-WP pontoon bridges over the Weser, thus delaying the fall of Bremen.

The West Germany battery commander glimpsed an aircraft's lizard patch underside for only a fraction of a second, but it was all he needed to convince the Sturmbannführer to unleash his mobile batteries of all-weather Rolands.

Three minutes later his crews reported that fifteen of the twenty-four missiles had found their mark, three of the nine misses being due to "circuit malfunction," a grab-all soldier's phrase covering everything from a slight wobble of the launch sleeve to a problem with the Roland's oxidizer. Sometimes the soft ground gave way beneath the trucks under the impact of the back-blast, and this could throw the rocket off course.

A short while later the Rolands' crews noticed that some of the debris coming down, wing segments, bore the distinctive red, white, and blue ball of the Royal Air Force. It was not long after this that the commander realized he had destroyed fifteen of twenty-four Canberra bombers.

CHAPTER FIFTY-THREE

North Atlantic

THE BIG CHINOOKS, making the distinctive *wokka wokka* sound that gave them their nickname, lived up to their reputation as one of the most reliable helicopters ever made. For William Spence, however, slipping between consciousness and black-outs from the pain, the noise of the rotors slicing their way over the heaving darkness of the Atlantic was a torture, the promise of deliverance fading by the second as he lay strapped into one of the six stretchers. How long he had been in the helicopter, minutes or hours, he did not know. At times the pain was merely a sore heaviness in his extremities, at others, so stilettolike that his body involuntarily heaved and jerked violently against the restraint straps. And always, added to the buffeting, in gusts of foul weather, there was the oppressive, stifling smell of oil. Lost in his delirium, he thought it was the smell of the Chinook's fuel, causing his nose to feel as if it were plugged with cotton wool, sinuses so blocked that all his breathing was by mouth, his lips cracked dry by the time the chopper was approaching the tiny slice of light that was the NATO-designated hospital ship, USS *Bahama Queen*, four hundred miles northeast of Newfoundland. The red-cross-painted landing pad was the stern platform of the twenty-thousand-tonner—a converted cruise ship, or "Love Boat," as they were called, which had quickly been pressed into service as a hospital relay ship by the U.S. government.

So hasty was the conversion that many of the plush luxury fittings were still aboard. And while nurses like Lana La Roche and her three colleagues had enjoyed walking on shag carpets down the passageways where private and semiprivate rooms now served as wards and a series of operating rooms, the senior surgeon had ordered all such carpets, drapes, and so forth removed in order to create a more aseptic environment. Normally it was a job that would have been done in the Halifax yards, but war and the fate of Convoy R-1 had meant making do with what was available, which, for Matron, given the perennial shortage of nurses, meant having to tolerate the "eager beavers," as she called Lana and what she saw as Lana's "clique." There was no clique, but Matron was convinced, as she told a disinterested nurse's aide, that the younger nurses congregated about the La Roche girl "no doubt because of her connections." Lana's and the other girls' bravery in volunteering for sea duty was not, in Matron's view, anything to be lauded—it was a nurse's duty to be where she should be in times of need—and she remained convinced that "the La Roche girl" was merely grandstanding.

"She'll be gone in six months," Matron declared confidently to those few she had selected as her "chums" in the head nurses' mess. "That type never stays," she pronounced, sipping her tea from one of the *Bahama Queen*'s Royal Doulton cups unlike the thick, institutional mugs she was used to. "Here one minute, girls like that. Gone the next. Novelty wears off."

"I would've thought," dared one of her chums, "that she'd be much better off with her husband. I mean, all that money. Servants probably."

Matron put her cup down, shaking her head at the other's naïveté. "That's just it, don't you see. Those kind of people with too much money get bored. Simply run out of things to do. 'Don't you know,' " she said, affecting an upper-class condescension.

"I would have thought she could have chosen something a bit less dangerous," another contested.

"Tosh!" said Matron. "Russians wouldn't dare attack a hospital ship. Oh no," she went on, seeing one of her chums about to interject. "I don't think it's because they wouldn't like to—that they're humane. Certainly not. People are capable of anything. I know that. But they won't because the brouhaha internationally would be very bad publicity. Not good propaganda at all. Ms. *Brentwood*, as she prefers to be called, knows that very well."

The red bar-shaped light came on. They were bringing in the first wounded, the most serious ones from the convoy; others, less seriously hurt, had been lifted to the nearest ships in the convoy to have their wounds attended to later in Halifax.

As the nurses hurried to the emergency bays, rubber soles squeaking along carpetless passageways, it struck Lana how well insulated the ship was, for the big Chinook was almost upon them and yet the sound of both its rotors was barely audible. Until she reached the floodlit stern. Opening the last door, she saw rain pelting through the air above the landing platform, a platform that only weeks before had been a swimming pool, tourists languidly drowsing their hours away en route to—where, it didn't matter, the journey itself being the occasion. She bowed her head in the face of the driving noise and spray, then withdrew, as ordered by the deck officer, to what had been a "pub" aboard the cruise liner. At least, she thought, there was no doubt about the destination of this ship, the purpose of one's life aboard her, as stretcher after stretcher was lowered and unloaded by seamen straining and slipping on the weather-wild deck.

Despite the stabilizers of the *Bahama Queen*, the ship was rising and falling enough to make unloading the Chinooks a precarious business as they hovered twenty feet above the deck like some great insects in the night, intravenous bottles swinging with even the slightest roll in the choppy sea, everyone anxious for the ship to be under way so she could get her nose into the wind and reduce the yaw.

Refusing the bright yellow wet-weather gear available to her, Matron was the first nurse to meet the incoming wounded on the blustery deck. She, like the officer of the deck, had ordered the others to wait inside, and watching her, several of the nurses warned that she'd catch her death of a cold, but Lana doubted it. "The cold would be too afraid," she told her friends.

Yet Lana admired "Battle-ax," as Matron had become known among the junior nurses, and when one of them commented, "It's a grandstand play, that's all. Wants to show how macho she is," Lana, looking out at the matron's stout frame, no more than a white woolen cardigan over her uniform, and rubber boots, which Matron insisted on calling "gum boots," to protect her from the elements, disagreed.

"I don't think she wants to look tough," said Lana.

"Well then, what's she doing it for?" asked Elizabeth. "She got a death wish?"

"I think she wants to be the first one the men see when they come aboard."

"Lordy," said Elizabeth, "that face ain't gonna cheer them up, honey. One look at her, they'll want to head back to war." The other nurses laughed except for Lana, her arms crossed, mood pensive as her group of four waited to be assigned one of the wounded.

"No, I think it's important to her," Lana continued, "that the wounded see a nurse's uniform. Know they're going to be looked after straightaway." Lana paused for a moment, trying to recall the words from one of the Halifax lectures: "Remember, emotional first aid in the first few hours is psychologically . . ."

Elizabeth chimed in, ". . . as important to recovery as hands-on medical treatment."

"Very good, Liz," clapped one of the others.

Elizabeth was nodding toward the deck as the last two stretchers were lowered, ropes on either end held taut by deck crew to stop them from spinning. "Talking 'bout 'hands-on treatment,' children, I'd like to get my hands on that bosun. I'd give that sweetheart all the first aid he wanted.''

"Elizabeth!" said Lana. "That's disgusting."

Elizabeth smiled across at her friend. "Lana, this war's just startin',' honey, and it's gonna get a whole lot worse 'fore it gets better.''

"If it can get better," put in one of the two Canadian nurses.

"Exactly," said Elizabeth. "So get it when you can, babe, before the mushrooms start sproutin'.''

"My God," said the other Canadian girl. "You really think some crazy'll push the button?''

Elizabeth, watching Matron trailing the last casualty into the reception area on the starboard side, where seamen were unbuckling the wounded, replied, "Some crazy started that, honey.''

All ten casualties now unloaded, the staff in the reception room, once the ship's stern deck casino, were now quickly returning the stretchers to the chopper.

The matron entered, soaking wet. "La Roche, your charge is Spence—first name William.'' She looked at the other two Canadian nurses. "You two prep him. I want a full sheet on him as soon as possible. La Roche?''

"Matron?'' Lana had given up on getting her to call her Miss Brentwood.

"Before he goes into surgery, check his tags for allergies,'' said Matron.

"Yes, Matron.''

The matron paused. "Have any of you seen an amputation before? I mean in theater, not in the training films.''

None of them had.

"Well, now's your chance. My guess is you'll be seeing lots of them, but there's no time like the present. There is no observation theater aboard, of course, but I could ask the surgeon.''

"I wouldn't mind, Matron,'' said Elizabeth.

"Very well.'' She looked challengingly at the other nurses, her eyes settling on Lana. "I think you should. More experience you get, more use you'll be. *If*, of course, you want to be useful.''

They were all shamed into it.

"Good,'' Matron said. "Theater Two in—'' she lifted the watch that was resting on her ample bosom—''oh four fifteen,'' she said, and let the watch go. "The captain will be heading into the wind to give us as much stability as possible, but there will, of course, be sliding from time to time. That's where you can be of some help. Pick the instrument up and into the sterilizer straightaway."

"Is it a bullet wound?" asked Elizabeth, in a tone that told them all she'd seen plenty of those.

"We're not sure, but both hands will have to come off. Hanging by mere threads."

"Oh, Matron—" One of the Canadian nurses turned the color of chalk; she had been feeling a little seasick since the ship had stopped, wallowing, more at the mercy of the sea than when under way.

"If only someone had packed them in ice," said the matron matter-of-factly. "But the helo had to pick up another four casualties." Lana was struck by the fact that the matron's British turn of speech became decidedly Americanized when referring to military matters, such as calling the choppers "helos" rather than "helicopters," as the two Canadian nurses did.

"My God," said the other Canadian. "Are there any ships left in that convoy?"

"I've no idea. Ours is not to reason why. All right, enough chatter. To work." She paused. "Oh, one more thing. After you scrub up, be sure *not* to use your hands to maintain your balance. It's the natural thing to do aboard ship and it's very unhygienic."

As they undressed the wounded Englishman, Lana carefully avoided looking anywhere else but at the man's face. It was the first shock she was to receive that night. He looked so young as to be no more than fifteen or sixteen—never mind that the ID tags said he was nineteen. There was no allergy code on the tags.

As they cut away the young sailor's oil- and water-sodden blue shirt, woolen sweater, and trousers, careful not to touch the raw lumps of mangled bone and flesh that had been his hands, the strong smell of oil persisted, and Lana, bending down, sniffing unselfconsciously like a bloodhound over a corpse, discovered that the oil smell seemed to be coming from his hair, and made a note of it. Elizabeth smelled it, too. "They start that bone saw up in there, they could have a fire."

Sometimes Elizabeth's callousness—or was it simply forthrightness or insensitivity?—shocked the other nurses, but after

Jay La Roche, it rarely bothered Lana. What did bother her in the OR was the surgeon's knife severing the tendons. She passed out and ended up with what Elizabeth called a "king-sized lump" on her forehead, which, Matron pronounced, was "La Roche's own silly fault."

When she had revived enough for two of the other nurses to walk her out of the theater, Matron had assailed her with the comment, "You almost struck the surgeon! How many times do I have to tell you? If you're going to pass out, please do so *away* from the table."

"Will the boy live?" Lana asked later.

"He's young," someone said.

Yes, thought Lana, praying for him as fervently as she had once prayed for herself in her war with Jay. Oh, Lord, she asked as outside, the Atlantic gale howled unabated about the lighted ship, let him live.

CHAPTER FIFTY-FOUR

North Atlantic

IN THE CLEAR, cold dawn, blue-white bergs calved by the land ice of the Greenland mass and floating south beyond Cape Farewell into the Labrador Current now approached Newfoundland with ominous majesty, towers of purest white as the sun climbed higher above a dark blue sea and against the aquamarine sky. The chop of the previous night had now abated over the Grand Banks, and the *Bahama Queen* sliced through the big swells as if the urgent nocturnal activity just hours before had never happened—as if, thought Lana, the world were at peace.

She was sitting in the intensive care unit referred to as the "deluxe suite," which it had been in its prewar days. The intravenous drip was regular, William Spence's heartbeat a little weak, but all other vital signs were indicating the nineteen-year-old had pulled through the double amputation. "The young mend fast," the surgeon had said, his scalpel indicating the severed hands, which, unknown to the ship's crew, would soon be dumped over the side with the other parts and garbage. "Healthy tissue," the surgeon had proclaimed, as one might refer to a horse worth buying. "If we'd gotten to him earlier—still, lucky to be alive."

"Lucky"—Lana wasn't so sure about that. Maybe one's future *was* written in the stars and we called it luck, but she still harbored the conviction that you made your own luck. But if so, had it been preordained that her marriage was to be a disaster? Was God then darkness as well as light? If so, why had she prayed so fervently for this young Englishman's life? Did she really believe? she wondered, or was she a fair-weather believer? Had her prayer accomplished anything or was prayer simply another way of so channeling, so concentrating your psychic energy that you did more than you normally would to help the situation to a successful conclusion? Especially when a young life hung by a thread.

As she watched the pristine beauty of the icebergs and the vastness of the sea, she thought of her brother Ray, who, their mother had written, was now entering yet another series of plastic surgery operations, and she realized that though she'd been unable to help her brother in the purposeless and directionless vacuum following her marriage, the young English sailor had become for her a kind of surrogate for her wanting to help, to do something, anything, and an escape from her personal ordeal, a chance to look after another victim of cruel circumstance, alone as she had been alone.

She took a lemon-glycerol swab from the kidney basin and gently daubed his lips. He was conscious now, mumbling what must have been a dozen questions, his eyes opening for a fraction of a second, then closing, the mumbling ceasing, replaced by a guttural, rumbling snore.

Elizabeth offered to sit with the wounded man to give Lana a break, but Lana said she'd stay on, and besides, there were more than enough nurses for the fifty-three cases now aboard.

"All right, honey," said Elizabeth, "but don't you go getting attached, hear? You know what happens."

"Elizabeth," smiled Lana, "you can rest easy on my account. Besides," she said sardonically, "I'm a married woman, remember?"

"That's what I mean," said Elizabeth in her usual forthright tone.

"He's just a boy," Lana replied.

"And you've got a soft streak in you a mile wide," said Elizabeth. "You're not his mama. You start with that and you'll be all burned-out 'fore Christmas."

Lana smiled, touched as always by Elizabeth's concern. "How come you know all about it, Miss Ryan?" she asked Elizabeth with mock formality.

Elizabeth looked down at her friend. "You always X-ray people?"

"Not if I can help it," Lana replied good-naturedly. "I just think this is the case of the pot calling the kettle—"

They cracked up, both laughing at the appropriateness, or was it the inappropriateness, of the cliché, Lana dropping her head against Elizabeth's stomach, still giggling. There was a sharp tapping noise on one of the glazed windows of the deluxe suite that looked out on the gigantic berg. Outside on the promenade deck, glaring at them through the glass with ill-concealed annoyance, was Matron. She was obviously saying something sharp to them, but the double glazing made it impossible to hear. Matron's voice coming out from the red face in a series of barklike sounds was quickly whipped away by the wind, her blue cape ballooning, quivering violently about her in the stiff Atlantic wind. Both Lana and Elizabeth tried their best to look properly contrite, but Elizabeth, a look of abject apology on her face, murmured, "Watch out you don't take off, you silly old bitch." That did it—they both doubled up laughing. The blue balloon was now moving down the deck against the roll of the ship, heading for the nearest door.

"I'm gone," said Elizabeth.

"*Elizabeth!*" Lana shouted. "Don't you leave—"

"Your shift, honey. I'm just visiting."

A minute later the whirlwind came in, hair wild. "If there's one thing I will not tolerate, Miss Brentwood, it's flagrant disregard for a patient's well-being."

"We were only sharing a joke, Matron."

"A *joke*? So you think war's a joke?"

Lana simply refused to be baited. "Excuse me," she said, and moved to check the IV drip.

"A joke!" repeated Matron, louder this time. The drip was a little slow and Lana unscrewed the clamp a touch.

"I fail to see—" began Matron, but then the PA system crackled to life. A man's voice.

"Matron, please report to OR Two. Matron to OR Two."

"I'll deal with you later," she said to Lana, her hair, still wind-strewn, looking like the wreck of the *Hesperus* as she bustled off with an air of officious efficiency.

When she entered the prep room for OR Two, Matron was still angry and realized she'd gowned up before she knew it. Hands held high, she backed through the swing doors into the OR. There was no one there.

She accused Elizabeth Ryan outright.

"Matron, I don't know a thing about it, honest."

"I'll find out," Matron declared. "I'll find out. I won't be made a fool of by anyone."

No one would own up to the false call, despite the captain initiating his own informal investigation, enjoying the joke as much as anyone.

As *Bahama Queen* continued down Newfoundland's coast, passing three more bergs, the dark line of Newfoundland's cliffs far off to starboard, Matron continued her investigation, her dislike for Ryan and La Roche having curdled to childish absurdity in its intensity, and would have been dismissed as such later on had it not been for its dire consequences, which, like the bulk of the icebergs, lay hidden far beneath the surface.

For the first groggy day William Spence thought his hands must be badly damaged, the phantom feeling that he still had his appendages persisting until the doctor had told him what they'd had to do.

When the reality hit, when his mind grasped what sensation still denied, Lana touching his shoulder gently, try as he might, he could not stop crying, his humiliation made worse by the mundane fact that unlike anyone else in the world, he couldn't wipe his tears away. His cumbersome attempt to do so with the bandaged stumps was at once grotesquely comic and awful to watch, his bumbling effort reflected in his sickroom mirror.

Lana knew she was caring too much but couldn't help it, for in return for her kindness, her attention to the smallest detail, William Spence, she knew, was falling in love with her. Once before there was a brief sojourn in Boston General during her six months' stint as a student nurse, and she had seen it many times and knew it was a common enough phenomenon the rapidity with which young men, especially, fell for the angels of mercy. But here, far from his home, the intensity of feeling for someone tending his every need, from a sip of water to a bedpan, who touched him, held his wrist gently for a minute, taking the beat of his heart, meant more than home. And for a young man, experiencing his first time at sea, his first time away, Lana became mother and sweetheart, so loved, so passionately desired, that the very hint of her perfume filled him with life those first few days, and when the coughing began, Lana held him during the violent episodes that would go on for minutes at a time until, the stumps of his arms throbbing with pain, he would collapse back onto the pile of pillows. Ironically, during these episodes, his face, flushed by the coughing, would take on a healthy-looking pink compared to the drained postoperative face of a few days before. And she was with him as he talked with the intimacy that only patients and airline passengers do, telling her of how different he was from his father, for whom everything had to be explained in terms of rational behavior. How they had difficulty in getting close to each other, though he clearly admired and loved his father a great deal.

"I suppose," said William tiredly, his mouth dry, the morphine barely holding the pain at bay, "all fathers are like that. And all sons."

"Not all," Lana replied, struck by the fact that he had so shyly avoided calling her by her first name. No one had told him she was separated, and in his naïveté, not seeing a wedding band or engagement ring, it hadn't occurred to him that she might have been married, a belief that only made his dreams of making love to her as the untouched "she" all the more erotic and insistent. Normally in the other hospitals it was the practice to assign male nurses to assist the incapacitated male patients in bathing, shaving, and so forth, but with most eligible males being called up for what some generals were now secretly saying would be a longer war, the overwhelming bulk of the nursing jobs fell to the female nurses and the nurses' aides.

The first morning, after they had sailed past the icebergs north of Newfoundland and were heading along the island's western coastline, Lana bathed him. He had been too groggy to notice and had kept falling back to sleep. The second morning, however, he was as aware of his erection as she, his face turning beet-red. She simply put a warm washcloth over it, saying, "It's what we in the trade call a 'tent.' Means you're getting better." He tried to think of anything else—cold showers, long walks— but nothing worked. The astonishing sexiness of her uniform that was not supposed to be sexy was too much for him, and he grew harder. Lana returned to the subject of his family, asking about his mother—Anne, wasn't it?—his sisters, Rosemary and Georgina.

"I think you said Rosemary was a teacher?" she said, beginning to make the bed, giving him an excuse to turn away.

"Ah, yes," he said, so tense, so ill at ease, he imagined that being interviewed for assistant chef at the Savoy couldn't be more anxiety-producing. "Yes," he answered. "Ah—Rosemary's a teacher. English—Shakespeare."

"Uh-huh," replied Lana. "In London?"

"Yes, er—I mean, no—ah—she teaches in a public school. Ah, what you'd call a—"

Lana gently raised his right arm.

"A private school," he raced on. "I don't think you have them in America."

"Private schools? Oh yes. I went to one before college—"

"Oh, I mean you don't call them public schools, you call them private. We call private 'public.' " He was in a torture of embarrassment and lust. "It's all very confusing, I'm afraid."

"Not really," she said, smiling. "You're just all mixed-up over there."

"Yes," he admitted freely. "Yes, we are rather."

There was an awkward silence as she fluffed the pillows behind him. Her perfume washed over him. Lana gently moved his arm.

"Would you like me to bring a tape recorder for you?"

Immediately he thought of a tape measure. Maybe his wasn't as big as others she'd seen. How many had she—the filthy bastards.

"Tape?" he asked, his voice raspy, cracking as if she'd said "snake."

"To send home to your family." She removed the facecloth softly, swiftly, pulling up the sheet. As it touched his loins, he was terrified he'd have an orgasm right there. It would be sheer bliss. Then he realized what she was telling him, ever so nicely. No, more than nicely—wonderfully. He smelled her perfume again, drinking it in. He loved her. The thought of doing it with her—he'd never had any woman—suddenly seemed rather dirty, unbecoming of her beauty, her goodness.

She had, he now realized, really been telling him ever so gently that while he could never write again, he could and should start thinking about other people—how his parents would be sick with worry after the rather brief message that had been sent to inform the War Office of his whereabouts and condition. It was her thoughtfulness that touched him now and stoked his need for her as mother and lover. But he could see no chance of the dream materializing.

While he was waiting for the tape, William Spence's brain raced with the things he had to tell his parents, for despite a persistent weakness, and the exhausting episodes of coughing, his spirits were buoyed by Lana's presence.

When she brought the tape, it was as if he'd suddenly developed stage fright—couldn't think of a single thing to say. Ever considerate, Lana had started the recorder and left, but the tape in motion only made it more difficult for him to think of something to say. Though he'd tried clumsily to push the "stop" button with his elbows, he sent the machine instead crashing to the deck. He was in such an agitated state that tears of exhaustion and rage streamed down his face and he started another bout of coughing, humiliation heaped on top of his frustration with his inability to perform such an elementary task. When she returned, Lana interpreted the signs at first as evidence of indescribable pain. But if he had such difficulty, why hadn't he pushed the buzzer? His elbow would easily— Then she saw that the cord, with its emergency push button, had swung away down the side of the cot as he'd made a grab for the recorder, which now lay on the deck, its plastic case cracked, the tape unraveling. She eased him back on the pillows and, taking a moistened towel, wiped his face, her voice as soothing as her touch. She told him not to worry about it; she would bring another and would think of something so that he could stop and start a recorder by himself. As she leaned over him, tucking in the bedclothes, her breasts touched his chest and he wanted to hold her, kiss her. It was an ache inside of him.

"I love you," he said simply.

She said nothing for a moment or two, straightening up the bedclothes, then stopped. "I know," she said softly, and bent down to collect the broken pieces of the tape recorder.

When she brought him a second recorder, the talking came easily, as now that he'd declared himself to her, all things were possible. He spoke of his older sister, Rosemary, with whom he shared a fond affection. He was not anywhere as intellectual or trained in the classics as she was, he told Lana, but Rosemary had never made him feel that not going on to university was a shame, unlike Georgina, who, as fiery as she was young and beautiful, had castigated him soundly for not having a "social conscience" and planning to do "something with your life." Georgina was in her third year at London School of Economics and Political Science, and her social conscience had embraced the left of the political spectrum—a "born protester," her father often lamented, claiming, not too far from the truth, that she seemed to have more causes then courses.

It was still difficult for William to speak to his father, even on the tape. He reminisced about the walks up on the chalk downs, the small, winding, tree-shaded lanes, and the holidays they had spent cruising the canals once so awfully neglected but now largely restored, like the two-barge-wide lock at Stoke Bruerne, where all the people had lined the canal to watch them squeeze through. Petrol must be rationed now.

When it came to the double amputation, he spoke quickly about the wonders of modern technology, how Nurse Brentwood had told him of the fantastic things they were doing with computer-controlled limbs these days, and how things you never realized—tape, for example—were a darn sight easier than having to write a long letter. "Never was much of a speller anyway, was I?" he asked his father. But he could not tell his father he loved him, no matter how much banter or reminiscence might have provided the opportunity to say it. He tried to say it outright near the tape's end, but the words choked and he ended with a cheery "Love to all," not realizing how many times he'd mentioned how kind and what a "wonderful person" Nurse Brentwood was, not using her first name, as for him to have done this would have somehow constituted an invasion of his love for her, cheapened it in some way. The name Lana had become something intensely private to him, something he carried with him.

By the time he had finished the tape, he was coughing again from the exertion, this time so badly that Lana took it upon herself to go and suggest to the senior medical officer that he authorize an X ray.

"That your boyfriend in two oh one?" the young medical captain asked.

"The patient," said Lana icily, "in two oh one. Yes."

"Why?" the captain pressed, reverting to rank after her rebuff.

"Well, I'm used to patients having that acetonelike breath—with a bad cold, flu, or something on top of what they've got already. But his breath smells of oil. I think he must have inhaled some."

"Quite possible, but we should have noticed if there was a problem before this."

"It's difficult to sometimes," proffered Lana. "Every man that came aboard—I mean that's been in the water lately—had it on him, hair, fingernails."

"Lipid pneumonia?" said the doctor, his tone more reasonable now. "Possible. Sometimes doesn't show up for twenty-four hours or so." Lana also remembered that it only needed a few milliliters of oil in the lungs for it to be fatal in some cases.

"All right," said the doctor. "Let's have a look at him. I wouldn't mention why, though."

"No—of course," said Lana.

"Nurse?"

"Yes?" She turned. The doctor was taking off his stethoscope, slipping it into his white coat. "Sorry about that 'boyfriend' crack. Uncalled for. I'm glad we've got nurses like you."

"Thank you," she said, and as she walked away, the captain watched her for a long time, his fingers drumming on the stethoscope. Coming into the room, Matron saw him watching Lana Brentwood as, against regulations, Lana went down the stairs between upper and main deck facing forward instead of backward for safety's sake. The captain was watching her backside through the very proper starch of her uniform.

"Doctor?"

He jumped. "Yes, Matron?"

"One oh eight. I think you should see him. I think he's a malingerer."

"Evidence?"

"Oh, all vital signs are normal. That's what first aroused my suspicion."

"Hmm, still," the captain countered, "never quite know with these possible neuro cases. We've done the pointed stick bit on him. Let me check." The doctor scanned the list of thirty-seven cases. "Far as I remember, he didn't feel a thing."

"He *said*," replied Matron, "that he didn't feel a thing. I'm not so sure. In any event, I've seen some you could stick a hat pin in and they wouldn't flinch. Not if it meant they could get out of the fighting."

"All right," said the MO. "Any suggestions?"

"I'd wait till he was asleep."

"My God, you're a sly one, Matron. I could be sued."

"In private practice, yes," said Matron evenly, "but not in the navy, Doctor. We're here to insure—"

"Yes, yes, all right. I'll have a look at—" he took the chart from matron, "—Johnson now. Is he asleep?"

"I'm afraid not."

"Perhaps," said the captain, "he needs a sedative. Settle him down a mite."

Matron said nothing. She didn't believe in unnecessary medication. "Well, when do you suggest?" pressed the doctor.

"Perhaps later this evening."

The doctor sighed. "All right, but I would have thought we've got better things to do than weed out—"

"Doctor—if I might remind you. We have a very specific directive about this from both Ottawa and Washington. We must weed them out. Set an example. Besides, once they get ashore, that type is quite likely to pack up and—"

"Yes, yes. All right, Matron. I'll jab him in the butt at midnight. How's that?"

"I don't enjoy this any more than you do, Doctor."

"No, of course not," he replied, but believing she did.

As she walked along the deck on her rounds, Matron saw the line far to the west that marked Newfoundland and the tiny dots out from it that were the fishing boats of the Grand Banks, and on the deck below she could see the La Roche girl pushing the Spence boy to X ray.

CHAPTER FIFTY-FIVE

THOUGH IT WAS the mass of Soviet and Warsaw Pact armor together with Soviet Yak ground-attack aircraft that had pressed home the first attacks on the NATO positions in southern, central, and northern Europe, once the battle was joined, Soviet and NATO tanks mixing it up together in the massive confusion of dust, smoke, and darkness, it was the awesome and concentrated fire of the Soviet artillery that drove home the breakthroughs. While the Allies had expected the twenty-three-mile-range, eighty-pound shells from the Soviets' self-propelled BM-27s to precede any Soviet breakthrough, it was an entirely new development that was now turning some NATO withdrawals into routes. This was especially true along the alluvial flats of the Danube, all the way from the Austrian-Czech border in the east, westward deep into southern Germany.

For Soviet artillery commanders, the most impressive and dangerous aspect of American arms was the ability of the Americans to move artillery batteries quickly and to reengage within minutes. Up until the late eighties the U.S. Army had also possessed the most accurate 155-millimeter in the world. The reason was political.

In a move that the South African ambassador in Ottawa had reported to Pretoria as being "typically Canadian" and a move Pretoria took immediate advantage of, the Canadian government had expressed no interest in a new artillery piece designed by a Canadian. The Canadian-invented gun was then built by the South Africans, a fact that made it ineligible for purchase by the U.S. Army, whose procurement was closely monitored by antiapartheid lobbies. The G-6, as the South Africans called the new 155-millimeter gun, was a self-propelled howitzer capable of firing a hundred-pound, HE bag-cartridge-propelled shell a distance of forty-two miles. It was the batteries of the Soviet-purchased G-6s that systematically destroyed the American and West German defensive lines and attacks.

The only comparable guns had been the enormous rail-mounted weapons of Nazi Germany. The G-6, however, was much lighter, mounted on a highly mobile vehicle which also carried forty-four rounds. It was not simply that the G-6 batteries, with their eighty-degree arc of fire and extraordinarily long range, could outreach the other Allied guns but that the G-6's accuracy was such that the S-WP forward observers could see it tearing apart not only the big concentrations of Leopards and M-1s, as if they were metal shacks, but individual tanks as well.

Not only this, but the forty-two-mile range of the G-6 meant that the Soviet guns were reaching beyond U.S. First and Second Armored and the West German Twelfth Armored, severing their vital supply lines, allowing the bulbous-eyed Soviet Hind choppers to be whisked forward, cutting off the trapped American and West German columns, picking them off with laser and infrared antitank missiles.

After ten days of savage fighting, the Soviet–Warsaw Pact's thrust along the Danube was beginning to pivot northward to join up with those forces that had broken through at Fulda and were now separating out into a "breaststroke" pincer in which the left flank was to drive south to meet up with those coming via the Danube, the northern prong to meet up with those forces who had smashed through the North German plain to Bremen.

The only real danger to the Soviet thrust remained Allied air superiority and the possibility, albeit a slim one, that Austria would throw its lot in with NATO. To thwart this, the Soviet commander of Southern Forces Europe, Marshal Gordayev, had SPETS teams ready near all river crossings from Innsbruck in Austria's southwest through Salzberg, and from Austria's north-western region all the way east to the Carpathians on the Austro-Czech border.

By now NATO intelligence was as stunned as Seoul had been by the extent of the sabotage by Soviet-trained sleepers put "in place" during the detente of the Gorbachev-Reagan-Bush years. As a result, NATO intelligence teams were, as one British Foreign Office official put it, "swarming all over the place like a plague of locusts.'' With the result, he reported, ''that our chaps,'' by which he meant MI6, ''are finding the situation terribly confusing. Amateurs are tripping all over one another during the night.'' In Vienna on the Reichsbröche Bridge, U.S. and British agents even exchanged shots, each believing the others were Russian operatives about to place an explosive charge midspan. The ''charge'' turned out to be a ''rather sturdy soap-box,'' reported the Foreign Office, left by some untidy fishermen.

''Of course,'' the Foreign Office official, an under secretary, went on to explain to his vexed minister in Whitehall, ''no one was hurt. Our chaps from MI6 were using revolvers which hadn't been fired, I daresay, since World War Two—and the Americans, getting dangerously close, from all accounts, luckily heard one of our chaps swearing. So vociferously, in fact, that the American recognized him as British.''

''There's no need for that,'' said the minister sternly. ''Absolutely no call for profanity. I thought we were recruiting a little higher off the shelf than that.''

''Quite, Minister. But I'm afraid you see Redbrick all over these days. Different class of people altogether. My God, you ought to see the written reports. Dangling participles positively *litter* the page.''

The minister sighed. The trouble was that this egalitarian notion that you were as good as any other man was too entrenched. It was enough to make one yearn nostalgically for the DOM—Days of Maggie—to return.

But while the Vienna mix-up was symptomatic of the early days of confusion, there was already a recognition by some of the Allied heads of stations in Austria that if ''Chocolate Eclair,'' as the Austrian president was known, did fall off the fence to the Allied side, then one of the ideal ways of sabotaging the bridges would be from one of the many barges plying the over-two-thousand-mile river. Accordingly, MI6 and CIA agents all along the east-west Vienna-Passau sector on the Austrian side were noting the large white numbers on the backs of the barge wheelhouses. Any barge that proceeded back and forth too often within a twenty-four-hour period became marked as a potential target, a potential SPETS sabotage platform. The Austrian river police, worried about offending either side, nevertheless inspected many of the barges for explosives, for, while frightened of offending the Soviet Union, Austria was equally concerned about the financial ramifications following an argument with the United States. No explosives had been found.

"Well, of course not," said the minister of war, dropping the MI6 report from his hand, glancing down at the traffic along Whitehall. "These days you only need a satchel of explosives, don't you?" he asked his secretary.

The secretary knew it wasn't as easy as that; charges had to be placed precisely.

"I think we're ready in any case, Minister," replied the secretary. "If a barge so much as pauses longer than thirty seconds under any of the crossings from Austria into Czechoslovakia, we have informed the Austrians they can expect to be fired upon."

"And what if our Austrian friends don't turn to us but decide to run with Ivan? He certainly has the best of the field so far."

"Then of course, Minister," said the official, smiling, "we will blow the bridges ourselves."

"Hmm." The minister was still looking out his window, the first time he had done so for twenty-four hours, a camping cot having been set up in the office annex, as he'd stayed behind working hard on the problem of transporting the oil to the Continent now that the Chunnel had been blocked. He was surprised to find it was already dusk as a flock of pigeons dived and whirled in unison toward the stately Houses of Parliament. In a gloomy moment he wondered whether the mother of parliaments would survive. He suspected not.

"I've only one problem with your bridge readiness plan, Hoskins."

"What's that, sir?"

The minister, a cup and saucer in hand, was looking at the distant smudge marks against the sky, vapor trails crisscrossing as the Royal Air Force fought for supremacy over the channel. Perhaps it had been a strategic error to bomb Berlin, but for the moment retaliation from the Soviet–Warsaw Pact air forces seemed out of the question, at least on any large scale, the AA defenses in southern England being equipped with American radar-linked missile systems that some said were more sophisticated than the AEGIS system aboard the most advanced American cruisers.

"Sir?"

"What—oh, yes," said the minister, taking a sip of his tea. "Don't you think the Russians have already thought of bridge patrols, barge spotting, and all that?"

"Unquestionably, sir, but one can only—"

The minister lurched forward. There was a clatter of broken china, tea splashing over the desk blotter, and Hoskins' lapel was warm, a mixture of blood and tea. Staring at the dead minister, Fitz backed away from the shattered window, the bullet hole's spider-web radiating from its center.

BBC and ITN were instructed, under threat of D notice, to report the minister's death as heart attack. Superintendent Favisham pointed out, however, that this might be a little "thin" as the minister was known as a "physical jerks man—jogging and all that. I suggest a stroke. Burst blood vessel. Can happen to any of us. Anytime."

"Thank you, Superintendent," said the minister's secretary. "A stroke will do nicely. Man under enormous pressure." He flashed a smile. "But aren't we all?"

The superintendent didn't like him—thought he was fruit.

"Tread carefully, Superintendent," said the secretary, "with ballistics and all that. You know what these reporters are like. There's enough anxiety already." He paused. "Any idea who it might be?"

"No, sir," said the super. "Interpol has a long list of potential head cases. Ready to pop any member."

"Indeed. Then I need hardly tell you, Superintendent, that this will chill every member of Cabinet. Despite the present security, they'll want to know whether we've taken all the appropriate precautions. They'll certainly be asking how it is that a minister of the Crown can be assassinated in broad daylight from government offices on the other side of the street."

The superintendent shook his head, unconsciously dropping his top denture at the same time, a habit the under secretary found as irritating as it was vulgar.

"I don't think so, sir. From across the street, I mean. Acute angle, top of his head, you see, blown right out. More like it came up from the street, I expect."

"In any case, I'll need a full report by—" The telephone light was blinking. PM's office.

"Yes, Prime Minister? Yes, sir. Of course." The secretary

put the phone down slowly, trying to put on as brave a face as possible. "We're withdrawing along the Danube. New defense line is from Regensburg south to the Black Forest."

The Superintendent's geography was a little rusty. He hadn't made it to Redbrick, let alone Oxford or Cambridge. The best he had was criminology at the Polytechnic.

The secretary explained. "A withdrawal of some eighty miles. We're trying to 'consolidate'—I think that's what we're calling it. From Regensburg down to Switzerland. Russians have some huge bloody guns up against us apparently. Trouble is, it's as far as we can go, I'm afraid. With the French right behind us but not behind us—if you follow my meaning."

The superintendent was trying to follow it on the map of Europe above the carpet stained where the minister had fallen.

It was foggy and nearing dawn in the Sea of Japan.

"I've told the troops," Captain Al Banks informed the general.

"How'd they take it?" asked Freeman, who was sitting at his desk surrounded by maps of the operation Washington had called off.

"Tell you the truth, General," said Banks, "I think some of them were pretty relieved."

Freeman nodded. "Natural. But it takes the edge off of 'em, Al. We've got to keep them wound up. Stop them moping about home, girlfriends, wives. Rome burns and we wait," said Freeman disgustedly.

"We could show them movies, sir. Got lots of them. Or go over the rehearsals again."

"No, not the rehearsals. Hell, we don't know yet whether we've got anything to attack. No, they've been over it enough times. Any more, they start thinking they know it all, get stale. That's dangerous. No, your idea about movies is a good one, Al. Keep morale up." The general sat down, pulled out his bifocals, as discreetly as possible, and peered up at the map toward Pyongyang, several "Firebird" high-altitude photographs showing there'd been no substantive changes in the AA positions around the North Korean capital in the last twenty-four hours. "Ah, I don't know, Al. Maybe they're just sending us this stuff to keep us quiet while they squash the whole idea in Washington. Goddamn it, the plan's right—all we need is the

weather to hold and we could give 'em such a kick in the ass—''
He took off the glasses and dropped them on the desk map. ''I
ever tell you about that airplane in Canada?''

''No, General.''

The general rubbed his forehead and sat back. ''Left Montreal for a place on the prairies. Ran out of gas halfway there at
forty thousand feet. Air Canada it was—one of the best safety
records in the world—''

''Then how—''

''Metric!'' explained Freeman, smiling sardonically up at the
captain. ''Took off in Montreal, checked the goddamn tanks
twice. Converted liters to gallons. Multiplied by the wrong conversion factor. Did it *twice*, once when it left Montreal, second
time after it landed in Ottawa. Then bang—no gas.''

''Jesus, what happened?''

''Started to fall like a goddamn rock, that's what happened,''
said Freeman. ''Only thing they had in their favor was the pilot—''and here the general was pointing at Al to drive home the
point—''one in a million—had been trained in gliders. Was able
to manhandle that brute and slide it toward an abandoned airstrip. Only problem is, he was headed for the wrong airstrip
until his copilot, who happened to have lived in the area years
before, knew where the right strip was. They brought it down.
Undercarriage had no power, so failed to lock. That saved
them—otherwise they'd have mowed down about fifty people
who used the airstrip as a Sunday runaround. Only casualties
occurred when they started going down the escape chutes—two
rear ones were too high off the ground. Lot of people got hurt.
No fatalities. And all because of metric, Al. *Metric.* I've been
triple-checking all of these figures for the attack. Less gasoline
we have to carry, more men and ammunition.''

''There a problem?''

Freeman handed him a Xerox of a logistics and supply sheet.
''Some ass in Washington did a metric on us. Here—look at
this. NATO liaison, you see. Liters instead of gallons. Would
have put all the Chinooks into Crap City on empty. How do you
like them apples?''

''For cryin' out loud—''

''I'm going to stay here until all these figures are checked,
just in case Washington deigns to let us know what the hell

they're doing. Are they fighting a war or are we going dancing with these bastards?''

"I think you could do with a movie yourself, General."

"No, but you go ahead, Al. And Al—"

"Yes, General?"

"We have any movies with that Fonda dame in 'em?"

"I—think so, sir."

"Throw 'em overboard."

"Sir, I—"

"I mean it, goddamn it," said Freeman, using his folded bifocals as a pointer, his frustration at waiting for word from Washington bubbling to the surface. "I won't have that *female* on my ship in any shape or form whatsoever. That understood?''

"Yes, sir." Al Banks didn't think it prudent to remind the general that though he was in command of the possible invasion force, it was not his ship to dispose of ship's property, movies or otherwise. But it did occur to Banks that perhaps he should advise the general of something else: that females, three, in fact, had been in the audience last night as he'd delivered his salty and somewhat profane speech to the troops.

Freeman was appalled. "My God—that's terrible. *Women*? In the audience?''

"Yes, sir."

Freeman's right hand ran through the graying hair, his left hand with the folded bifocals spreading out in a gesture of utter surprise. "I'd no more swear in front of ladies than—" His head shot up at Al Banks. "What the hell were they doing there?''

"Combat roles, General. Navy choppers. MGUs."

"What's that?"

"Mixed-gender units. Supreme Court decision that—"

"By God," Freeman said, shaking his head, "those old farts have got a lot to answer for, Al."

"Yes, sir."

The general was pacing again, hitching up his belt now and then. "What kind of woman would want to fight? I don't understand it. Delicate creatures that—"

"I don't understand it either, General, but some of them aren't so delicate."

"Huh—well, I suppose you and I qualify as cavemen?" He paused. "Well, as long as they're in support roles, suppose we have to accept the fact. Long as I don't have to put up with them in tanks.''

There was silence, the vibrations of the ship noticeable now, the LPH's roll increasing.

"By God," said Freeman, his eyes narrowing at Banks, "you're not going to tell me we lost that one, too? The tanks? Not in the goddamned tanks?''

"Afraid so, General. Supreme Court previously backed off on some combat roles, but they said in a time of national emergency—''

"Good Christ!" bellowed Freeman. "I told Wexler—I don't want my men riding around with pussy in the front seat. They've got enough to look after.''

"I don't think we have too much to worry about, General. It just came through in the last week. Armored units haven't got—''

"No," said Freeman, "and I'll tell you what. They won't have. Now, I want you to find out who those females were that I, unknowingly, addressed last night and tell them to come to my cabin." Freeman grabbed his cap. "On second thought, I'll go to them. Probably a goddamned Supreme Court decision about them reporting to their commander's cabin. I'll be charged with molesting pussy on the high seas!''

"Ladies," began Freeman, "I've come here this evening to apologize for any profanity I might have used—I, ah, certainly hope I never made any disparaging remarks about your—the opposite sex. Women in general.''

All of them could have been his daughters. For the first time in years he was tongue-tied. "I—uh—that is to say, I never have, never will support the use of, uh, inappropriate language in front of, uh, women or seek to embarrass, uh—I'm sorry. That's all.'' With that, Freeman turned, leaving Al Banks, who barely managed to get in a wink at one of the women before he, too, was gone, trying to catch up to the general.

Inside, the three women were looking at one another in astonishment.

"What the fuck was that all about?'' one of them asked.

"Don't ask me," said another. "I think he's just old-fashioned. A commander chauvinist pig.''

"Oh," put in the third chopper pilot, a young, sandy-headed

girl with a bachelor of science degree out of Penn State, "I think he's kinda cute. Besides, I prefer old-fashioned men."

"*Cute!*" said one of the other girls. "After his speech? Nothing cute about that, sweetie. He's probably one of those guys who thinks his prick is a gun."

"Oh," said the sandy one, "he's not that type."

"Yeah, I know, Sandy. God, flag, the wife and kids. See the wedding band with the West Point ring?"

"That's what I mean. I like men who have values."

"It's one way you could get promoted, I suppose. Or get the clap."

"Well, I'm glad he's in command. They say he's a stickler for details," said Sandy, turning and looking at the barometer still falling. "And I don't like the thought of driving through this lot with some young whiz kid directing me from back on ship. I'll be quite happy to have Freeman up front."

"If we're lucky, honey, you won't have to drive anywhere. Washington'll kill it before it kills us. I joined to see the world."

"This *is* the world."

"I don't like it. Not this part anyhow."

"Who does?" asked Sandy.

"Freeman, for one. He's busting to go."

"You really think Washington'll cancel it?"

"I don't know," said one of the others. "We're just the gofers around here."

In Schönbühel, Austria, the Danube, rarely blue, more often green with pollution, was winding its way slowly along its two-thousand-mile course through some of the most beautiful country in Europe, by castles and patchwork fields, on through Vienna to Bratislava in Czechoslovakia, beneath the ultramodern span of the bridge of the Slovak national uprising, past the great spires of Buda and Pest, over the great Hungarian plain, down through Yugoslavia to Belgrade, and on through Bulgaria to the river's great delta in the Black Sea.

If you had to bail out, the NATO pilots said it was best along the Danube—but not on Ulm's spire. Hundreds of pilots did bail out in the first few weeks, a third of them falling into enemy hands, but the rest, except for a dozen or so who met with misadventure, hung up in the woods or drowned before they could break free or reach shore, found their way back to their units within three or four days along the verdant plain. It wasn't simply a matter of friendly populations, not yet overrun by the S-WP juggernaut, helping so many fliers, but the superb "rescue" facilities of the NATO units, who, even under the severest weather conditions sweeping down from the Carpathians and the Bavarian Alps, would do everything possible to pick up a downed flier.

For the civilians in southern Germany, caught between the two sides, it was a dangerous business, for the advance units of the Soviet–Warsaw Pact attack were almost exclusively Russian, the latter holding the Czech brigade only as support, their fighting ability held in contempt by the Soviets. This meant that if a NATO flier was caught by the enemy, it would most likely be a Russian unit, which had a clear and brutal rule: anyone having helped the flier was executed along with his or her entire family, and the nearest village razed to the ground, the girls and women delivered to the Soviet troops. Despite this terror, NATO fliers still came back, aided by civilians who understood that if the cost to the Russians of advancing every mile in Germany was not maintained, the great "rollover" convoys now en route from New York, Boston, and Halifax—trying to learn from the experience of Convoy R-1—might well arrive to find nothing to reinforce.

CHAPTER FIFTY-SIX

OF ALL THE surface ships operating in the North Atlantic, the minesweeper was now king.

Due to the experience of Convoy R-1, wood- and fiberglass-hulled vessels suddenly emerged from dull, mundane, and in many cases outright pitied, existence in the backwaters of the navy into the exalted ranks of leaders in the age of nuclear-powered missile ships. The admiral in command of the hitherto ugly duckling minesweeper fleet of forty ships that were to accompany the first three convoys was told not to gloat but merely to do the job and to do it quickly. The minesweepers did the job as quickly as conditions in the Atlantic allowed, approaching it with the zeal of newfound importance. They also gloated. Oh, how they gloated, flagging, as it was a court-martial offense to break radio silence, the warships behind them with messages such as "Follow the leader—Compliments of MS-190," or "We'll tell you when it's safe." It became an unofficial competition between British and American sweepers as to who could be the most insolent and get away with it.

"The HMAS *Gordon* will be happy to show you the way."

"By God!" fumed a Royal Navy destroyer captain. "They're cheeky bastards."

To make it worse, the Canadian and U.S. minesweepers were among those "combat 'support ships' allowed to have women aboard. One of the American ships, USS *Twin Forks*, was skippered by a woman, and on the third day out for one of the massive four-hundred ship-convoys on "rollover" to Europe, a pair of women's lacy briefs was hoisted to the masthead, "Compliments of 'rollover' leader." This moved a U.S. admiral to issue a firm rebuke by semaphore to the minesweeper, but as the message came through, the panties were gone, the minesweeper's captain nonplussed and assured by the crew that the lookouts on the other ship must have been seeing things. It was a brief, light relief in what was otherwise a grim business.

Four hundred and seventy miles north of Newfoundland, another convoy was attacked by ten Soviet Hunter/Killer subs. All but two of the nuclear subs were destroyed, but not before there had been a "run-under" torpedo assault by all ten subs, resulting in thirty-seven Allied ships, twenty-eight of these merchantmen, sunk. When the convoy had re-formed in a defensive diamond east of the sub, it found itself unwittingly driven, or "bloody well herded," as one British frigate captain put it, into a minefield laid about them by six Backfire bombers. Three of the Backfires were shot down on their way into "egg laying," but their deadly cargoes landed intact.

Here, once again, the Russian numbers pointed to only one conclusion—that if the Allies could not reduce the rate of loss, whatever supplies and men did arrive in Europe would be insufficient to replace the men and matériel already lost, let alone to reinforce NATO. In this case the unrelenting Soviet–Warsaw Pact land offense would decide the issue. A further complication for NATO was the fact that with so many towns and cities in the Russians' path, the S-WP attacks sent millions of civilians fleeing westward, tying up the vitally needed West European road and rail systems.

By now the USS SN/BN *Roosevelt* was going up for its second attempt at receiving a VLF burst message. This time additional aerial was extruded from the stern, like some great worm from the belly of a whale.

"Start the count!" ordered Robert Brentwood.

"Counting . . . five minutes . . ."

At the three-minute mark Brentwood knew he was not going to get a message. "Okay," he said evenly at the five-minute mark. "Reel her in."

"Reeling in, sir."

"Very well. Mr. Zeldman, resume zigzag pattern for Holy Loch. ETA?"

The first officer glanced at the computer as Brentwood on the periscope island ordered, "Up search scope."

"Up search."

There was a quiet hum as the oil-mirrored scope slid up inside the master sheath housing several other periscopes as well.

"ETA Holy Loch," Zeldman reported, "six hours approximate."

"Exactly, Ex."

"Six hours, three minutes, forty seconds, sir."

"Very well."

Brentwood knew that if the Wisconsin aerial "farm" was out and TACAMO aircraft had failed to overlap sufficiently to contact the *Roosevelt*, then the United States would have notified U.K. control to beam out a VLF signal. If not, it meant the Soviets were jamming satellite bounce-off signals between the States and Britain, or the British aerials at Holy Loch were knocked out, or Holy Loch itself was in the hands of the Soviets. Brentwood knew he had only three choices: stay where he was; head for Holy Loch and risk a trap of acoustic/pressure mines at the entrance—some keyed to *Roosevelt*'s specific signature; or run for cover and head back to the United States.

He reversed his cap, eye glued to the scope. He flicked on the control room monitor relay so the men on watch could see the same infrared images he saw. Nothing but gray waves stacked all about them, creased with white lines of bioluminescence.

"MOSS in tubes one and two."

"MOSS in tubes one and two, sir."

"Very well. ETA Holy Loch?"

"ETA five hours, fifty-seven minutes."

"Speed?"

"Thirty knots."

"Increase to forty-five."

"Increasing to forty-five."

One of the planesmen glanced over from his steering column at the operator on trim, rolling his eyes heavenward. "Watch the dials, sailor," said Zeldman sharply.

"Yes, sir."

"Revised ETA Holy Loch?" asked Robert Brentwood.

"Four hours," answered Zeldman. "Including corrections for currents plus or minus fifteen minutes."

"Very well. No active sonar. Passive only."

"Passive only."

"Call me when we're ten miles off."

"Yes, sir."

As Brentwood left the redded-out control center, Zeldman heard one of the sailors whisper to another. "What's Bing up to?"

Zeldman let it go as if the scratchy noise of the ocean had drowned out the whisper.

"Don't know," answered another of the men on watch. "Probably wants to get his book."

Zeldman still held off saying anything. Now and then you had to let the rein loose a tad—up too tight, they were as apt to make a mistake as they were when too relaxed.

CHAPTER FIFTY-SEVEN

THE MIDNIGHT MOON was bright, and a mile or so beyond the *Bahama Queen*, one of the bergs, a silver tower in the moon glow, split, its cracking sending hundreds of birds rising above it in panic like falling confetti, the wave running toward the hospital ship as the main berg and its calf righted themselves. For a moment, broadside to the moon, they looked dazzling white on the gunmetal sea.

"When will I know?" William Spence had asked her, and she had to wait until he had finished coughing before she replied, asking, "Know about what?" She knew what he meant but didn't want him to worry unnecessarily.

"The X rays."

"Oh—tomorrow morning, I guess."

"I'm coming apart," he said, the violent coughing starting up again, so that she got up and slipped the elastic about his head, placing the plastic mask over his nose and mouth, altering the rate of oxygen flow until the small, black plastic marble was unseated, jumping up and down inside the flow indicator.

Spence was perspiring so much, the sheet was clammy about his chest, and Lana could tell the other pain, from the amputations, was also tormenting him, the medication wearing off again, the pain boring into him again. But she knew she couldn't give him any more morphine for half an hour. If it were up to her, she would have given it to him now—it wasn't going to make much difference. The lungs in the X rays had been a diaphanous white. He was so weak that his so-called "walk" to the washroom had degenerated during the last twelve hours to nothing more than a shuffle. They had performed a miracle of modern surgery in keeping him alive after the trauma of the evacuation from the *Peregrine*, but now the killer of more ship-wrecked sailors than torpedoes or shells, oil, had lain in wait in the lungs, threatening to drown him slowly. With only a cough to announce it, the lipid pneumonia had come upon him swiftly, the final quietude of pneumonic death in any hospital called "the old person's friend."

While holding his cough-wracked body, Lana recalled the X-ray technician as he had stood looking gloomily at the film, watching it rock to and fro with the action of the ship, the very motion somehow an obscene mockery of real life.

"There will," Matron had told her matter-of-factly, "be moments of serenity, even reverie. In the end they're quite content."

"With a double amputation?" Lana had asked tartly.

"You'd be surprised, my dear," Matron had replied.

No, thought Lana, you'll be surprised. This boy is going to fight with everything he's got.

"The X ray doesn't tell us the whole story, nurse," the MO had advised her in a more understanding tone. "Even so, I'm surprised the prednisone didn't help—I'd thought there was definitely an allergic component that the prednisone would deal with. Well, all we can do now is watch him. Could be a turn-around before we reach Halifax."

She had been with him eight hours straight, and now in the calm following the wracking coughs, every one of which she had felt like a blow to her own body, Lana leaned over him and with a cool, white facecloth, as white as the ice, he thought, she dabbed his body cool, gently patting him dry. She saw him smile, or rather his eyes moving suddenly, full of life, the rest of his face covered by the semitransparent green mask. "What are you grinning about, Mr. Spence?" she asked with playful severity.

"I can't help it," he said, his voice sounding nasal from behind the mask. "I don't need this mask anymore."

"You sure?"

"Yes." Closing his eyes, he lifted his left arm and lowered it gently onto her hand. She did not take it away but reached up with her left hand to turn off the oxygen flow, its soft hiss fading, the sound of the waves from the calved berg still slapping the ship's flanks. She was standing by him, her right hand still beneath his bandaged arm. He looked up at her and wordlessly she sat down by him, her hand still beneath his arm, her other hand gently stroking him and seeing the miracle of the pain not gone but momentarily defeated by her gift of touch. She raised her left hand higher, kissed her finger, and gently stroked him again. He groaned in ecstasy, his head beginning to move slowly, joyously, from side to side, and in that moment, out of all the pain and the evil of Jay La Roche, Lana emerged as gentle as a virgin, but knowing much more, lowering her head, her long, soft hair falling on him, then she kissed him there, the firm but pliant wetness of her lips encasing him, drawing him into her, her tongue sliding hard and fast and then slowly, lovingly, as he groaned, his whole body beginning to arch and rock, arching again, then arched as if frozen in time, shuddering before he collapsed against the bed, bathed in sweat, his eyes glistening with life, looking at her, then slowly filling with such calm that they said nothing until the pain, like a vindictive husband wanting to kill, attacked again.

Quickly, alarmed, she looked at the clock, rearranging her clothes and hair. It was still ten minutes to go before the next injection. Now the flush of love in his face left him, like a red curtain torn aside, his face stunned with ferocious pain, white, as pale as moonlight. She took the hypodermic, injected him, and knelt by him, ready with the mask should the coughing return. It never did, and as she told him she loved him, he went into a deep sleep, a tiny spot of blood seeping through on the stump of his right hand, as scarlet against the bandages, she thought, as a rose against hard snow.

She pressed the buzzer and the cardiac arrest team arrived. He revived on the second "jump," but later that night the oscilloscope's hiccuping green sine wave went flat, and in place of the lively "bips," there was a long, steady tone.

"In all my career," Matron fumed before the chief surgeon and the ship's chief medical officer, "I have never seen such a flagrant violation of procedure."

The MO, the young captain who had referred to Spence as Lana's "boyfriend" a couple of days ago, could see the pain in Lana's face, and for his part, the morphine shot she'd given the patient too early would hardly have made any difference. He told the chief surgeon so. And in his view it certainly didn't warrant a court-martial, as Matron was pressing for.

The matron's head shot up, looking over at the surgeon.

"It's hardly the morphine I'm concerned about, Mr. Reilly." Even now she insisted on the British convention of referring to chief surgeons as "Mister," its usage conveying a higher status than "Doctor." "Though giving the patient the injection earlier is, in my professional opinion, also thoroughly reprehensible."

"Then what is this all about?" asked the surgeon, nonplussed.

"Ah—perhaps," the MO interjected, "Ms. La Roche would care to step out for—"

"No," said Lana.

"All right, then, Matron, I think you'd better go ahead," said the MO.

"The sheet, sir . . . it's . . . it's filthy."

"Filthy sheets?" said the surgeon, pushing the question back at her and looking at the MO for clarification.

"*She . . .*" began Matron archly, "did things to him."

"Oh—" said the surgeon. "*Oh—*" He paused. They could hear the ship's foghorn as it entered the area off Cape Race. "This is a very severe charge, Matron. I would advise you—"

"I don't deny it," said Lana.

Matron glanced quickly at the surgeon, making it quite plain she expected the maximum punishment for such unprofessional conduct and would not rest until she got it.

"Ms. La Roche," the chief surgeon began, "you must realize how serious this is."

"Yes, sir."

Now there was a silence in which the captain noticed for the first time that he could hear the clock in the cabin ticking very distinctly. He shifted a few pens on his desk pad. "I'm afraid I'll have to refer this to HQ. Are there any—mitigating circumstances you'd like to add—"

"The boy was dying," said Lana.

There was silence again, Matron staring at her. Finally the surgeon, doodling uncomfortably on the blotter, said, "That doesn't excuse it."

"*Exactly!*" said the matron.

"All right," said the surgeon. "That's all."

* * *

Out on the deck, where the chilly fog now came tumbling through in gusts, Matron paused before taking the steps down to her cabin deck. "If you think I've done it because I don't like you, that's not true."

"Oh really?" said Lana.

"The point is, my girl, that we have to set an example for the others who come after us."

"Yes," said Lana. "Imagine if every nurse did it, *and*," she added sarcastically, "right in the middle of a war."

"Don't be insolent! You don't seem to have realized something, young lady."

"And what's that?"

The matron stood very close to her, and Lana could smell her bad breath as she began to speak. "*You* might have killed him. A shock to the heart like that."

"He was dying."

"You don't know that."

"Didn't you see the X rays, Matron?"

"I've seen more X rays and more deaths than you, young lady. He might well have recovered—*if* he had been left alone."

"I don't think so," replied Lana, but Matron thought she saw a glimpse of fear in the young woman's eyes and she pressed home her advantage.

"You'll never know, will you?"

The temptation of guilt, of a hundred letters to Ann Landers about unprofessional conduct of nurses, flashed through Lana's mind, doubt flickering in her eyes for a moment, and then it was all gone, rejected utterly, as if her whole being had irrevocably changed at that precise moment in her life. "I gave the boy love."

"Is that what you call it?" sneered Matron.

"Yes," said Lana, "and I'm sorry for you."

"*You!* Sorry for *me*!"

"Yes," Lana said softly, pulling her cape tightly about her, the fog from the Grand Banks colder by the moment, enveloping them both and hiding the bergs. "I'm sorry for anyone," said Lana, "who hasn't had love. It shrivels your heart to nothing." Lana turned and walked slowly away along the deck, past the dim outlines of the lifeboats.

* * *

The young medical officer managed to get Matron to strike out some of the more hostile adjectives in her report about Lana La Roche, and while, informally, he convinced the surgeon not to recommend a court-martial, he could not prevent transfer, to the Matron's delight, to a forward hospital—in what the nurses called "America's Siberia": the Aleutians.

"As godforsaken a place on this earth as you could imagine," the medical officer informed Matron in return for her retraction of the prejudicial adjectives. "And," he added hastily, "under strict supervision."

"She should be drummed out of the service," Matron retorted. "She has no place—absolutely no place—in—"

"Well, Matron, you won't be bothered by her anymore."

The matron, however, was barely appeased. "She's a bad penny, that one. Mark my words, Captain. She'll turn up again."

The MO made no comment on her prediction but did tell Matron it was the best he could do.

Before she packed, the *Bahama Queen* passing by McNab's Island, through the narrows, past the Imperial Oil refinery into Canadian Forces Base Halifax, Lana sat down and wrote a letter to Mr. and Mrs. Richard Spence in Oxshott, England. Trusting the highly reliable fleet mail service—much of it sent electronically from base to base—more than she trusted the civilian post, Lana addressed the letter care of her brother at Holy Loch, with a covering note to him just as she had done with the tape that William Spence had made a few days before.

Dear Mr. and Mrs. Spence:

My name is Lana La Roche and I was a nurse on your son's ward aboard the hospital ship *Bahama Queen*. Although we've never met, I feel I know something about you, as William talked quite a lot about his family. I know he sent a tape, and though, of course, I don't know what he said about his wounds and the major surgery he had—I'm guessing he didn't say much at all about this and so I thought it might be of some help if I could tell you a little about the circumstances as I know that by now you will have received official notification from the DOD of his death and that after what appeared to be the chance of a good recovery, his passing must be a terrible shock. It was a combination of things, mostly the fact

that he had contracted pneumonia from oil he had inhaled
while trying to rescue shipmates trapped in the engine room
of his ship, and while we were concentrating so much on the
severe wounds to both arms, the wretched pneumonia, as it
so often does, was already forming in his lungs. By the time
it was detected, I'm afraid that plus the amputation proved
too much. He was a wonderful young man and, though weakened by his ordeal, quietly brave—not only on the *Peregrine*
but on the *Bahama Queen* as well, where I think he knew the
end was near.

All I can say is that he clearly loved you all very much and
told me so, and that helped him a great deal. We buried him
at sea yesterday, as regulations call for. It was a very simple
but moving ceremony. I asked the ship's first mate to mark
the spot as near as he could on a chart of the area, a copy of
which I've enclosed. I will keep the original for another time
to send as I'm forwarding this by Fleet SAT Post—electronic
mail. I don't know if this will help any, but the first officer
told me the location of the burial is remarkably exact as they
take bearings from Loran and satellite.

I've addressed this letter care of my brother Robert, as I
did the tape I sent on for William, asking Robert to pass it on
also. I hope we can meet someday. Please don't bother to
reply, but if you wish to write sometime, and I can tell you
any more about William's time on the ship, please write me
care of the address in Virginia on the envelope and they'll
relay the letter to me.

Sincerely yours,

Lana La Roche

In the covering note to Robert she told him he could read the
letter, as it would fill him in on the news, "if that's what you
can call it," and also reminded him that, as she'd mentioned in
her earlier note, whenever he got back to base and received the
tape and the letter she sent him, he should advance the tape for
a minute or two until William started talking—"otherwise the
boy's parents might think there was something wrong with it."

The censor patched out *"Bahama Queen"* and went over it
again, making sure there was no mention of the USS *Roosevelt*,
on which her brother Robert Brentwood served, and whited out

the time and place of the burial, giving another point hundreds of miles away so as not to give away any more information about the hospital ship's location.

CHAPTER FIFTY-EIGHT

STANDING AT THE bow, his tank commander's collar pulled up, hands gripping the rail like a Roman tribune aboard his chariot, Douglas Freeman went up to the *Saipan*'s starboard flying bridge as the LPH headed north in the Sea of Japan, having reduced her speed from fifteen to a mere five knots. When Al Banks found the general, he immediately told him he had good news and bad news and asked him which he wanted to hear first.

"Both barrels," said Freeman, adding grumpily, "They've turned it down."

"No, sir."

Freeman pushed himself back from the rail. "Don't say they've approved?"

"Hold on a minute, General. Good news is, the Taiwanese have told Washington they'll back off for the moment. They're reinforcing the offshore islands of Matsu and Quemoy. That's provocative, but as long as they don't start shelling the mainland or launch an amphibious assault, Washington believes China will stay out of it. They haven't withdrawn all the way to Taiwan, but at least they've withdrawn to their side of the dividing line in the Taiwan Strait."

"All right, so China's off our back. Quid pro quo. They won't resupply the NKA in return for us keeping the Kuomintang on a leash. I don't like it, but right now I don't care what those lily-livered bastards in Foggy Bottom have arranged. What's the bad news? Have we got the go signal or not?''

''Monsoon, General. Moving down from the Sea of Okhotsk.''

''It figures,'' said Freeman. There was disgust in his voice, but it was a calm disgust at receiving the news of yet another delay, and the blowup that Banks had expected never materialized.

''Now,'' said the general calmly, ''God's on their side.'' He turned back to face the sea. ''When will it hit us?''

''Ten hours—maybe less, the met boys say.''

''Why?'' asked the general slowly, darkly, in a tone so controlled, so ominous, that Banks would have preferred some of the general's feigned short-fuse rhetoric. ''Why is he doing this to me?'' repeated the general.

Freeman leaned against the rail, one foot up on the lower rung.

''All right, Al,'' he said wearily, without looking around at his aide. ''Let me know when the monsoon's expected to be over. 'Course,'' he sighed, ''by that time Washington'll probably have changed their minds.'' Now he turned around from the rail. ''The Yosu perimeter?''

''Half what it was, General. Intelligence estimates it can hold four days at most.''

Freeman had turned away again, looking up this time at the starless sky. ''God, I hate nights like this.'' There was a long silence—uncomfortable for Banks.

''Anything else, General?''

''No, Al. Keep me updated on that damned monsoon.''

''Yes, sir. Good night.''

As impatient as General Freeman was, alone on the upper deck of LPH *Saipan*, below in the oil-smelling belly of the LPH's hangars, David Brentwood, one of the two thousand marines and airborne troops still waiting, put his camouflage pack against the bulkhead as a pillow, loosening his load-bearing vest and taking off the Kevlar helmet, placing his protective eyeglasses inside it, and unsnapping his belt. He was reading the note on the postcard again.

''Hey, Stumble-ass, don't worry about it,'' said Thelman. ''She'll be there when you get back.''

"Then," said another marine, "keep off twenty-six. Gunner kept those cold pills. He'll be about as fast as Tim Conway."

The thing David couldn't work out was why on earth Melissa had sent a postcard of herself—one of those camera machine shots—without an envelope. It was the flirt in her, he thought—she'd know very well all the other guys would see it. And there was nothing really in the note—just a "wish you were here" scrawl, a faint tropic isle background, and Stacy, the note said, had passed his friggin' IR major with straight A's. Got a deferment from the national call-up because IR was "hot" right now.

"Hey, man," said Thelman. " 'Tain't a Dear John."

David wasn't so sure.

Across the world, an ocean and continent away, the mood in the control room of the USS *Roosevelt* was as tense as First Officer Peter Zeldman could remember it. What would "Bing," Robert Brentwood, do? They were going up to a hundred feet for the third attempt to receive a burst message.

"Five hundred—four fifty—four hundred—three fifty . . ."

Robert Brentwood stood on the raised periscope island, both hands behind him on the guard rail, legs crossed, looking with quiet confidence across at the sonar, a lot of fresh sea noise coming in, but nothing that stood out in the clutter.

"One hundred, sir."

"Very well. Roll out VLF aerial sixteen fifty."

"VLF to sixteen fifty."

They watched the VLF registering on the stern monitor relay screen as a white trace as the aerial, ready to receive anything on the three-to-thirty kilohertz band, extruded from *Roosevelt*'s stern pod, the long wire rising on buoys until it lay extended approximately eighteen to twenty feet below the surface of the sea.

"Begin the count," ordered Brentwood.

"Five minutes and counting . . . four minutes thirty seconds . . ."

At ten seconds to go, the tense mood in control and throughout the sub had changed to a palpable gloom. Brentwood and his crew knew that in five or six hours they would be no more than sixty to eighty miles out of Holy Loch and would have to go up for a burst message. If none was received then, they would have to stay out with the assumption that the war had gone nuclear and that the *Roosevelt*, on Brentwood's authority alone, would launch its six eight-warhead-tipped MX missiles aft of the sail at its preselected forty-eight targets in the Soviet Union.

"Wind her in," said Brentwood calmly.

"Sir," proffered Zeldman, "should we try HF?"

"Negative."

Brentwood had already figured that one out; using the high-frequency aerial would mean taking the sub up farther so it could literally poke its stick HF aerial above the surface, the chance of receiving messages being much greater but more dangerous, as one penetration of the sea-air interface by the aerial could be recognized by Soviet satellite intelligence, if not intercepted by another Soviet sub. "Sticking your head out the front door" is what the crew called it, whereas with VLF you stayed submerged, not letting the enemy know where you were—the only potential giveaway a noisy VLF reel retraction or any other noise on the sub that might be picked up by the Russian subs or by their SOSUS networks. For this reason no radios were allowed without headphone attachments, and the cook could not even chop onions during silent running.

"Take her down to one thousand," ordered Robert Brentwood.

"One thousand, sir."

"Carry on, Ex. Resume zigzag."

"Resume zigzag. Yes, sir."

Brentwood asked the electronics officer if the failure to receive a VLF burst message could have anything to do with their own VLF aerial.

"I've checked it out, sir. It's fine."

Brentwood thanked him and quietly, calmly, addressed the whole watch, not using the PA system so as not to risk the slightest hull reverberation; the sound of the aerial retraction was risk enough.

"Right, you all know the situation," Brentwood began. "Next attempt to receive burst message verification will be our last. If we do not receive any—if it has been a case of land farm knockout—we should be close enough to the Scottish coast on the final BMV station to elicit TACAMO recognition signals. But if not, then we must assume the Soviets have taken out both Holy Loch and Wisconsin transmitters and possibly the floating dry dock in the Loch itself. If we do not receive TACAMO, we will proceed north—get as close to Kola Peninsula as we can—and unload. Any questions?''

In the red glow, a blue denim sleeve, looking darker than it was, rose. It was one of the operators on planes control. "Turbulence during aerial layout indicates bad weather topside, Skipper. TACAMO aircraft might be grounded—or being jammed.''

"Believe we would have picked up jamming,'' said Brentwood, turning to Sonar for confirmation. "That right?''

"Yes, sir. No jamming apparent.''

"All right,'' continued Brentwood, "we'll be close enough in so that even a relatively weak TACAMO signal or farm transmission should reach us. We might risk a quick HF exposure. But if we do not receive any signal, then our course of action is set.''

"Yes, sir,'' said the planesman, but his face told Robert Brentwood he wasn't happy.

No one was.

The salt night wind was stiff in his face, the smell of monsoon rain approaching, a peculiarly clean smell that he'd never experienced before, as Douglas Freeman started to berate himself, talking as if he were dressing down a besieged prizefighter between rounds. "Well—you saying God's on their side. That's as self-pitying a piece of shit as I've heard in a long time, Freeman. You need a good kick in the ass.'' He gripped the railing, which narrowed sharply at the bow, the sea splitting open beneath him. "God helps those,'' he told himself, "who help themselves, *General*.''

When he reached Banks's cabin, his aide saw the sparkling impishness of a child in the general's eyes. "Banks. God helps those who help themselves!''

"Yes, sir.''

"Al,'' said the general, a mood of such excitement coming over him that the tenor of his voice grew more stentorian by the moment as he invited his aide out on the deck. The sea was becoming rougher all the time, but Freeman seemed oblivious to all around him as, leaning into the wind, he recalled what he termed "the fields of honor—Agincourt, the first English conquest of Normandy in the early fourteen hundreds, Waterloo, Gettysburg, the Somme, Normandy, Stirling's brilliance with the long-range desert group, the eccentricity of Wingate, Patton,

Heinz Guderian, Liddell Hart, Lawrence, Napoleon, Mountbatten." He mentioned Mountbatten and Burma several times, and each war, he told Banks, was different, though the terror just the same. "Worst mistake you can make, Al," he told Banks, "is to seduce yourself with history. Technology changes everything. Only the men are the same, and with new technology, even they change. You realize in the first two hours of the Arab-Israeli tank battles in Sinai, over ten percent were disabled. Men have never before been under such strain—not even in trench warfare. The sheer hitting power, mobility, means there's nowhere to hide. No time for respite. You have to fight or die. Come with me, I've got something to show you."

Captain Al Banks listened to the general's plan carefully before he spoke. After hearing it, he thought that either General Freeman had a screw loose or was some kind of genius.

"You seem to know a lot about Pyongyang, General."

"Yes I do," said Freeman, in no mood for false modesty. "Blindfolded, I could take you through Pyongyang." He paused. "You think I'm nuts?"

"General, I just hope you're right."

Freeman was grimacing as a huge cloud of spray enveloped the ship. "So do I, Al. So do I."

CHAPTER FIFTY-NINE

"SONAR CONTACT!" *ROOSEVELT*'S operator said. "Good bubbles, sir," by which he meant the cavitation of the unidentified ship's prop was very definite on the sound graph.

"Trawler?" suggested Brentwood as he grabbed the extra set of earphones.

There was no reply from the operator.

"Closing?" asked Brentwood

"Think so, sir. Slowly. Think they might be trawlers."

Brentwood turned to Zeldman. One of the burst messages they had received before the "Zippos" had started was about the trawlers who'd attacked R-1. "Cut to one-third."

"One-third, sir."

Now the sonar hydrophones would have reduced interference from the *Roosevelt* itself.

"Soon as you can, Sonar," Brentwood enjoined.

"Yes, sir."

Zeldman, his face in the light of the control room taking on a deep sunburned color, asked, "Herring boats?"

"Possibly," answered Brentwood.

"Closing, sir," came Sonar's voice. "Fifteen knots."

"Tad high for a fish boat, isn't it?" put in Brentwood.

Sonar made a face that said six of one and half a dozen of the other. " 'Bout right if they're hurrying to close in a net, sir."

"Or attacking," said Zeldman. "Doing a Norwegian."

"Torpedo status report?"

"All eight loaded, sir. One to six inclusive Mark-48s. Seven and eight MOSS."

"Very well. Lock in seven and eight to sonar feed."

Two seconds elapsed.

"Seven and eight MOSS locked to sonar, sir."

"Very well," Brentwood said, speaking on the intercom to the lieutenant in charge of the torpedo room up forward. "Stand by." Brentwood now switched his attention to the screen readout. "Have we got a signature match for friendly?"

"Negative," answered Sonar.

In his mind's eye Brentwood could see the convoys of well over three hundred ships en route to Europe behind him.

"Sir. Two other contacts coming in. I'd say range six miles and closing."

"Signatures?"

"None of ours. But trawler class."

"Very well. Designate targets Alfa, Bravo, Charlie."

"Designated Alfa, Bravo, Charlie, sir."

"Very well. Alfa bearing?"

"Zero seven four," said Sonar.

"Mark. Fire seven."

"Seven fired and running."

Zeldman instinctively looked above, as if trying to see through the sub's pressure hull, as the torpedo, running at forty-five knots, the sub's attack speed, streamed away from the sub, giving off the signature of the *Roosevelt*. At seventy-six seconds after release of the torpedo, Sonar reported, "Alfa—violent pulse."

All faces in the *Roosevelt*'s control room were locked on to the dials, the loudest sound the men's breathing, all knowing that "Bing" had elected to roll dice with the MOSS. To fire another MOSS now or another torpedo might give the deception away. The price for firing the MOSS was, you had to stay absolutely still or you might end up being hit instead of, or as well as, the MOSS. Zeldman and Brentwood watched the sonar screen intently, seeing the luminescent dot on the top of the green screen signifying that something had been fired from the first trawler.

"Track the splash," ordered Brentwood dispassionately. "Designation Tango."

"So designated," answered Sonar. "Ah—sir—Bravo Charlie, fading."

"Frightened them off?" wondered Zeldman, his tone half wishful thinking, half doubt. Brentwood said nothing. Target designated Alfa was also turning now, its echo fading. But whatever it had dumped was still falling and, worse, spreading.

"Mines!" proffered Zeldman.

"Sonar?" asked Brentwood, lending his authority to the question. The operator looked puzzled.

"What's up?" asked Brentwood.

"Don't know, sir. Fuzzy."

Brentwood held up the earpiece. It *was* fuzzy, a soft sound, like milk simmering, then more diffuse. The one big chunk of sound definitely broken up, as a pile of mines would.

"Bearing?"

"Zero eight niner."

"Mark. Fire one and two." The sub felt a gentle push.

"One and two fired and running, sir."

Again Zeldman looked up, the Mk-48s running off to his right from the sub's bow at ninety miles an hour, uncoiling their thin guide wire until the torpedo would be close enough for the target's sound to activate the Mk-48s' own sound sensors, which would then cut off from their wire feed control and home in by themselves.

At four minutes plus three seconds, the sonar screen went into what looked like power failure whiteout, as a screen goes blank when hit by a high voltage without a surge control bar, the screen about to die, but the green and amber tints started to bleed back. There were loud cheers and excited congratulations all around.

"Shut up!" It was Brentwood, as sharp as they'd ever heard him. "Sonar—readout on Alfa, Bravo, Charlie?"

"Still fading, sir."

The control room remained quiet, no one moving, waiting to see if the Alfa, Bravo, or Charlie fish boats were returning or whether passive was picking up anything on them at all.

"Distance to Holy Loch?" Brentwood asked Zeldman.

"A hundred miles, sir."

"Very well. We'll wait ten minutes, then proceed. Plot a new course to the loch."

"Yes, sir."

Brentwood glanced around at one of the spare watch-standers, another "Bugs Bunny," or sonar operator, with whom he could replace the present sonar operator, whom Brentwood now asked to report to his cabin in fifteen minutes. The sound wave from the Mk-48 explosions now hit them, the sub rolling slightly, then righting itself.

The sonar operator didn't know what to make of it but took his leave to go down in his slippered feet toward the gallery for a much-needed coffee with liberal helpings of sugar.

Two of the other watch-standers in control, out of Brentwood's view because of the scope island, made "what's with him?" faces at each other.

After ten minutes, Brentwood turned over the watch to Zeldman. But before he brushed the heavy green curtains of the combat control center aside, he looked around at the men on watch, from sonar to planesmen. "If I ever hear an outburst like that again, I'll cancel all liberty on this pig boat, That clear?"

Muffled acknowledgment.

Brentwood did not raise his voice—indeed, it was softer than the usual tone he used issuing orders—but the repetition of his question now took on a more clipped tone. "Is that *clear*? Everyone?"

"Yes, sir," came back the affirmation.

"In the two seconds it took you to sound off, an enemy surface ship could have fired an ASROC-SN-12, which travels—at what speed? Wilkes?"

The planesman frowned, the glistening worry lines emphasized by the red glow. "Can't say exactly, sir."

"Ex?" Brentwood didn't want to show any favorites.

"Eight hundred meters a second, sir," answered Zeldman.

"Correct. Twice as fast as our best ASROC, gentlemen. *And* sixteen times faster than our torpedoes. Message received?"

"Loud and clear, sir," said the Trim petty officer.

"Very well. Carry on, Ex."

"Aye aye, sir."

After Brentwood left control, there was a sigh of relief in the room, and Zeldman's presence, though clearly felt, was more like that of an assistant coach—the big tension was off, which Brentwood could feel as he stepped out of the control room and which is what made him worried. The most difficult thing at sea on such long six-month patrols, was to keep the men on the razor's edge. A second or two had always been important in war, but never so much as now.

"What's bugging Bing?" Brentwood heard a seaman ask as he passed the galley, his antistatic moccasins moving noiselessly on the shiny spill of red light bouncing off the polished decks and bulkheads.

"His book's probably out of print," answered someone else, and Brentwood heard them laughing.

When he got to his cabin, the sonar operator was waiting outside apprehensively. Brentwood took off his cap, pulled the narrow green drapes across the cabin entrance, and waved the operator into his cabin. "Come in, Burns."

"Yes, sir."

Robert Brentwood yawned, excused himself, and motioned the seaman to a chair, but the seaman declined.

"You did well up there, Burns."

"Thank you, sir."

Brentwood noticed the boy had a prominent Adam's apple, just like his kid brother Ray. "But there's something you did I don't want you to repeat."

"Sir?"

Brentwood pinched the bridge of his nose between his thumb

and forefinger, the tiredness in his eyes now more pronounced under the regular cabin lighting. "You hesitated on giving me two readings.'Ah—something.' I don't want any 'ah—somethings.' It's at least a second's delay, and it could be enough for a target designation of us by an enemy sub. We've got enough acronyms—letters for symbols—as it is. Forget about 'ah'—sounds like 'R.' ''

"Yes, sir."

"One more thing, Burns. What the heck you think we blew up? I would have expected more of a bang. I've heard forty-eights go off against targets. Usually make a lot more noise than that.''

"We might have hit fish, sir."

Brentwood looked up. "Fish?"

"Yes, sir. The way I figure it—the 'splash' from Alfa we got could have been them ditching their catch—to get away."

Brentwood sat back against the bulkhead, thinking about it. "When they heard our torpedoes?"

"Yes, sir. Coming in on their fish sonar—or Fathometers, as the Brits call them."

"Okay, Burns. You can go. You did a good job."

"Sorry about the slipup—"

"Just chalk it up," said Brentwood. "We learn from the mistakes. Or we should."

"Yes, sir."

Robert Brentwood thought about it for a few seconds after Burns had left but then rang control.

"Yes, sir?" answered Zeldman.

"Pete? Why do they call me 'Bing'?"

"Sir?"

"They call me Bing."

"Don't know, sir. Maybe you mentioned him once or twice."

"I'm sure I didn't. That woman at the admiralty party last time at Holy Loch—she asked—wait a minute, I asked *you* about this Bing character."

"Ah, yes, sir, believe you did."

"You son of a bitch, I'll get you for this."

"Yes, sir."

"Carry on. Call me half an hour before our last—I mean our next scheduled BMV/TACAMO station."

"Will do, Captain."

Robert Brentwood lay back on his bunk, his gaze focusing on the Pacific Northwest calendar he'd taped up on the bulkhead showing a radiantly pink Mount Shasta, its volcanic cone rising majestically into the sky. Half the days were crossed out by black marker pen, the rest, after the "hump" of the patrol, in red. The book on Crosby should be at Marriage's by now, waiting for him. Was there something wrong with Crosby—not liking women or something? He pulled out the store's card as he couldn't remember its address, but soon he was wondering about what he would do if *Roosevelt* didn't receive a burst message from either land farm or TACAMO aerials. He might never see a book again. He loved reading—could never understand people who were retired saying they got bored. There was so much yet to do, so much to learn. It was something that became more and more apparent after a long six-month patrol around the Atlantic, and all the time on radio silence. One thing you realized was how people ashore took their luxuries for granted, everything from a daily newspaper, a TV or radio report that told them what was going on in the world, to their freedom to go and do as they pleased. On radio silence ever since they'd received the first few bursts alerting them to the fact that the United States was at war with the Soviets, they had no idea of the extent, and beyond the convoys, the intensity, of the war, and knew next to nothing about the titanic land battle that must now be under way in Europe. Unless it had already gone nuclear, which would explain why *Roosevelt* hadn't received burst messages as scheduled in the last three listening slots.

He thought about the family, Lana, whether she and her husband had sorted things out, or had she gone ahead and joined the Waves? And he wondered if young David had been caught up in the war. But soon his sense of isolation that is the submariner's lot in time of war gave way to concern about how he was going to fill out the required "firing report," to explain to COMSUB-LANT—Commander Subatlantic Pacific—why he had just fired off two million dollars' worth of the American taxpayers' money. At fish. The truth was that sometimes, with all the electronic gadgetry, you still didn't know—you had to make a judgment call. He hated doing the paperwork, but dutifully started. All the while, however, he was aware of a growing sense of unease. If he received no message on the next TACAMO station, it would be up to him—his decision and his alone whether or not to go to the "nuclear mode." To attack his predesignated targets in the USSR might mean turning what probably up to now had been a more or less conventional war into a nuclear one. If so, would his men be ready? He believed they were. If he had to do it, speed and smooth operation would be everything before the enemy could get a fix on him.

He finished with his paperwork and lay down on the bunk, clipped on his Walkman earphones and, hands behind his head, thought of London and New York and the trips he'd thought of taking on the long furlough, as this time they would be in dry dock in Holy Loch for a new coat of anechoic paint. If Holy Loch was still there.

But something told him he would not be going to Holy Loch. Or was his natural sense of caution giving way to an uncharacteristic strain of morbidity at the end of the long war patrol? He closed his eyes and tried not to think about it, tried not to dwell on the fact that the *Roosevelt* alone carried more explosive power than all the wars in history combined. He switched off the lamp, his right foot tapping to the deep, authoritative lament of Johnny Cash and "Sunday mornin' comin' down."

CHAPTER SIXTY

IN THE SEA of Japan it was the middle of the night, and Al Banks had grave doubts about the general's plan. "I don't know, General. It's unprecedented."

"Al," said Freeman, looking up at his aide, "you're a fine officer. You've got initiative, too. I don't know of any staff officer that could have coordinated the logistics of this operation on such short notice. You've done a superb job."

"But—" interjected Banks philosophically, "what's the qualifier, General?"

"You know squat-all about history. Hell—" The general pushed back in his chair, holding the bunk edge to stop it rolling hard forward again, his right hand smacking the operational map of Korea draped over his desk. "Hannibal didn't let a bit of bad weather stop him going over the Alps."

"He had elephants, General. If I remember my history. We only have choppers."

"We flew night missions in Nam. Gunships were one of the success stories, Al."

"Yes, sir, but not in monsoons. That's like flying into a car wash."

"It's not that tough," said the general, tapping the map now with his bifocals.

"Pretty tough, General."

Freeman was up pacing again, legs bracing against the increasing roll and pitch. "You know, when the Japanese drove south in World War Two, everything fell before them. Singapore, Hong Kong, Philippines, Dutch East Indies. God, they were on Australia's doorstep before you could say uncle. They were good. Damned good." He paused and looked over at Banks. "Hope they still are, Al. How many have we got? Two support companies?"

"Yes, sir. Washington thought that *if* they could be brought onto our side—in some kind of support role—it would give an international flavor to the attack. Would make—"

"My God—" scoffed Freeman, "—'international flavor'—is that what they actually said?"

"Yes, sir."

"Well, anyway, point I'm making is that those Japanese in WW Two were only stopped when the British and us and the Aussies finally stopped pissing ourselves from fright. Now, Mountbatten told 'em the super Japanese were no more 'super' or 'natural' jungle fighters than anyone else. Hell—they're still the most urban country in the world. Training's what did it, Al. Training, training and morale. And that's how we beat them in the end. But before that, we had to have a success, to break the aura of invincibility they were gathering about themselves. That's what this mission's all about. To go in and break the goddamned mold of invincibility these jokers are being cast in. Even, I might add, by our own newspapers." His hand flew out toward the big wall map, striking east of Korea in the middle of the Sea of Japan, then moved quickly left toward the North Korean capital. "I'll tell you another thing. I've been reading the intelligence reports from way back on this runt Kim Jong Il. There's one hell of a lot of dissent in that country. Under cover, of course. They shoot you in that place for smiling." The general got up, walking over to one of the manila folders he'd spread out across his bunk, and picked up one with a red secret diagonal stripe across it. "Take a look at this," he instructed Banks.

"His old man. Not much secret about that, General."

"That's not the point. Those fools in Washington would classify their laundry lists if they could." Here Freeman's eyes narrowed like an infantry squad leader under attack, voice lowered, whispering to his men about the best way through the wire. "You know what Daddy did? Built a gold statue of himself. Sixty-six feet high. Egomaniac. Everyone's scared shitless of him, but look at him— a pudgy-faced dictator. No courage—no vision, except a squint-eyed Commie view of the world that not even the Russians or Chinese believe in anymore." Freeman walked over to the wall map. "Al, I've got a gut feeling that if we can nail this bunch—if we can catch 'em with their pants down—" he swung about at Banks "—we'll not only do a Doolittle—give everybody at home something to yell and whoop about—we might just pull the whole rug from under him."

"I'm not so sure, General. First we have to get there."

"That's why I called you here. I want to see all chopper pilots now."

"Now, General?" said Banks, looking at his watch—it was just after 3:00 A.M. "They'll all be sacked out."

"You get them up and ready. In the dungeon back there. I'm going to tell them about Mountbatten. Came the monsoon, every son of a bitch used to dig in till spring. Including the Nips. Mountbatten turned it around—had our side attacking in the monsoon."

"But our troops would be in the air, General. Taebek range is over five thousand feet high."

" 'Freeman's variation,' we'll call it," said Freeman, smiling.

"Sir. Can I speak frankly?"

"Only way, Al. Shoot."

"General, ordering your men to deliberately fly into a hurricane—well, sir, nobody, and I mean *nobody*'s, going to like it."

"What the hell's the matter with you? I don't want them to *like* it. I want them to *do* it." He paused. "I know what I'm asking these men. But, Al, if I didn't think that, God willing, we could do it, I wouldn't be pressing ahead with it." Freeman paused for a moment, staring back at the map. "Don't worry, Al—everyone's going to volunteer."

"I wouldn't bet on that, General."

"I would," answered Freeman, turning, grinning and taking Al's shoulder, steering his aide toward the door. "You go and get all those sky jockeys in that briefing area back there and I'll show you. By the way," he asked Al, "you think those young ladies were appeased by my apology for my—er—the language I used? Goddamn," said the general, shaking his head, "the very thought of using foul language in front of the fair sex fills me—"

"I'm sure they're not losing sleep over it, General. Tell you the truth, I think they were rather flattered."

"Think so?"

"From what I've heard. Word is, they think you're 'cute.' "

"Cute? I'll settle for that. At least not everybody hates my guts for this operation."

"The day is young, General," said Banks. Freeman grunted. Banks was running his eyes down the list of chopper pilots and cabin numbers on the roster. "If I could suggest, General, it might be as well to remember they'll be called to this impromptu meeting as well. So with all due respect, sir, I'd watch my . . ."

"I will. Appreciate the advice, Al. I'll speak to them first— it'll be clean as a Bible meeting."

As Al Banks closed the door to the general's cabin and made his way forward along the seemingly never-ending cream passageway, with aquamarine trim—some psychologist at the Pentagon had said "pastels" increased morale—Banks knew it was going to take a lot more than pastels to get the chopper pilots to go along with the general's dangerous plan. He made a bet with himself that they'd object loud and clear. The marines, of course, weren't included in the bet. They would fly into hell if ordered. They thought a "request" was some kind of fatal disease. No, it would definitely be the helo pilots who'd balk—after all, they'd be the ones who would have to navigate and fly through it. Then Freeman would be forced to either order them aloft into the monsoon's fury or wait till it had passed.

The forty chopper pilots filed in, sleepy-eyed, resentful at being turned out at such an hour.

"Atten-*hun*!" called Banks. There was scraping of chairs and Banks saluted as the general, clean-shaven, immaculately trimmed, his general's star bright on the drab camouflage background of his battle dress uniform, took the podium.

"Be seated," said the general. It was easy for the audience to do, most of the forty-odd pilots still half-asleep.

"We are going into battle at last!" added Freeman dramatically. It had its intended effect—waking up any of those in danger of nodding off.

"Washington's given us the green light. As you know, there's a storm—monsoon—heading our way. Now, I don't know about you, but I'm tired of sitting around on this bucket—" Freeman, seeing the look of alarm sweeping across Al Banks's face, quickly held up his hand. "No offense to a worthy ship or those who sail aboard her. Fine ship. Now," said the general, arms akimbo, left hand resting on the holster, a pugnacious set of his chin telling everyone that *he* was ready, "I hear the calmest place in the world is in the eye of the storm." He flashed a grin.

"How 'bout getting out again, General?"

Banks turned his head, frowning reproachfully at the interjection. Had to be a regular army jockey; a marine pilot would never have interrupted an officer like this. But, surprisingly to Banks, Freeman, with his fast-spreading reputation aboard *Saipan* as a stickler for discipline, didn't seem to mind the question. The general's camouflage Kevlar helmet, its wider, much less rounded contour so different from the old steel U.S. helmet, rose slightly as Freeman's eyes sought out the questioner way in the back.

"You fly out, son. Same way as you go in."

Many of the pilots were shifting uneasily in their chairs.

"Of course," continued Freeman, "it will be strictly voluntary. No one will think any worse of you, except *me*, if you don't." Then in a flash of an eye Freeman fixed his gaze on the three women pilots in the front row. "McMurtry—how about you? Game for it?"

"Ah—ah—yes, *sir*!"

"Outstanding," answered Freeman. "I'll fly in with you. Chopper One. Anybody else?" He looked up, smiling, as if getting ready for a picnic, wondering who'd volunteer to bring the hot dogs.

The two other women were putting up their hands, followed by every pilot, including the "driver" of He-26, who was shaking his head even as he was volunteering. "That son of a bitch," he whispered to his copilot. "Fuck—we've got a mad general, a fucking monsoon, and fucking Dopey on the chain gun. What more could you ask for?"

"Hey, buddy—if you can't take a joke—"

He-26's pilot nodded. "Yeah, yeah, I shouldn'a joined. I *didn't.* I was drafted. This gung ho son of a bitch is gonna get us all killed."

"You volunteered, my friend," said a tall Negro pilot to his left, the LPH's resident slam-dunker at the basket.

"Well," said He-26, "Whatcha gonna do? Three pussies put up their mitt—'yes, sir, I'll go and get killed with you.' " He-26 turned to the basketball player. "Whatcha gonna do?" he repeated.

"Then," said the basketball player, all smiles, "he isn't such a dumb son of a bitch after all."

"Well—he's a son of a bitch."

"That goes with the star," said Basketball.

"I guess. Jesus, I could do with a Bud."

Down below in what the marines had dubbed the "dungeon," the assembly deck, marines were buckling up, getting ready to go after hearing the first news of the pilots' "volunteering." The marines were glad to be on their feet; at least they were *doing* something. The regular army platoons drafted from Freeman's infantry support company were grouching, tabbing the three women pilots who had started the "mass hysteria," as they called it, "Lippy, Hippy, and Titty." Despite the grumbling, most of them, like the marines, were glad to be heading off shortly from the rolling, vomit-stinking LPH, quite a few of the men having been sickened by the constant yawing and dipping of the ship and by breathing in fuel exhaust as choppers were warmed up, the wind gusting so badly that often fumes were driven back down the stacks.

"Wait until you get airborne," said one of the marines, clicking on his Kevlar flak vest. "Make this rockin' and rollin' down here seems like kids' stuff."

"Bullshit!" said one of the regulars, taking his place on the parallel chalk lines, his platoon assembling. "I've been in a chopper before. Think I'm green or something?"

"A mite green about the gills, I'd say," said the marine.

"Bullshit!" the man repeated. "I've been airborne before."

"In a *monsoon*?" asked the marine, now checking his MIRE freeze-dried, ready-to-eat meal pack. "Care for some beef stew?"

"Animals," said the regular. "They're animals, those marines," he told his buddy.

"Yeah," said the other regular. "Have to be, where we're goin'."

Suddenly they heard a boom, like some huge doors on a warehouse opening, and a Klaxon alarm—the elevator warning—the ship's platform ready to take up the first load of Apache gunships. The transporter choppers would follow.

CHAPTER SIXTY-ONE

THE TWIN-ENGINED PROWLERS, six of them, with Tomcats covering, were now crossing the North Korean coast, their ECM jammers in the wing pods and in bulging rear fin tips ready to do battle with "Charlie's"—in this case, the North Koreans'—beams, airborne or ground.

It was hoped that by the time the subsonic two-seater Prowlers with their Tomcats cover were approaching Pyongyang, pulsing down their countermeasure beams from the ALQ-99 jammers, Shirer's wave of Tomcats would be ready to "clear the lanes" of any MiGs that might try to intercept *Saipan*'s chopper force.

Hauling field-pack 155 howitzers, two 125-ton Galaxies were being escorted across the sea of Japan by six Phantoms, each carrying a "buddy" refueling pod.

At sixteen thousand feet above the weather, the first diamond of Tomcats, riding shotgun for the unarmed Prowlers at nine thousand feet, were on radio silence. Seventy-one miles in from the North Korean coast, over Changdo-ri, the Tomcat leader saw a blip, one of the twelve Prowlers dribbling to the right off his NEPRA—his nonemitting passive-mode radar—screen. There was no call for help from any of the twelve electronic counter-measures aircrafts, for that would have meant breaking the radio silence, and the Tomcat leader assumed, correctly, that the lone Prowler in the monsoon was experiencing mechanical difficulties. Its gap in the Prowler wedge was filled, the rest of the Prowlers closing up as if guided by some invisible hand. It gave the Tomcat leader a quiet sense of pride in the professionalism of the carrier's family of pilots. The Prowler might be forced to ditch, but rather than emit a giveaway signal, it had simply turned off into the monsoon-torn night alone, any call for help calmly stifled until the plane returned to the carrier's patrol zone—if it got that far.

Despite all the alarms aboard his F-14, the Tomcat pilot's eyes kept monitoring the instruments, moving from altimeter, bottom left of the HUD assembly, to the banks of dials below the compass on his right.

Aboard the Prowlers it was rough going, the black torrential downpour shot through with pockets of less dense air, the unarmed planes having a bumpy ride that irritated the two ECM officers in the rear compartment, for as good as the ALQ-99 jammers were, the turbulence didn't help.

A hundred miles behind them, coming westward over the Sea of Japan, was the small armada of forty Chinook choppers, led by an arrowhead formation of five fierce-eyed Apache helos. Each of the helos sprouted wing pods of nineteen 2.75-inch rockets apiece, an infrared TV masthead sight, and laser range finder for eight Hellfire missiles and the Hughes chain gun.

As the lead Apache rose from a thousand feet above sea level, its copilot saw the red light go on above the "check on systems" display and heard the accompanying buzzer warning them they had insufficient power for the steep climb over the Taebek Mountain range, invisible in the rain but not more than five miles away. The pilot glanced across at the terrain-contouring-map video display confirming they were getting too close to the peaks around Konjin to make any shallower-angled approach, which, in any case, would require breaking formation.

"Lose the port Sidewinder!" ordered the pilot.

"Done," replied the copilot.

"That was quick."

"Red light's still on."

"Lose the other one," said the pilot. The light went off, the chopper's rate of climb increasing. "You were next," he told the copilot.

"Thanks."

They had just sacrificed their two best antiaircraft defenses in order to better protect the troops following them in the Chinooks.

Ten minutes from Pyongyang on the Tomcat's NEPRA screen, a blip was appearing on the far right. Very fast. Then another. Three more—the dots heading for the eleven Prowlers.

The Tomcat leader switched on his air-to-air Sidewinders, heard their growl, and called to the flight, "Tomcat leader. Five Bogeys, maybe more, one o'clock—twelve thousand. Strikers go!"

These were the six Tomcats in front of the diamond, now going down behind the Prowlers, who were already starting to pour down their rain of powerful beams to overwhelm the SAM radars, and dropping chaff as well.

In the semicircle of twelve three-missile-apiece SAM sites east of the city, NKA operators hit the siren buttons as their radars suddenly turned to snow, the eleven Prowlers coming down guided by their red TERCOM—terrain-contour-guided radars. With a constant video feed of mountains, dips, and rivers flashing by, the pilots and crew in their blinkered canopies, windshield wipers on overtime and no use at all; the planes effectively flew themselves. This allowed the two EWOs in the rear section of the plane to direct their jamming beams at any energy source distinct enough to look as if it might be trying to "burn" its way through the heavy-duty beam screen of the Prowlers. Three of the Prowlers were destroyed in twelve seconds, balls of curling orange as the MiGs, now twenty in all, screamed in from the west, another seven from the north, the squadron of MiGs on the ground at Pyongyang Airport all but wiped out by three of the Tomcats striking with two-thousand-pound laser slide "walleye" bombs.

A MiG, unable to come out of the turn, smashed into the flatland west of Turu Islet on the bottom left-hand stretch of the S made up of the Potong River and the much wider Taedong. Halfway up the S, where the river straightened between two islands and flowed under Taedong and Okryu bridges, it passed Kim Il Sung Square. Beyond the square was the wing-tipped Grand People's Study House, and near the riverbank, framing the square, the Korean Art Gallery to the south, the History Museum to the north. But now none of this was visible except as sharp angular shapes on the helos' video displays. Several SAM sites sprang to life, firing blindly, radars jammed but hoping to bring down the "American pirates," as a hysterical Pyongyang Polly was describing them on state radio before it, too, went dead.

"Think they'll expect troops, General?" Lt. Sandy McMurtry asked Freeman in the lead Chinook.

Freeman tightened his helmet's chin strap, smacking her affectionately on the shoulder. "It's all right, Lieutenant—they just think it's a bombing mission." He pointed at the Chinook's radar. "Moment those helo gunships break for perimeter defense, you take me right on down where I told you."

McMurtry had already keyed in the square that she and the others had gone over so often in their minds during the pre-op discussions on the LPH while Washington had whiled away the time, or so it seemed to them, making up its mind. For a second McMurtry saw the distinctive shape of the ninety-foot-high Arch of Triumph, a slavishly brutal imitation of the Parisian original, and south of it the outline of the Chollima, the famed winged horse of Korean legend. Then momentarily everything was lost as stalks of searchlights exploded from the defensive circle ringing the city and now crisscrossed the sky, reaching up, feeling the rainy darkness for the enemy bombers who, Pyongyang Polly had said, were trying to pollute the sacred birthplace of "our dear beloved leader."

Amid the chaotic sound of rain, intermixed with antiaircraft fire and the never-ending electronic beeping of warning and centering indicators, McMurtry's earphones were nothing but a garble of noise as Prowlers and the NKA AA batteries engaged in a war of the beams, for without targets, the huge twelve-finned Soviet SAMs were useless.

In the torrential downpour of the monsoon, which lowered visibility to zero, it was all instrument flying and landing for the helos, and here the American know-how was overwhelmingly superior, the Prowlers "frying" the NKA's radar screens clean of any targetable image, allowing the Tomcats streaking in behind to drop their five-hundred-pounders with a devastating accuracy not seen since the Vietnam War.

"Bogey on your tail," yelled a wingman, the striker leader, his bomb gone, hitting the button, going from air-to-ground to air-to-air in milliseconds, screaming up deep into the monsoon, and gone in a crimson flash, a collision with a MiG on the cross vector.

"Aw shit!" said the wingman, going into a roll, locking on to a MiG's afterburner and engaging his own, his Tomcat now on full war power, its fuel consumption ten times its pre-afterburner phase, heading into the three-minute zone in which he'd use up a third of his total fuel, his wing automatically sweeping back now that the bomb load had been dropped from the more stable wings-out position. He saw the MiG in his sights, pressed the cannon. The MiG was gone—not hit—quickly reducing speed, the American overflying him so that now he was up again behind the American, his air-to-air Aphid waiting for the growl, not sure whether he heard it in the confusion, waiting for the light. The American broke, so did he, both into scissors at the same time, their reaction times to this point exactly on par. The American popped four incendiary flares and dropped, the Aphid catching one of the flares, exploding. The wingman looked for the MiG, but he'd vanished, another missile, American or Russian, he couldn't tell, passing well ahead.

As the Prowlers completed their turn south, one of the city's searchlights, having given them up, lucked out on one of the Apaches. Suddenly all the searchlights converged on the Apache. The helo pilot tapped down his sun visor, put the Apache's nose down, and fired both pods: thirty-eight 2.75-inch rockets. Four of the beams died to pale yellow, then nothing. The Apache was still coming down fast in the strange white-black river of night and searchlights, the copilot picking up a SAM site in a searchlight's spill.

"Let 'em go!" said the chopper pilot, the copilot firing all eight Hellfires, the helo's underchin chain gun spitting a long, bluish-white tongue down at the NKA's SAM site. The SAM site exploded, the chopper, its rear rotor's pitch-change spider damaged by AA fire, canting crazily to seventy degrees, the small blades chopping into the tail drive gearbox. The pilot glanced across at the copilot—he was dead, head slumped, lolling in the turbulence of low air currents, the rain so hard it sounded like a hose on the fuselage as the Apache's pilot braced for the crash. The helo smashed into the dark blanket that was Changsan Park on the city's northern outskirts. Its explosion terrified the well-to-do Party administrators celebrating the imminent victory, in a day or so at the most—when General Kim's *komtŏt*—"bear trap"—would clang shut on the Yosu/Pusan pocket. They had just drunk to the extermination of the *mikuk chapnomtŭl*—"American bastards"—and their ROK lackeys.

The thirty-seven remaining troop transports, three taken out by MiGs, were now chopping air, settling down on the big square, using infrared, Freeman having already selected the huge concrete bulks of the Art Gallery and History Museum as flank protection, as well as the thickly treed parks about the square, which would give added protection from the small-arms fire that was bound to open up.

Though he would never know it, the pilot of the downed Apache, in panicking the second and third searchlight batteries, allowed fourteen of the thirty-seven big Chinooks following Freeman's to land in Kim Il Sung Square virtually unseen.

The American troops, to the utter astonishment of only a few janitors and museum night watchmen, poured out from the long, black shapes in a circle of machine-gun, rifle, and other small-arms fire that was quite audible, even over the rolling thunder of the aerial combat high above. And it was at that moment, with the SAM sites impotent because of the still-falling parachute chaff from the striker Tomcats, and the city militia excited and startled, that Freeman's plan saw its first success: an unopposed landing of his troops.

Many of his men were busy pushing the big Chinooks clear of the central area so that others could come down to unload.

For a totally unexpected and eerie moment, all the streetlights came on as one of the Tomcats' five-hundred-pounders hit the Pyongyang thermal power plant, ironically switching on lights that were meant to be off during the curfew. But the surge of power was too much, and the next moment the plant and city were once again in darkness. In those fleeting seconds almost every man had frozen or dropped to the ground in Kim Il Sung Square, three of the half-dozen cameramen Freeman had insisted accompany him recording the moment on tape, getting three of the Chinooks in the process of unloading their troops, the cameramen not realizing its significance until much later. After the Chinooks had unloaded, they moved off with others as part of the twenty Freeman had detailed to proceed across the city to the Pyongyang Airport. Here they would hopefully be met by the airborne regiment in the two Galaxies with the four self-propelled 155-millimeter howitzers.

Despite the blackout of the city, some of the SAM sites did receive the extra surges of electricity from emergency generators, but it was not enough. It was as if the Americans had drawn an impenetrable canopy through the rain-laden sky over the entire city, a canopy in which signals were either soaked up or bounced back as *chapsori*—"rubbish." And when fourteen SAMs were fired, their long, red tails and back-blast illuminating the immaculately clean and deserted streets around Potonggang Station, what their jubilant ground crews didn't realize was that the blips they had momentarily picked up and fired upon were not F-14s at all but F-14 simulators of the "box of tricks," as the Tomcat pilots called them, which fell through the sky, steadied by spring-loaded fins. Eight of the SAMs hit the decoys, exploding, turning the rain to vapor in the immediate area, debris falling down to the elated NKA crews. It was not until an hour or so later that they discovered their error when puzzled Party officials, racing out for propaganda displays of shot-down American wreckage, could find the remains of only one F-14 amid the litter of SAM casings.

The top floor of the Grand People's Study House, Freeman had told the marines, would command a sweeping panorama of the city through infrared binoculars, and it was taken by a squad of marines without opposition as the helos kept landing and the remaining Apaches, loaded with antitank and thousands of small antipersonnel mines, made what they called, in General Freeman's lexicon, a "ring around the craphouse," using Kim Il Sung Square as the center aim point. From the twenty-two-storied Kim Il Sung University on the northern outskirts to Pyongyang Station on the south side through the Victorious Fatherland Liberation War Museum far to the west of Kim Il Sung Square and back around East Pyongyang Stadium, the Apaches led two of the big Chinooks, which laid a string of explosives, while in the square, four of the eight American Motors Hummers, or Humvees, as the troops called them, came down, slung under the last XC Chinooks. Once unhooked, the Humvees, equipped with a .50 machine gun and infrared swivel antitank launcher in the back, were quickly manned by driver, codriver, and six men armed with SAWs—squad automatic weapons— and demolition charges. Two of the four Hummers that had not made it to the square were totaled, their parent Chinook striking a tree and overhead wires near the History Museum, sliding and tumbling down the embankment into the Taedong River. The other two Hummers had been aboard a Chinook when, only forty feet above the ground in front of the Grand People's Study House, it collapsed in a sudden wind shear. In the occasionally flare-lit air, it looked like some great, exotic brown cucumber broken in the middle, its quiet *poof* of flame starting to spread quickly. A marine sergeant thrust his M-16 at the nearest man, went in under the wreckage, crawling into the Hummer's cabin. Slithering across the rain-slicked vinyl seat, he was unable to raise his head any higher than the steering wheel because of some part of the chopper's fuselage sticking in through the driver's window.

"Where's he goin'?" shouted another marine.

"Fishing!" another shouted, his mood of bonhomie the result of having passed from sheer bowel-freezing terror into a reverie of relief at still being alive.

The Humvee came to life, jerking out from the wreckage, dragging pieces of fuselage with it.

"On his honeymoon," someone else shouted. The buoyed mood of the men was caused not simply by the lack of any determined resistance on the ground, evidence of the fact that so far Freeman's gamble of surprise had paid off, but because of the absence of any vehicular traffic that might be bearing NKA. The magnificently spacious streets around the square were deserted, a possibility that Freeman had privately entertained from the SATINT he'd studied so closely aboard the *Saipan*. But it was a hope that he knew could be ended any moment by a sudden convoy of infantry coming up from the south or, if the Chinese were still supplying the NKA through Manchuria, troops from the north. Nevertheless, for the moment it was a surprise that helped mitigate the loss of the three Hummers and the crews of the downed Chinooks. Once he was sure the perimeter from the river up past the museum, around the People's Study House and back to the art gallery, was secure, Freeman sent out three Humvees to complete the next phase of Operation "Trojan."

One of the Humvees, its nine men all wearing infrared goggles and hunkering down, except for the machine gunner and the ATGM operator, headed north from the square along Sungni Street, swinging left on Mansudae Street. In the last of the flares dropped by the Tomcats, who were low on fuel and returning to sea, their position taken up by Shirer and the second wave, the marines could see the dim outline of the Arch of Triumph half a mile or so away. But their interest centered on the sixty-six-foot-high brown statue of Kim Il Sung in front of the Museum of the Korean Revolution. Off to their right they could see Chollima statue, the winged black horse, peasants joyously riding it, Marxist holy book held aloft, the book invisible in the rain.

It took the demolition team four and a half minutes to place the plastic hose cylinders around the dear and respected leader in front of the Museum of the Korean Revolution and another two minutes to insert the wire and run it back off the spool, several hundred yards to where the Humvee had been stationed as an advance guard.

"I don't like this," said one marine. "Too fucking quiet. Where're all the people?"

"Inside, you dummy. Where would you be?"

"Come on—hurry it up," cut in the corporal as they hoisted the spool aboard the Hummer and drove slowly in the direction of the trees that hid the Grecian facade of the Pyongyang Art Troupe Theatre across the wide boulevard.

"Okay," said the corporal, "let's do it."

There was a dull thud, the ground trembled, and the blast rustled the wet ginkgo trees, water coming off them in a spray, and the air filled with dust that quickly fell in the rain. Kim Il Sung was no more.

The driver of the second Hummer lost his way, his navigator rifleman giving wrong directions, so that now they were headed toward the Pyongyang Seafood Direct Sales Shop several blocks up from the square by the river.

"Where the fuck are we?" someone shouted.

"Gooks—dead ahead!" A police car, its Klaxon squawking, its blue light flashing urgently, was tearing down Okryu Street, wet leaves flying up behind it, orange sparks seeming to come from its interior. Small-arms fire.

The Humvee's .50 Browning stuttered, hot casings steaming through the rain. The police car wobbled, then careered wildly, ran across the street, struck the curb, rolled, ending up on its side, wheels still spinning. A man came scurrying up from the cabin like someone trying to escape from a submarine. The Browning stuttered again and he slumped back, arms caught in the door in a V, the fire licking at the rear wheels.

The marines had another look at the map. "Christ, you're nowhere near it, Smithy." It was a gross exaggeration—in fact, the driver had only overshot a right turn past the seafood building by less than a hundred yards.

"Back 'er up," the corporal ordered, and after thirty seconds ordered, "Now turn right and straight ahead."

It was another couple of minutes and they were on the western side of the east-west Okryu Bridge, which spanned the wide Taedong River and was one of the two main bridges by which any counterattack from the east would most likely come. The driver still felt spooked by the apparently deserted city, which till now had not offered any resistance on the ground to the landings in its main core, even though the sound and fury of the air battle was enough to awaken the dead, sonic booms rolling overhead—at times so loud, they were mistaken for the monsoon's crashes of thunder. To the north, the marines could see forked lightning reaching right down to the hills.

The third Humvee had already reached Taedong Bridge, half a dozen blocks or so to the south of Kim Il Sung Square, and the demolition team had started laying their charges when the first of three NKA armored cars started across the old wide span, the armored vehicles' ghostly outline visible for only a second in the light of a flare. The antitank rocket fired from atop the Hummer—the distance to the armored cars no more than three hundred yards—exploded against the bridge railing. The armored cars kept coming, their machine guns now spitting fire and finding their mark, the marine driver and machine gunner thrown back hard against the canopy, dead, the antitank missile operator behind them taking second aim. The lead armored car's machine gun opened up again, and the AT operator fired. The lead armored car burst into flame, followed by a sound like falling pots and pans as the vehicle stopped. Without hesitation the second armored car behind the first broke out and took up the attack. The third armored car braked, using the first car's wreckage as good cover, barely showing its main gun. The new lead car fired its main gun and the Hummer leapt into the air, the AT man dead.

Beneath the bridge, which lead onto Mansudae Street, the demo team kept working along the slippery embankment with the extraordinary concentration of sappers, whom Freeman had always held to be among the bravest of the brave. The rain was still heavy and the lone American marine on the bridge took cover behind the burning U.S. truck, not seeing the cupola of the lead car open until its top-mounted .76 began raking the Humvee, pieces of metal and upholstery flying through the air.

"How long?" the marine called to the sappers.

"Two minutes max!"

"They're on top of us."

"Hold 'em, Arnie!"

Arnie dashed from the big stanchion near the end of the bridge across the traffic lanes behind the burning wreckage of the Hummer, hearing a faint gurgling sound coming from it. Going low, catching a quick look at the armored car, he saw the NKA car commander, his leather World-War-II-type helmet striking the marine as old-fashioned as hell. The marine gave him a full burst. The man flung his arms back before he slumped over the right side of the cupola. The marine heard a lot of shouting coming from inside the armored car, but still it kept coming, turning now to ram the Hummer. Arnie dropped his heavy automatic squad gun, ran left to a gap of about three feet between the Hummer's rear wheels and bridge rail, saw the armored car, now only six feet away, going straight for the front of the American truck. It took him one, two—three steps, up on the wheel guard flange, and two grenades down the cupola, conscious of a stringent, unpleasant odor: the dead man's breath as he lolled on the cupola. The second car veered and hit Arnie so hard, the demolition team, running the wire back and slipping on the grass, heard their buddy's ribs snap like sticks. Now the second NKA car was blocked by the V formed by the wreckage of the Humvee and the other armored car. It backed up and suddenly its searchlight penciled out along the embankment. The cupola opened and the gunner, the .76 coaxial slaved to the searchlight, sprayed straight down the approach to the bridge. Out of nowhere, a MiG flashed low, canisters falling, the armored truck enveloped in napalm, the pilot having mistaken the three armored vehicles in radar clutter as American.

One of the demolition team cranked, and the other pushed the plunger. They felt a slight tremor, heard a thud, then a louder claplike noise. The approach to the bridge had collapsed only six or seven feet, but until it was fixed, nothing would be coming across to Mansudae Street.

By now the unarmed Prowlers had been gone from Pyongyang twenty minutes, though it seemed much longer to some of the men on the ground. Still the possibility loomed—was it possible that Freeman could make his "Doolittle" hit-and-run and get out virtually unscathed?

Shirer told half of his remaining twelve F-14s who had made up the second wave to drop chaff and go for railyards on the city's south side, and he designated three strikers to take out the six bridges across the Potong, particularly Chungsong Bridge on the southwest side, where reinforcements might be rushed from the port of Nampo twenty miles to the south.

Laser-guided bombs took out three of the bridges, but Chungsong in particular was an elusive target, its span running over the island of Suksom pleasure ground, bisecting the target, making it more difficult to get at. What made the situation worse all of a sudden was that the chaff jamming over the city was now coming to an end, even as Shirer could see MiGs, at least twenty of them, coming from the west, another seven from northern airfields, perhaps in Manchuria.

Until the Tomcats were relieved in five minutes, they would have to leave the ground force to its own devices. The AA batteries were opening up again now that the jamming was weakening. Shirer half hoped the MiGs would reach them before his and other Tomcats' fuel dictated a withdrawal, for in the mixed-up blips of Tomcats and MiGs in aerial combat on their screens, the NKA batteries, including the remaining SAMs, would be more discerning lest they hit one of their own.

"Outstanding! Excellent!" were repeated so often in the first half hour by Douglas Freeman as he saw the thousand men secure the perimeter around the square that he began worrying it was all going too well, a suspicion now reinforced by the Tomcats' leader telling him that though the next wave of Tomcats hadn't arrived, he would have to take his flight back for refueling.

Then, on the PSC-3 Manpak satellite-bounce radio he was using, Freeman heard the eight hundred airborne troops from the two Galaxies were pinned down at the airport. One of the big planes was forced to stand, soaking up small-arms fire as its men unloaded, and one of the Phantoms that had escorted it was shot down as, low on fuel; it turned back with the second Galaxy, which had delivered its load of four howitzers.

The howitzers and their ammunition had come in on pallets from the Galaxy now heading back, but one of the drag chutes had failed to open, so that the guns were now at the outer flooded edge of the airport rather than on it, and in the darkness a fight raged between the U.S. Airborne and NKA militia for possession of the guns.

Freeman knew it would all come apart if the six tanks that the Airborne's Colonel Menzies was now sighting managed to reach the airport before the Airborne could get the 105-millimeters into action.

"You can take care of them, Rick," Freeman told his Airborne commander. "Their goddamned rattletraps come apart if you fart. Over."

"They're our tanks, General. Captured M-60s."

The general paused. "Then knock 'em off with the howitzers."

"When we get—" There was an explosion in the background, drowning Menzie's voice. When he came back on the radio, he told Freeman, "General, there's a good chance we're going to lose the Galaxy. It's one mother of a target—even in the dark. I'm concerned about my men, General. If that big bird goes . . ."

"Then our empty cargo Chinooks can take you out. . . . How long do you think you can hold?"

"Not a matter of holding, General. We can hold all day, but it's no good if we can't get out."

"An hour's all I need, Rick. You hold."

"Yes, sir."

A minute later the sky over the airport went yellow, followed by an explosion—the Galaxy going up in flames, illuminating the Airborne better than any flare.

Freeman turned to Al Banks and a marine major. "I'm going on to Mansudae Hall." He said it as if he were going over to the PX for a moment. Perhaps, thought the major, the general's enormous self-confidence came from the long hours of preparation, of poring over the SATINT and Japanese intelligence reports. But then, anyone could read a map. There was more to it. Freeman's élan had spread through all the men, now digging in around Kim Il Sung Square, readying for the inevitable NKA counterattack with three of the bridges on the west side still intact.

"By God," said Freeman, "what I wouldn't give for an M-1."

"Hey, General. You don't need a tank, sir. You got us."

They were hunkering down close to the Hummer.

"Where you from, son?"

"Brooklyn, sir."

"You stay by me. I need a man like you."

"Where we goin', General?"

"We're going to start a fire, son, right in that runt's scat of government. By God, those Commies talk about ten days that shook the world. We'll do it in ten minutes!"

There was a shuffling sound—the boy's buddy hitting the cement.

"Down!" bellowed Freeman. There was another shot, but they couldn't see where the sniper was.

"Medic!" called the boy from Brooklyn. The stretcher bearers ran over, crouching. There was a flash in the darkness south of the square from the Haebangsan Hotel. The Humvee's machine gun roared to life, illuminating the rain and several marines nearby.

"Got him!" shouted the gunner.

As the Medevac team were lifting the downed marine onto a stretcher, Freeman touched Brooklyn's arm. "C'mon, son. Work to do."

"Yes, sir." But now the boy's voice was cracked with emotion as one of the medics, seeing his colleague was stripping open an emergency field dressing, reached over and stopped him, then pulled the marine's poncho over his face, the rain bouncing off it with a drumming sound.

"Let's go," said Freeman, leaping into the Humvee, a squad automatic weapon with him and grenade vest packs in his right hand.

He turned to the marine major in charge of holding the square. "Give us forty-five minutes, Major. We're not back, you go ahead with the withdrawal."

"We'll stay as long as we can, sir."

"You'll stay forty-five minutes and get your ass out of here. Second Tomcat wing'll have enough to do with those MiGs without baby-sitting us. That means I want choppers in the air at oh six thirty. You hear me?"

"Loud and clear, sir."

The major's biggest worry wasn't whether the general would get back or not but how best to protect the Chinooks, scattered all around. So far the general's plan was working well, despite the airport and the sporadic fire of some Home Guard and militia troops working their way up Sunji Street, the marines now in the process of blocking it off. While this was happening, the major saw that the men from Freeman's infantry were pushing more of the Chinooks to the west end of the square between the big protective blocks of the Art Gallery on the square's south side, the History Museum on the north, and the river directly behind them to the east. Meanwhile on the top floor of the six-storied Grand People's Study House, marines with spotting scopes took up positions.

"What's your name, son?" Freeman asked the Humvee's marksman/squad leader next to him in the cabin.

"Brentwood, sir."

"All right, Brentwood—we're going to visit Mansudae Hall. Ever heard of it?"

"On the map, sir. Aboard the LPH when we were going over the—"

"Well, son, you're going to see it up close. You ready?"

"Yes, sir," said Brentwood.

The general knew he wasn't. No one was, before their baptism of fire.

"So far," Freeman had told Al Banks back at the square, "on the ground we've only had chicken-shit resistance."

"I think that's about to change," Banks had cautioned.

"By God, it's the curfew," Freeman had proclaimed in a

moment of revelation, standing in the pouring rain, arms akimbo. "Thought everyone was staying inside from fright."

In the Hummer, Freeman could hardly breathe, so excited was he by the prospect—a vision of glory so powerful—heaven so clearly on his side with the curfew and the rain and the monsoon, together with *his*, Douglas Freeman's, idea to attack when no one else would, that the general found it impossible to contain his exhilaration. "Hot damn!" He smacked the dash, the startled driver almost driving into the curb and having to hastily readjust his infrared goggles as they swung right at the Grand People's Study House, rushing the four blocks to the Mansudae Assembly Hall.

In the ice-cold depths of the North Atlantic, eighty-three miles west of Scotland, the USS *Roosevelt*'s executive officer, Peter Zeldman, gave the skipper his wake-up call. "Captain. Message station coming up."

"Okay, Pete. Be right there." Robert Brentwood pressed the "stop" button on Johnny Cash's "Don't Take Your Guns To Town," swung his feet off the bunk, and made his way over to the washbasin to wake himself fully, his eyes and throat dry as parchment. He made a mental note to tell the chief engineer to turn the switch up on the air/water content control. Brushing his teeth, his mirrored image looking better than he felt, he was struck by how the public face—the face of duty—so often and so convincingly hid the deepest fears of the inner shadows. He glanced at his watch. Three-twelve P.M. Back home—that other planet—his mom would most likely be having her morning coffee, his father at the New York Port Authority, pushing paper and moving ships, cutting corners where he could and sucking Tums where he couldn't. And what about Lana? Had she and La Roche patched it up? For the life of him, he couldn't think why a man would want to break with a beautiful girl like that. Maybe it wasn't all La Roche's fault. It took "two to tango," as his mother never tired of saying. Anyway, hopefully, if the burst message did come in, he'd be in Holy Loch tied up within two to three hours and there'd be lots of mail for everyone. Maybe a letter from young David, though that was too much to hope for, knowing his younger brother's "allergy" to writing anyone. Well, hell, thought Robert, replacing the toothbrush, you're no letter writer yourself, pal. You ready, Brentwood? he asked himself. Ready. And willing?

No—but ready.

The moment he began the walk toward the control room, Brentwood felt every sailor he passed watching him, wondering. He nodded to most and stopped at the galley.

"What's on, Cook?"

"Roast lamb and mint sauce, sir."

"Gravy?"

"Yes, sir."

"Trying to make me fat?"

"No, sir."

"I'll have to start training like Wilson."

The cook grinned, hoping the skipper wouldn't notice he wasn't wearing his chef's hat.

"Wilson's down there now, sir." He nodded back toward the missile bays. "Doing his laps."

"Hope he's wearing sneakers," Brentwood said, half in jest, half seriously—the "on station" behavior code forbidding anything that would make a noise loud enough to be picked up by an enemy's towed array.

"Good," said Brentwood, about to move on. "And Cook?"

"Sir?"

"Get that hat on."

"Yes, sir."

Stepping into control, the sub still rigged for red, Brentwood could feel the tension.

"Depth?" he asked Zeldman.

"Five hundred, sir."

"Very well. Take her to one hundred."

"To one hundred," confirmed the diving officer, standing behind the plane and trim operators, their half-wheel steering columns moving gently with hydraulic grace.

"Four fifty . . . three hundred . . . three fifty . . ."

Brentwood pressed the intercom for all sections, from torpedo room up forward through "Sherwood Forest," the missile bays, to the reactor, to call in for status reports.

"Three hundred . . . one fifty . . . one hundred, sir."

"Very well. Roll out VLF."

"Roll VLF."

The sub shifted slightly.

"Upwelling, sir," commented the diving officer, noting the sudden change in salinity and water temperature.

"Stop VLF," commanded Brentwood.

"Stop VLF."

Brentwood watched the depth gauge, its needle moving slightly, up again, then down. The sub shifted a little more. Last thing he needed was an inversion layer, a sudden change in water density that could suck the sub down before enough ballast could be blown to regain neutral boyancy, driving the boat down, hitting the bottom at 150 miles per hour. The needle moved down again and back.

"Retract VLF."

"Retracting, sir."

Brentwood was now receiving status reports from all the sections. Everything A1. "Pete, let's take her on a mile or so. Get her away from this upwelling nonsense."

"Yes, sir."

Robert Brentwood looked at the steering computer's clock—at an easy twenty-five knots they should reach a new position in plus or minus four minutes, depending on local sea current/ salinity/temperature variations. It would mean running out the VLF a little faster and risking a little more noise for *Roosevelt* to hopefully clear the upwelling and still have time for a ten-minute wait—but this should be no problem.

In itself, the fact that the old Cold War rule of Soviet–Warsaw Pact armies forbade anyone under the rank of lieutenant to possess military maps did not seem particularly significant. While it was something that had astonished the Americans and British in the long-gone days when NATO had invited Soviet–Warsaw Pact officers to observe NATO maneuvers, it had not occurred to anyone that the antiquated rule, buried in the bureaucracy of the Soviet army, would have much significance. After all, even platoon officers didn't require maps, their particular tasks, such as taking a farmhouse, a ditch, or a hill, "microrated," in the jargon of the strategists and tactical warfare experts, a small piece in the vast jigsaw of war. Most combat troops, only 25 percent of any army doing the actual fighting, rarely knew or cared about the wider battles. All that mattered was for them to survive, to take the particular objective on any given day with minimum casualties, not knowing till it was over, sometimes for months, even years, what part they might have played in the overall scheme of things, whether they had won or lost or had merely come to a bloody draw.

But on this October day, while the USS SN/BN *Roosevelt* approached message station, and another Brentwood, half a world away, was running up the stairs of the granite and marble Mansudae Assembly Hall, now quickly having been reinforced by North Korean regulars and the "black-pajamaed" militia of Pyongyang, something decisive was about to happen on NATO's central front.

In the southwestern corner of West Germany, a weary Margaret Ford, the young lieutenant of twenty-six, and her crew of three were about to launch another Lance missile with conventional warhead. Ford's crew was one of twelve out of the original twenty-three that had been located for "shoot and scoot" counterbattery fire in the German Black Forest. A light shower had fallen, but now the sun was trying to come out as the rain clouds passed over the forest into France, and Margaret Ford, though she did not know it, was about to change the course of history.

An advance Soviet mobile observation post, a camouflage-painted, fourteen-ton Russian BMP—armored personnel carrier—traveling at thirty miles per hour, carrying ten troops and armed with the standard 7.62-millimeter machine gun and antitank missiles, stopped on a side road, twenty miles from the Black Forest, now a blue smudge through the hazy autumn mist. The greenish-yellow poplars were turning and flickering in the midafternoon sun, and the sudden warmth made the fields steam for miles around.

The men aboard the BMP had been on hard rations for forty-eight hours, with little sleep, nerves jangled by a brief but spirited American counterattack which had taken place behind them on the big bend in the Danube as it curved south from Regensburg. And when they had lost their officer, whom they all liked, during the fierce American 155-millimeter barrage of high explosives and armor-piercing shells, the Russian crew's morale had taken a bad body blow. Had it not been for the sergeant's initiative in keeping the men going, they would probably have called for a break earlier on, but now the sergeant didn't have a choice as one of the men *ema stalo plokho*—"was feeling carsick"—in the personnel carrier, not surprising in the suffocating heat from the engine and the sun combining in the coffinlike interior. The jolting, jarring, and lack of any sense of direction for the men inside was guaranteed to leave even the toughest reeling after three or four hours in battle conditions. The sergeant had yelled for the driver to stop earlier, but the BMP was so noisy, he couldn't be heard. But with the smell of the man's vomit filling the already dirty and stale air inside, the driver finally got the message.

As the sick man rested with a few comrades, the sergeant and four others took the BMP over to a green hill nearby from which they could see a farm about two or three kilometers away and had a good view of the retreating American 155s, their flashers, like pieces of a shattered orange mirror, visible at the edge of the dense forest. The sergeant and his comrades could also see that the shells from their own guns roaring away five miles behind them were falling short of the West German and American batteries.

Suddenly a Lance missile could be seen streaking from the blur of the woods twenty miles away, but the elevation of the Russian guns was still obviously too acute and the sergeant was reporting on his radio, "Too short! Too short!" until the Soviet battery commander, in an effort to save valuable rounds at the end of the already overextended supply lines from Czechoslovakia and Poland and Russia, ordered the fifty big G-6s to "down." New salvos came screaming lower, over the crew of the BMP, the G-6s' twenty-five-mile range six miles longer than the American 155s'. The Russian shells not only tore into the American and West German positions in the Black Forest, but over four hundred of the HE shells crashed into the forest on the French side.

Several of the Allied TV, print, and radio reporters covering the war from the Rhineland—on the supposedly "safe" French side—were hit. Two of them, both French, were killed, another, British, from ITN, badly wounded.

A shell or two amiss was to be expected, inevitable perhaps, but not salvo after salvo—and all because the Russian sergeant, not knowing precisely where he was, only in front of American and West German guns, had simply done his job.

When West German TV and British networks bounced the signals via satellite throughout Europe, particularly into the homes throughout France, "the balloon," as the newly sworn-in British minister of war said, with barely concealed satisfaction, "went up."

"*Vive la France*, gentlemen," he said. "She is now in the war. Which means, gentlemen, we have our ports."

"So far," said an assistant who foresaw something in Gallic disposition within NATO that either the minister did not want to admit to or did not appreciate being referred to, especially not by a junior member of his department.

"*So far*' will do quite nicely," said the minister icily. "If it won't do, Parks," the minister continued, holding his glass out for a refill without looking at the steward, "I suspect you could be called up. Yes?"

The NKA militia approaching the dark, four-storied monolith of concrete and granite that was the Mansudae Assembly Hall were terrified when they saw the Americans. It wasn't merely the rolling thunder of the overhead battle amid the monsoon that so unnerved them—it was the Americans' goggles.

Freeman and the nine men in the Humvee, having taken infrared and "starlite" goggle training in stride, could not know that, for all the wrong reasons, they conformed to everything the militia had been told about the U.S. imperialist warmongers—like the banner of the Sinchon "Museum of American Imperialist Atrocities," which depicted socialist toddlers joyously shooting the U.S. monster, in the form of a wicked-eyed "Uncle Sam." It was part illusion, reinforced by the reality of the size of the marines' packs, the big SAWs—squad automatic weapons—and the hideous-looking robotic eyes of the infrared and starlite goggles, the gray plastic lenses protruding from a base the size of a quarter, tapering to a dime-sized lens, giving the Americans an even more terrifying, unblinking appearance to the heavily armed NKA militia and police now defending the North Korean Assembly Hall. The two men on the Humvee's ATGM-mount and .50 machine gun created a murderous firefront, the remaining seven, including Freeman, split into two teams, Freeman with his SAW leading the three in front, the probe team, while the four others, all equipped with SAWs in support, were moving in reaction to the lead team's situation.

The big doors were closing as Freeman, Brentwood, and Brooklyn started up the stairs—a burst of fire from about twenty feet away to their left clipping Brentwood's helmet, the three of them going down hard on the cold marble steps, Freeman yelling back at the Humvee to "take out that—"

There was a "whoosh" of flame only feet above their heads and the loudest explosion David Brentwood had ever heard, as if someone had let off two massive firecrackers strapped to his head, the noise added to by the reverberations of the huge door, now agape, not unhinged but licked by the yellow flame of the antitank rocket that left a large, jagged section blown out from the door's right panel, smoke bleeding from it like dry ice, and part of the lower hinge torn, curled back as neatly as a pop can tab—two dead NKA militiamen, another crawling away from the door.

"Let's go!" Freeman shouted, got up and led the probe team, Brentwood on his left, Brooklyn to his right, up the remainder of the long, wide marble stairs. From the sides of the building and from two third-floor offices either side of the draped NKA flag, from where the dear and beloved leader had issued some of his most famous edicts, flashlights winked in the power outage, then went out themselves.

Almost to the door, night became day, and the three Americans saw two groups of black-trousered militia coming from both corners of the building. David Brentwood to Freeman's left returned the fire.

"Come on!" yelled the general. "We're in the sack."

It didn't make any sense to David, but he was only too happy to obey the order. Once inside the door, his ears still ringing from the noise of the antitank missile hitting the big doors together with the din of the machine gun raking the militia outside, he realized what the general had meant. The two groups of militia coming from either side of the building couldn't fire at the Americans without fear of hitting one another in a "fire sack"—the realization bringing a flashback to David of his instructors at Camp Lejeune.

Inside the building an emergency battery light created monstrous shadows. The infrared goggles proved to be of limited use and the three men quickly took them off. While the goggles had allowed them a clear picture of the enormous spotted marble columns with massive sculptures of revolutionary workers and peasants clustered about their bases looking heavenward, they robbed the three Americans of peripheral vision. It was a trade-off—wider vision but less distinct images. David could smell strong wax polish and hear the tinkling of chandeliers, then echoes of boots coming up the marble steps outside the door.

One burst and General Freeman had taken out the emergency light, the foyer plunged into darkness, illuminated only by the eerie light of the NKA flares going up outside over the Assembly Hall and Kim Il Sung Square. There were two muffled explosions and the general knew two of his Chinooks had gone, crimson flames leaping high in the rain. As they advanced down either side of the foyer, Brentwood taking the inside of the left column, Freeman fired a "draw" burst. There was no response.

"Bastards are upstairs," he said in a hoarse whisper. How he wished he'd been able to bring in a dozen trucks, or even the three other Hummers that had been destroyed, for this place, Freeman knew better than any of his troops and even fewer of his officers, sustained the power and majesty of Communist North Korean power. The runt's Brandenburg Gate, as he had called it.

There was a loud shout from somewhere upstairs. Though he knew no Korean, to Freeman it sounded more like a revolutionary slogan than an order. The general knelt and unclipped his PRC "satbounce" walkie-talkie radio. "Freeman to square. You reading, Al?"

"Yes, sir, loud and clear."

"Those two Hummers back from the bridges yet?"

"Only one, sir," replied Banks.

"Get an HM squad up here fast, Al—start firing two hundred yards back and lay 'em right on the top. Synchronizing?"

"Go, General."

Freeman flicked the cover of his watch dial up. "Oh five thirty-seven. Now."

"Got it. Ten minutes."

"Affirmative."

There was an orange flash, and a sound hitting iron. The ATGM launcher on the Humvee outside had been hit. The machine gun kept going, defying all logic, in a continual burst. Then, through the warped rectangle of the door, David Brentwood saw the soft glow of the burning Hummer, the two Americans slumped over the canopy, the machine gun still firing as at least twenty black-pajamaed militia appeared about the flames. The machine gun stopped, the gunner's body collapsing into the pyre. David could now hear someone on one of the twin staircases that descended either side of the foyer. Freeman was on his PRC again. There was a crackle of static. "Banks?"

"Sir?"

"You left the square with that HM yet?"

"Negative."

"Then pack your Humvee with every man you can get in. Mortar crew's going to need as much covering fire as we are. The bastards are all round the building. And Banks—" Freeman's voice faded for a moment, then came back. "I want a photographer and a flamethrower."

There was another surge of static on the line and Banks needed to confirm. "Photographer . . . flamethrower, sir."

"Fast as you can."

"We . . ."

Freeman didn't hear the rest—young Brentwood had opened up with his SAW, taking out the first two militia to make it through the door, a third tripping over them, the rest breaking either side, lost to the darkness behind the huge marble columns.

"Brentwood—Brooklyn?" Freeman's voice took on the tone of a basso profundo, its echo bouncing off the nearest statue of a hero worker. "Alternate fire! I'll start the next one. Brentwood?"

"Sir?"

"You got a PRC?"

"Yes, sir!"

Suddenly the darkness was split by the flash and telltale staccato of AK-47s, glass breaking behind Freeman and Brentwood, bullets singing as they struck marble. Brentwood squeezed off a burst, quickly moving to the next column, every nerve raw, not knowing how far up the hall they'd gone.

"When our boys start moving up those steps," Freeman yelled, moving toward the balustrade, "we go up to the first floor. Got it?"

"Yes, sir."

"Brooklyn?"

There was a squeaky reply—Brooklyn so terrified, he could hardly find voice.

"More gooks, General," Brentwood called out, and fired at the door, seeing there were too many for alternate fire if they were to stop them. He thought he got one or two, but the rest had disappeared like the first group, left and right of the cavernous foyer, behind the columns.

His ears still ringing, heart thumping, David crossed the hall, letting off another burst as he reached a marble column close to the balustrade. He felt something stinging him—his left leg—momentarily wondering whether or not it was a bullet but having no time to dwell on it. Soon, he knew, the enemy, now over the initial surprise, must figure out a rough plan of fire without hitting one another.

Now the Koreans were shouting instructions to one another, adding to the sense of increasing chaos. A second later two grenades shattered the air with purplish white. There was a scream, and in the flash Freeman glimpsed three militia coming up his side, rolled a fragmentation grenade, turned about the nearest column, and let off a quick burst from the SAW. Next instant he was on his back, Kevlar helmet hitting the marble floor, a needlelike pain down his neck, the NKA troops shooting wildly, hitting windows and turning the water-slicked floor that had caused Freeman to fall into a gallery of elusive running shadows.

The moment he saw the brilliant light, David, his training overcoming instinct, froze as another militiaman fifty feet away fired a second flare inside the building. David knew what they were looking for would be movement, not shapes, as he pressed himself hard against the pedestal of a peasant woman at harvest.

Brooklyn forgot his training, swung out between the columns with his SAW, and crashed to the ground as a dozen militia cut him down. The flare now fizzing in the far corner, Brentwood snapped into the prone position and swept the floor with a full magazine, hitting four of the militia, sending the others racing back behind the columns toward the door, the general getting one man silhouetted in the penumbra of the flare's light.

"Brentwood!" he yelled. "Go for the balustrade. Back of it there's the auditorium. Give 'em a burst and head back!"

As the general fired the covering burst up the stairs, David ran between the columns, heading beyond the foyer toward the faint outline of the auditorium door, plate glass collapsing from the windows either side of the assembly hall from ricochets. When he reached the auditorium he turned hard right inside, sweeping the SAW in front of him—astonished to see the emer-

gency lights down by the stage were on, casting a soft glow over
the two thousand seats that smelled like a new car's upholstery.

The door burst open and the general came in, almost taking
Brentwood with him, the hot barrel of the gun striking David's
flak vest, the general swearing, his SAW's sling having got caught
in the breech, jamming the gun. He yanked hard at it, but it
wouldn't budge. His PRC surged to life, Freeman still uttering
oaths, cursing himself now for having left the volume switch
up. "Forty-dollar fine," he said to Brentwood, who tried to
smile but couldn't. It was all he could do to get enough saliva
to swallow. Freeman turned the volume down and heard Banks.

"General, this is square one."

"Reading you," said the general. He was disappointed it
wasn't his mortar crew outside.

"General," Banks went on in an excited voice, "one of our
ROK interpreters has plugged into Charlie traffic—seems—"
Banks's voice rose and fell in waves of interference, and Freeman could hear the gunfire around the square. "Seems, General
. . . the runt's in Mansudae Hall."

"For Christ's sake!" hissed the general. "Why the fuck you
think we're here? Intelligence confirmed he's been holed up here
since . . ." There was more static, but this time it seemed like
the tearing sound of a machine gun in the background.

"That all?" said the general.

"Yes, sir. Mortar crew should be there soon."

The general turned the volume switch off. It had suddenly
become very quiet. They heard the patter of sandaled feet. He
yanked at the SAW strap again, but it wouldn't come free. As
David drew his bayonet from its scabbard, handing it to the
general to cut the strap loose, Freeman saw the boy's hands were
shaking uncontrollably.

"Don't you worry about it, son," the general said, in a barely
audible voice, his breathing slowing for the first time since they'd
entered the great hall. "You're doing just fine. We'll get the son
of a bitch." Brentwood began to speak, but Freeman held up
his hand, motioning above with his thumb. "Some of the monkeys are going up the stairs. Good."

David guessed there'd been about half a dozen or so, and
when they didn't find any Americans upstairs, they'd be coming
back down. His apprehensive gaze upward conveyed his fear to
the general. "Don't worry," said Freeman, smiling. "We'll be

all right.'' He nodded his head down toward the stage. ''You like the front seats or the mezzanine?''

David couldn't think straight, let alone respond to a joke. All he knew for certain was that he was down to his second to last clip and that whenever anyone told him everything was going to be all right—it never was.

At latitude fifty-six degrees north and longitude seven degrees ten minutes west, a hundred feet below the sea's hard blue, USS *Roosevelt* was eighty miles west of Scotland and thirty miles north of Ireland.

''Any upwelling here?'' the captain asked.

''No, sir. Salinity, temperature look fine.''

''*Look* or *are*?''

''They're normal, sir.''

''Very well. Ahead five knots, roll out VLF to two thousand.''

''VLF rolling, sir.''

At the Sorbonne in Paris, over five thousand leftist students, some of them anarchists, were rioting, fighting police, protesting France's decision to ''defend the borders.''

In Whitehall, the new minister of war was on the scrambler to 10 Downing Street.

''Agreed, Prime Minister, it's not a declaration of war *per se*. But I should have thought that a 'defense of one's borders' means . . .'' The minister grimaced. ''No, Prime Minister. Yes, it is possible. Very well. Yes. Right away, Prime Minister.''

When the minister of war put down the phone, his hand went to his forehead in an effort to remember what he'd been saying to Under Secretary Hoskins. But his mind was still on the prime minister's unsettling reservation about the French action. ''PM's office can't seem to understand,'' began the minister, ''that while the French response means we can't use their rapid deployment force in NATO as yet, we will be able to the moment any foreign troops violate French soil. And—'' he looked across at his secretary ''—that has to happen—otherwise what's the bloody point of the Russians fighting the bloody war? If we have the French ports for resupply, we still stand a chance. We don't have the Chunnel, and if we don't have the ports and the Bolshies continue to hold Holland, Hamburg, and Bremen, and take Rotterdam, then I should think we're in very deep. Wouldn't you agree, Hoskins?''

"Yes, sir."

"I mean to say—" the minister's right hand reached out for an elusive word as if he were answering an opposition question in the House "—the bloody French *have to* come in." He paused for a moment, thinking, hands behind his back, moving toward the now armor-plated window. "I mean, all that lovely wine. It's unthinkable."

"Yes, Minister."

But the reports from JIB—Joint Intelligence Bureau—indicated that the Russian surge had lost much of its wallop. There had been early snow, and the war of mobility was grinding down. It didn't go nuclear, as all the experts had told the minister it would, and "Pray God it won't," said the minister, adding, "But I think, Hoskins, contrary to what we all thought—I should say, what all the *experts* thought—that we're in for a long war."

"No VLF signal," Zeldman reported to the captain.

Captain Robert Brentwood nodded and gave orders to the executive officer of the USS *Roosevelt* to map a course for the next twenty-four hours that would place them in the deepest part of the Norwegian Sea—within comfortable launching range of targets on the Kola Peninsula and beyond, guaranteeing the *Roosevelt*'s missiles the minimum possible CEP—circular error of probability—when striking all twenty-four of the sub's designated targets.

In "Sherwood Forest," aft of the sub's sail, where the six missiles in two rows of three stood ready in their gleaming forty-foot-high, seven-foot-wide tubes, Raymond Wilson, one of the off-duty RCOs, or reactor control operators, was jogging, the steady hum of the forty-ton ventilators washing comfortably overhead like the pleasantly reassuring noise of a summer breeze in high timber. Wilson, the man whom Captain Brentwood and the cook had been joking about earlier, was in his workaday blue cotton and polyester jumpsuit and quiet matching canvas-sole shoes, the blue in stark contrast to the smooth, creamy white color of the missile tubes. He sat on one of the narrow flip-down benches near the bulkhead, taking his pulse, his breathing slowing, whole body relaxed, yet his senses acute, missing nothing, the odor of the sub like that of a sparkling clean showroom—a world away from what he'd been told were the stink-holes of the old World War II diesels. He felt good— fit, confident he'd live to be a hundred.

The second wave of refueled Tomcats had now penetrated North Korean airspace once again, and a swarm of thirty MiGs attacked, the Tomcats breaking, fifteen to do battle with the MiGs, the other fourteen Strikers racing for Pyongyang.

Then everything happened astonishingly fast.

In the Mansudae Hall, General Freeman and David Brentwood ducked automatically, for no amount of training could steel a man's nerves against the instinctive reaction to seek cover from high explosive. There was a high, whistling noise, then the next explosion shaking the building, plaster flaking off the auditorium's walls. Another crash, more plaster, a lot of yelling from the floor above. The mortar crews were doing their job. Freeman and Brentwood heard confused firing upstairs, then shouting. As the North Koreans started back down the stairs, Brentwood and Freeman stepped out, firing two long bursts into the bunch of figures, killing several, sending the rest scuttling back up the stairs. Freeman was calling the mortar crew on the PRC. "Cease firing!" There was one more explosion.

"Let's go!" said Freeman. "You take the left stairs."

They made it up to the second floor without incident, but on the third they saw a small group of NKA, who suddenly retreated, their silhouettes clearly outlined in the fires that had been started by the mortar shelling, toxic smoke already rising from the burning red carpets. The NKA squad could now be heard above on the fourth floor, and Brentwood and Freeman followed, the din of firing beneath them on the ground floor telling them the marine reinforcements from the Humvee were coming in.

On the fourth floor, the NKA, unable to go farther, the exit door stairwell filled with smoke, suddenly split into two groups, three of them, or so it looked to Brentwood, melting into an office on the left, the others disappearing through a door on the right side of the hallway.

"They've got the runt!" yelled Freeman. "That's why they're trying to hightail it. You take the left."

David Brentwood stood, back against the hallway wall, his

SAW pointed at the office door, the general doing the same across the hall, raising their guns, butts positioned for eye protection, blowing out the locks, then spraying the doors. There was return fire on Brentwood's side, hitting his Kelvar vest, slamming him across the hallway—bullet holes peppering the stone-finished walls high above the opposite door, sending bits of marble flake whistling through the air as Brentwood, down on the floor, emptied the rest of the magazine into the door.

There had been no return fire from the general's side. He went to fire again, but the SAW jammed. He threw it down, pulled a grenade, kicked the door in, his hand a blur in and out, the grenade's explosion sending a cloud of dust and paper floating gently to the red carpet. He went in low with his Beretta, right hand arcing, his left cupping for support. There was silence, a lot of paper still falling, the room thick with plaster dust. By now, David Brentwood was back on his feet, SAW blazing, going into the room from which he'd taken fire. In the office across the hall, Freeman, debris still settling about him, saw four figures: two stunned officials in green Mao suits, one of them a small, pudgy man with glasses, the other covered in fallen plaster, a streak of blood on his face, and two NKA officers, one of whom, a lieutenant, was dead in front of the desk, the other, a major, on the floor to the right, his uniform in tatters from the explosion, moaning and clutching his stomach, rolling in debris beneath a picture of the "dear and respected leader." The picture was amazingly intact, not even its glass broken, but the grenade the general had thrown in was meant mainly to terrify and stun—which it had clearly done.

Inside the other office Brentwood saw that both men, militia, were dead, one still holding his AK-47, staring up through the light given off by the advancing flames. David whipped about as he heard heavy firing down on the main floor, sounding like marine SAWs. He saw his magazine was empty and reached for the last one.

The general waved the two officials away from the desk with his Beretta. "Over there, Comrades!"

Brentwood was walking over, having taken the finished magazine out and seeing the toxic smoke that was billowing at the blocked end of the hallway and moving quickly toward them. "General, we'd better—" He saw the one groaning on the ground rolling a grenade at the general's feet. Knocking the general farther into the room, Brentwood scooped the grenade up, throwing it down the hallway. Its blast took out three neon light fixtures and blew a door in, the NKA's Major Rhee coming at them, a knife in his hand. The general fired four times and now the picture of the dear and respected leader tilted sharply, its frame shattered, the leader solemn at a ridiculous angle.

"Move!" yelled Freeman at the other two, "before I shoot you, too, you goddamned Commie rats!"

A marine at the far end almost fired a burst before he saw the other marine, Brentwood, and General Freeman.

The general tore off the NKA soldiers' dog tags, at the same time trying to apply pressure on where the knife had cut him on the left arm.

The Tomcats were again too good for the MiGs, the American jets' fly-by-wire technology far superior to the Russian- and Chinese-made controls when it came to using circuits instead of ailerons. And while the monsoon was abating, the rain was still so heavy that flying by instrumentation alone put the Americans still further ahead, the final toll in this sortie, four Tomcats lost to fifteen MiGs. And what Freeman had expected to be the worst of it, the fighting withdrawal, went far better than expected. Ironically, his decision to attack in the monsoon, when flying by instrument was the only way, had been the best decision about the use of an air force since the world war had begun. Within hours its implications were revolutionizing NATO strategy, giving new hope to the exhausted and outnumbered NATO pilots in the European theater that the bad weather of winter might promise to give them a decided edge against the Soviet and Warsaw Pact planes.

Success on the ground at Pyongyang, where Freeman's chopper had been the last to leave, was not due solely to the Tomcats' superb ability to keep the MiGs off the Chinooks, but was largely due to the three remaining Apache helos, which, rearmed from free supply Chinooks and lighter because of the jettisoned extra fuel pods they had had coming in, rose from the square like angry gnats and attacked the NKA armored column approaching from Nampo, able to come down directly above the tanks' turret tops, the latter being the most vulnerable armored section of any battle tank. The Hellfire missiles set the first six PT-76s afire, the bigger, heavier tanks, including most of the captured

American M-60s, having already been sent south days before for Kim's final push on the Yosu perimeter.

Even so, when Freeman returned to the *Saipan*, he was a disappointed man. The little pudgy official with the glasses, interrogated aboard the LPH, was not the "runt" after all but a senior official with the NKA's ministry of supply, merely working late at Mansudae Hall.

"Where the hell is he then?" thundered Freeman, drained and tired.

"They say he's well outside the city in a bunker," Al Banks informed the general. "Apparently, first radar alert they had, or rather when our Prowlers started scrambling, they got him out."

"By God," Freeman said disgustedly, "he's a goddamned coward as well."

"General, sir, you did a magnificent job. Washington expected us to take sixty—eighty—percent casualties. We got out with less than fourteen percent."

"Well, Al," said the general, who kept moving around in the sick bay, the SB attendant trying to clean the deep knife wound, "fourteen percent is two hundred and eighty men, Al. And not to get that Commie bastard is a—it's—"

"General," interjected Banks, his relief at getting back alive infusing him with the same excitement as it had the media types now filing their stories via the fleet communications center aboard the *Salt Lake City* south of them, "we, *you*, got into the North Korean capital—in the worst weather imaginable—shook the hell out of them, and came out. General, our boys in the Yosu perimeter are counterattacking like you wouldn't believe."

The general started to simmer down. "The other two attacks on Taegu and Taejon—how are they doing?"

"Proceeding as planned, General—knocking the hell out of their supply line, and the NKA air force has shot its wad. *Salt Lake City* tells us our attack sucked up most of the MiGs from the south away from the perimeter. Our reinforcements are unloading at Pusan now."

"Well," said the general grumpily, but clearly bolstered by the news, "that was worth it. But it sticks in my craw that we never got that turd." His arm was still, the sick bay attendant working fast, but there was a lot of dirt in the wound and grease around it, and the attendant was wondering whether he should remind the general or not about the importance of cleaning a wound, in combat or anywhere else, as quickly as possible. He decided to take the plunge.

"Sir, you should get a tetanus booster."

"What—oh, all right, son. If you say so."

The general turned to Banks again, who was looking a bit out of it, thrown off balance by the LPH's sharp turn and a long roll that sent the medic's kidney dish clanking on the steel deck. "You'd better sit down, Al."

"I'll be okay, General. Sorry, I forgot what we were—"

"That other bozo in the Mao suit. Who's he?"

"Ministry of supply's secretary—or so he says. We could run it through the Pentagon computer link if you like?"

"No. Waste of time."

"The dog tags you got, General. Both NKA officers. One a lieutenant. The other was a Major Rhee. Intelligence. He—"

"Yes—the son of a bitch tried to kill me. Sneaky bastard. Well, that young marine—Brentwood—and I. We put pay to that. Damn knife wound." The general held his arm up. "Looks like I've been in a whorehouse brawl."

"Well, General, you did better with that major than you think."

Banks turned to the ROK captain, who had been sitting quietly by the sick bay's centrifuge. "That right, Dae?" asked Banks.

The ROK captain turned to Freeman. "Sir, the major was from General Kim's personal staff. Intelligence. We found these on him—" The ROK officer handed Freeman a bunch of creased papers, a lot of Korean characters on them that the general didn't understand but recognized as the outline of military areas V and VII. It was the Yosu-Pusan perimeter. The ROK officer leaned forward, pointing to various Korean markings outside the perimeter on the Nam River, which flowed down from Taegu, breaching the perimeter at Chinju fifteen miles north of Sachon. "Disposition of U.S.-ROK forces, General." He pointed next to the cluster of Korean characters above the arc of the perimeter. Next to each character were rectangles and squares of NKA troop and all battery dispositions. "Kim's strategy for his attack on Yosu, day after next, General."

"*What!*"

"Don't worry, sir," Banks quickly interjected. "We had the

information coded and SAT-bounced to Pusan HQ, Washington, and Tokyo before your chopper and the last Tomcats crossed the coast.''

''Hot damn!'' said Freeman. ''We didn't do so bad after all, Al.''

''That's what I've been trying to tell you, General.''

''Dumb bastards. Should've burned 'em,'' said Freeman.

''We didn't give them much time, General,'' responded Banks, sharing the general's excitement. ''We were in and out of Crap City under two hours.''

''Seemed like two days,'' confessed Freeman.

''Most of the men feel the same, sir.''

Suddenly Freeman fell silent. ''We get all our dead out?''

''No, sir. Airborne over at the airfield took the worst of it. It was the howitzers that the NKA were really after—thought it was a major breakthrough from the South. They wanted Rick Menzies' big guns.''

''Rick get out?''

''No, sir. He was spiking the guns last I heard, but that's only hearsay—until we get confirmation from his two IC. LPH got a bit overcrowded here with everybody coming in. Some of the choppers went on to the carrier.''

''We don't need to get it straight from anyone,'' said Freeman. ''He was a pro to the core.'' The general winced as the medic touched his arm with the alcohol, then, seeing what it was, only cotton batting, looked embarrassed. ''Talking of pros, I want that Brentwood boy and that other man—'' The general tried to remember his name. ''Boy from Brooklyn—''

''We'll trace him, sir.''

The general was still avoiding the sight of the tetanus needle. ''Owe my life to that Brentwood. Make a note of it, Al. Silver Star.''

''Happy to, sir.'' The general grimaced as the cold steel pushed into his arm and he felt the antitetanus serum injected into him. ''Al—our photographers. Did they get out?''

''Four of the six, General.''

Freeman nodded. ''They get pictures of us all over Crap City?''

Banks was so tired that for a second he thought Freeman meant pictures somehow being spread all over Pyongyang like propaganda leaflets.

"They get shots of us?" pressed the general, the first time the ROK officer had seen anything like apprehensiveness, even fear, in the general's eyes.

"Oh, yes, General," answered Banks. "Two of them were up on the People's Study House. Top floor. Infrared shots mostly, around the square and of the howitzers firing. I think they got most of it, far as I know."

"You haven't seen any yet?" asked the general.

"No, sir. I—"

"By God, Al. We've got to get those pictures stateside right away. The president'll want to see them. American people need to know—"

Banks began to tell him that the news of the American raid via SAT signals was already burning the wires hot to half a dozen news agencies throughout the world. "It'll be headline news all around the world, General."

"*Pictures*, Al. Goddamn it—we need to get the pictures out. Know what the Chinese say. Picture is worth a thousand words."

"I'll check it out, sir."

"Do it now."

"Yes, General," said Banks wearily, getting up and feeling a little light-headed, making toward the sick bay door: "I'll get right onto it."

"Al—"

Banks turned slowly, trying not to look as fatigued as he felt. "Yes, General?"

"Al. There's only one son of a bitch in this world with a bigger ego than that runt!"

Banks looked puzzled.

"Me," the general said, and winked.

Banks was correct. Dawn now, 6:00 P.M. the night before in Washington and 11:00 P.M. in London—too late for the early evening news in America and pushing it for the midnight BBC broadcast, the news nevertheless took the world by storm. All programs in progress were put on hold as announcers cut in with the news flash of the American attack, the video pictures showing Kim Il Sung's enormous statue now a rubble on the ground, his body badly cracked yet clearly recognizable, the head split and nothing left standing but the hump of the pedestal.

Then came the biggest shock of all: cuts of Pyongyang Polly, picked up by satellite, announcing to the slow accompaniment of funereal music that "our dear and respected leader" had been killed in the American raid and that the new people's provisional revolutionary government was being led by "our dearly revered president, Choi Yunshik," formerly a vice martial in the Democratic People's Republic of Korea.

The CIA at Langley knew nothing more about Choi than that he was a middle-of-the-roader who had opposed the hitherto unheard-of Communist "succession" of Kim Jong Il taking over his father's title.

General Kim, it was announced by Pyongyang Polly, was being relieved, "due to ill health."

With the precious time and intelligence gained by Freeman's attack, the pressure on the Yosu/Pusan perimeter was immediately reduced. Kim's overextended supply line severed by the "Freeman-style" attacks, as the press was calling them, on Taejon and Taegu had only added to General Kim's problems.

Everything had come unglued for Kim, due in no small measure to the capture of Kim's plan of attack on Pusan from Major Rhee, who, after interrogating Tae at Uijongbu, had been given the plan by Kim to take to Pyongyang for the blessing of the NKA's general staff.

In Beijing, behind the highly lacquered bloodred gates of Zhongnanhai Compound on Changan Avenue, the North Korean ambassador reported that Pyongyang wished to "discuss the situation" with the United States, and as there was no official representative of the Democratic People's Republic of North Korea "in the imperialist warmonger's capital," the government in Pyongyang representing the "freedom-loving people of North Korea" requested that their comrades of the People's Republic of China might intercede on their behalf.

The ambassador's request was not well received by Premier Lin, who reminded the Korean that their late dear leader, Kim Il Sung, father of Kim Jong Il, whom Chinese intelligence knew was not dead but under house arrest, had once referred to the Chinese as "American puppets."

The ambassador was shocked, and said, with great respect, that he did not recall this.

"It was," said Premier Lin coldly, "in February 1989." With

that, Lin rose, indicating the meeting was at an end. Pyongyang would be informed of the Central Committee's decision.

The ambassador of the Democratic People's Republic of Korea bowed as low as his back condition would permit.

In three days Pyongyang, seeing their exhausted troops now reeling back from the Yosu/Pusan perimeter as far as Taegu, saw what President Mayne referred to rather mundanely as "the writing on the wall," at least as far as the Korean theater was concerned.

The CIA was receiving messages from Beijing's Bureau of Public Safety, the Chinese equivalent of the FBI, that "certain overtures" had been made from Pyongyang. These confirmed the CIA's suspicion, gained from Japanese businessmen who had visited North Korea prior to the war, that, despite the loyal public displays of affection, the wearing of sixteen different pins of their dear and respected leader, the most secretive dictatorship on earth had within it a boiling discontent. The people, in consequence of the economic cost of Kim Il Sung's lavish self-idolatry, and that of his son, were experiencing the lowest standard of living in the Communist world, it being estimated that over 20 percent of the country's GNP was going to the military.

General Freeman did not know any of this as he was in the throes of a violent allergic reaction to the tetanus shot. Nothing on his record sheet indicated any such reaction, it being hypothesized that the original vaccination given him years before had been made from a different serum. The knife wound had become badly infected, and in Tokyo's U.S. military hospital, to which he was transferred, surgeons were discussing the possibility that they might have to amputate.

At the moment the United States Congress rose in unison upon hearing that the heretofore little-known General Douglas Freeman, U.S. Army, was to be the recipient fo the first Congressional Medal of Honor in the Asian theater, Captain Robert Brentwood, U.S. Navy, was in the redded-out control room on the top level of the four-tiered sub off the Kola Peninsula. He was about to take the USS *Roosevelt* up for a last attempt to receive a burst message via the VLF. No message came.

"Retract VLF," he ordered. "Ready HF." This was a whip

aerial that would slide up from the periscope cluster to receive on the higher-frequency channels, but its appearance above the sea's surface could prove fatal if picked up by enemy SATRE-CON—satellite reconnaissance.

"Five minutes only, Pete," instructed Brentwood. "Then retract."

"Understood. Five minutes. Counting."

At three minutes fifty-seven seconds there was an electronic burp, the receiving screen registering digitized transmission from a TACAMO aircraft out of Reykjavík, Iceland.

There was a collective sigh.

"Jesus!" said one of the planesmen, too relieved not to break the silence order. Brentwood let it pass, relieved himself. From the computer room an operator handed him the computer-converted number-for-word message to USS *Roosevelt*: "Battle Stations Missiles."

There was no Klaxon or alarm chime as, following strict procedure, Executive Officer Peter Zeldman calmly announced on the mike, "Now hear this . . ." as he stood on the attack center's raised podium about the search-and-attack periscopes.

Next, Captain Brentwood ordered, "Set Condition One SQ"—the nuclear sub's highest alert.

"Set Condition One SQ. Aye, aye, sir," repeated Zeldman, then with all compartments "punching in" on the electronic state-of-readiness board, Zeldman confirmed, "Condition One SQ all set."

"Very well," acknowledged Brentwood. The USS *Roosevelt*, containing more explosive power than all the wars in history, each of its missiles forty times more powerful than the bombs dropped on Hiroshima and Nagasaki, was ready.

Leaving the attack center, Brentwood walked briskly forward of the BBQ sonar console, nicknamed "Barbecue" by the crew, past the NAVSTAR navigation console to the radio room, where he was joined by Zeldman and two other officers. As Brentwood and Zeldman watched, the other two officers from the sub's strategic missile division opened the two small green combination safes, extracting a black plastic capsule from both. The code phrase in each was the same—in this case "Anna Belle"—the fact that both capsules contained the same name confirming the Pentagon's order for *Roosevelt* to "fire all missiles."

"Neutral trim," ordered Brentwood solemnly.

"In neutral trim now, sir."

"Very well. Prepare to spin. Stand by to flood outer tubes."

"Standing by."

"Very well. Flood tubes one, two, three, and four." The outer doors of the torpedo tubes opened, followed by the hissing sound of air under pressure expelling water from the tubes, the four Mk-48 torpedoes sliding forward from their rail-tracked dollies into the tubes, ready to fire at any enemy sub or ship that might try to run interference with the missile launch.

In missile control the weapons officer, his gold submariner's dolphins insignia a bloody red in the light, began feeding the local orientating corrections for Kola Peninsula into the warheads' computers, aligning them to true north—insuring bull's-eye trajectories for the forty-eight reentry vehicle warheads atop the six missiles. "Spin-up complete," he announced, inserting and turning the circuit key he carried at all times on a lanyard about his neck. His assistant, a junior officer, walked, headphone wire trailing, along the narrow "Blood Alley," the redded-out corridor of tall, lean computers, ticking off each missile's status, verifying for the weapons officer that every one of the Trident-Cs was ready to pass through the last of its four prelaunch modes.

"Prepare for ripple fire," instructed Brentwood, his order calmly informing the weapons officer that *all* missiles were to be fired, the *Roosevelt* now hovering in neutral buoyancy at launch depth, a hundred feet below the surface. In ripple fire sequence, each of the six thirty-ton, eight-warhead Tridents could be launched with enough water above the sub to prevent serious "blast-off" damage to the hull's carbon steel fairing aft of the sail. It would also allow the missile to obtain optional launch from the moment steam pressure blasted each six-thousand-mile-range missile from its four-story silo. To thwart the danger of the sub yawing violently each time it lost the sixty-seven thousand pounds of each missile, the emptied tube immediately replaced by rushing water, the firing sequence would be staggered—in ripple fire—so that missile one would be followed by missile six and so on, maintaining the sub's trim.

His hands holding the highly polished brass rail that girded the control room's attack center, Robert Brentwood's lean frame bent forward, his deep-set brown eyes concentrating on the com-

puter screen directly in front of him. Checking that all missiles
were ready for launch, he held his key ready to click into the
MK-98 firing control system, the weapons officer waiting below,
the black flexi-hose trailing snakelike behind him from the plastic red firing grip in his hand, his thumb now on the transparent
protector cap, ready to flip it up and depress the red button. Six
times.

Only Brentwood, his executive officer, weapons officer, and
three vitally positioned crewmen could now tell, from the last
number-for-letter variation in the code, that this time Brentwood
would not have to insert the key and complete the circuits, it
being judged absolutely imperative by the president and the chief
of naval operations that a crew should not know when it was a
WSRT—weapons systems readiness test drill—until the final
seconds, if they were to maintain the razor-edge efficiency
needed to defend their country in the time of "maximum peril."

The trouble with this, as Robert Brentwood had often discussed with his younger brother Ray, was that after a high alert,
the natural reaction for the crew was to relax. This could be the
greatest danger of all.

CHAPTER SIXTY-TWO

THE RED ROW houses flashing by, their brick a contrasting
blur against autumn-stripped trees and fields of England's farmland, combined with the rolling rhythm of the Glasgow-London
express, lulled Robert Brentwood to sleep.

After the tension-filled months of war, he found it a joy to
simply sit back and watch the countryside and the towns of
England passing by. There was the ever-present danger of a
Russian fighter/bomber surge trying to break through the British

and American circle, but their main targets were the big ports down on the Channel coast. If the Russians did break through, the Royal Air Force's inner defenses were augmented, like the U.S. Marine Corps, with the Harrier, originally built as a close-support and reconnaissance aircraft, but which, since the first few weeks of war, had played such a vital role as defender and ground support in Europe that its status had now gone beyond its post-Falklands reputation as a good all-around aircraft. Its very name now elicited near-awed response, from pilots and civilians alike, a status that had been accorded only the Spitfire and Hurricane in World War II and, in the 1950s, the American Sabre in Korea.

But while the success of the Harrier against the Russian–Warsaw Pact air forces was now being discovered and talked about by the British public during a period in which both sides were digging in and resupplying, the plane's success had long been predicted by a "difficult," by which the English mean "eccentric," fifty-year-old classics teacher. Guy Knowlton, Ph.D., of Balliol College, Oxford, had also predicted, after his excavations during the summer "hols" before the war, that the probability of a modern war going on longer than anyone had predicted was indeed very high. Masses of men, their psyches savaged by the speed and devastation of high-tech mobile war, said Dr. Knowlton, would simply be unable to sustain the momentum. As they dug in, waiting for overextended supply lines to catch up with them, the trenches, said Knowlton, would become "a coveted place." The soldiers, as soldiers had done since the beginning of time, would discover anew that a trench, quite apart from being far more preferable than open-ground warfare, was a place where the hitherto unobtainable luxury of a hot meal, instead of C rations on the move, settled into a *predictable* routine. It was something the generals abhorred, for wherever men began putting up signs such as "No Vacancy," "Pete's Place," and mile markers to their homes, from Scotland to New York, troops became increasingly reluctant to get up and leave.

No NATO commander was foolish enough to think the war would remain static very long—that there could be any return to the kind of massive, wasteful trench warfare of the First World War. But the longer the trenches remained lived-in, the more difficult it would be to move men quickly when the present fall-off in hostilities heated up again. It was rumored, as Robert Brentwood had heard in Holy Loch, that a "deal" had been struck through Swiss mediators between the USSR and NATO to the effect that no nuclear weapons would be used. Whether this was true or wishful thinking, no one was sure. If it was true, then given the enormous gain in territory by the Russians at NATO's expense—almost all of Germany, northern Holland, and the low countries—it was inconceivable that NATO would now simply return to a cease-fire if the Russians did not agree to give the captured territories back.

CHAPTER SIXTY-THREE

"ANYONE WHO BELIEVES the Communists will let us have any of that territory back is a dreamer," General Freeman proclaimed to the clutch of White House press photographers and reporters crowding around him after the president had pinned on the general's Medal of Honor. The general saluted solemnly then raised both arms in a victory sign to show his well-wishers that his wounded arm was back in service.

Harold Schuman, as the president's national security adviser, was not pleased with Freeman's off-the-cuff remarks. The Medal of Honor, in his view, brought you respect—it didn't make you an authority on the delicate matter of diplomatic maneuvering, especially when Moscow and Beijing might interpret the general's words, at such a high-profile event with the president, as official U.S. policy.

"But it is our policy, isn't it?" the president challenged Schuman when the general had left. "I certainly don't intend spilling American blood to defend Germany, then turn around and tell Moscow it can have whatever it overruns. I'm certainly in no mood to 'stabilize' the position 'as is,' as someone at State said last night. The United States alone,'' Mayne reminded Schuman, ''has lost twenty-five thousand troops in this war. The very worst thing to do in my view is to give any impression to the Russians, or anyone else for that matter, that we're about to seriously consider redrawing the map of Europe on their terms. Why—it would make a mockery of what we've been through. Those boys would haunt us from their graves.''

''I couldn't agree more,'' put in Trainor. ''These are bullyboy tactics from beginning to end. Carve up half of Europe and then say you're willing to talk. Personally, I'd tell them to shove it.''

''I've no doubt you would,'' said Schuman tartly. ''But we must never close the door to negotiations.''

''Agreed,'' said the president, ''but this isn't the time, Harry. First we want the NKA north of the old DMZ, where they belong. And we need the European border where it was.''

''Well,'' mused Schuman, ''as far as Korea goes, it seems now we're in better shape than anyone had a right to expect.''

''*Because,*'' interjected Trainor, ''we gave Freeman—if you'll pardon the pun—a free hand there, Mr. Schuman. And State ought to realize that. Only thing those jokers understand in the Kremlin is the fist.''

''You're beginning to sound like General Freeman,'' said Schuman in a slightly disapproving tone. ''I hope it isn't contagious.''

''Well, he did one hell of a job over there, Mr. Schuman. You can't deny that. We could do with a few more like him in Europe.''

''It's a much different war in Europe,'' said Schuman.

''How?'' Trainor challenged him, suspecting that the national security adviser's comments about Freeman were motivated more from envy of the general's sudden celebrity than from any sound military consideration.

''We don't need cowboys in Europe, Mr. Trainor.''

The president held up his hand for an end to the disagreement. He was due for a meeting with the Joint Chiefs, and he intended to bring the matter of Freeman up there. Formerly the president's title of commander in chief was viewed by the vast majority of American people as a more or less honorary title until he was actually involved in the direction of some military action. In having taken the responsibility of giving the green light for the Pyongyang raid, his stocks were now high, and he intended using them as bargaining chips with the Joint Chiefs.

In the president's view, Freeman's raid had done infinitely more than turn around the position in Korea and raise American morale and status all around the world in perhaps its darkest time since the Cuban crisis.

Freeman, as far as the president was concerned, had overnight established a new battlefield code of conduct, showing what Mayne called "armchair video" commanders that in a rapid and highly fluid, high-tech mobile war, perhaps more than ever a commanding officer needed HUMINT—human intelligence—to get away from HQ and "go on the point." Precisely *because* of all the gizmology available, a commander ought to get out of the claustrophobic, noisy push-button world of divisional HQ tents and get into the thick of the fighting himself, just as some of the fighter pilots had found out that for all the benefits of instrument flying, sometimes it was necessary to simply shut off all the "incoming" buzz and use their eyes.

On this October day, however, with southern England flashing by, Robert Brentwood was one commanding officer who wanted to forget the war, and had it not been for Lana's letters and the tape she had sent to him from the Spence boy, he might have succeeded. He certainly would not have been on the 10:00 A.M. Glasgow to Victoria Station had he not read her letters, beginning with the last one she had posted. Now she had been posted to some "godforsaken rock," as she called it, the name of the rock carefully erased by the censor. It could have been anywhere, from Gibraltar to the Galápagos.

Robert was struck by the change in her tone. The self-centeredness of the beautiful coed and the bitterness of her failed marriage alike were conspicuous now by their absence. Instead she talked to her older brother about the terrible ordeal of Ray, of the Spence boy and how it had brought her closer to her three brothers. The war, she wrote, had not diminished her own worries, which she'd hoped it would, for despite common wisdom, she'd found that other people's troubles, worse though they may be, had not helped put her own "into perspective." That kind of thing, she discovered, was only a "short fix." Talking of fixes, she asked Robert whether it was true that many of the pilots were being given—the censor had crossed the word out, but she obviously meant amphetamines. "Yes, they are," would have been his answer.

After a long letter about Ray, her next had been almost exclusively about the Spence boy, not as a lover, Robert could see, though in matters of the heart he regarded himself as woefully deficient. She went on to tell him that if the war had taught her anything, it was that morale was often more potent than penicillin, that with a purpose before you, you could brave all kinds of horrors that normally would prove too much. Which is what had surprised her so much about William Spence's death. Unlike some of the smart-ass profs who were against the war and were 4-Fs and knew they wouldn't be called up, the young sailor had recognized that this was a war NATO and the United States had to win, that at the very best, it was good against evil. At the very worst, no matter what the deficiencies of NATO countries, there was a vast difference between a regime that could knock your door down and take you away in the early hours of the morning and a regime that was required to show good cause. Which was why, she told Robert, she had thought that William Spence, filled with old-fashioned love of country, family, would pull through. But then no one, including herself, had seen that ''old hag,'' pneumonia, creeping up, just sitting there, knitting, patiently rocking in the savage corner. Waiting.

It had made her even more worried about Ray. Apparently he'd gone into a funk until some admiral from La Jolla had visited him and told him straight that if he was going to go into a damn sulk over it and not see his kids, he might as well make himself useful—OD and clear the bed for somebody who needed it.

Some of the fighter pilots, she said, who were coming in were experiencing what they call ''electronics burn'' the result of an intense spitting kind of fire that came from all the high-tech, lightweight, but highly inflammable consoles they'd stuffed into the cockpits. Anyway, apparently Ray had had his sixth plastic surgery operation and had seen the kids. Everyone had a good cry, ''according to Mom,'' and Ray had started to make noises about sea duty, though that would certainly be a long way off if not out of the question. Maybe some form of support ship, a tender, spare parts or something. Mom was all in a flap because she'd just heard that young David was in for some kind of decoration.

Lana didn't know, though, whether it was such a good idea for Ray to try and get back in the navy. ''Knowing Ray,'' she'd written, "he'll probably be worse than—" she couldn't think of the name "—the man played by James Cagney in 'Mr. Roberts'—you know, the old grump who kept losing his palm trees."

"Anyway, Bob," she ended, "if there's one thing I've learned in the last two years, from Hong Kong to this godforsaken rock, it's that love is all that matters and you should give it wherever you can. Hopefully you'll get some back before we're all blown to Kingdom Come."

Robert Brentwood had read all the letters at Holy Loch and, as per her instructions, had run the tape forward a little so Mr. and Mrs. Spence wouldn't fret about not hearing anything for the first few minutes. Robert had waited to push the "stop" button, not wanting to intrude in any way on the boy's private thoughts to his family. But the tape was silent. It had come via fleet mail quickly enough, and Brentwood guessed the security and bomb people had done their job—the package going through X ray, and with it, the dead boy's last message to his folks.

Under the circumstances, and seeing that he had two weeks to fill, Robert thought that the least he could do was visit the boy's folks. Before going to Waterloo and catching a train down to Surrey, he had called into Marriage's bookstore. The same manager in Harris tweed who had taken his order several months before had just finished serving a customer when he looked up and saw Brentwood walking in. The manager beamed. "Welcome back, Captain."

Suddenly throughout the whole store, from the paperback section atop the old spiral iron staircase down to the hardcovers on the main floor, the staff and customers broke into applause. Brentwood looked around to see who they were all clapping for, and blushed like an afterburner when he realized it was him. It was the convoys—they were starting to get through, the convoys without which the British would die, let alone Europe, and the guardians of the convoys were in good standing with the people of Britain.

The manager, Mr. Harris, was quite definite about refusing payment, handing Brentwood a mint-new copy of *Bing*.

"No, look, I'd like to—" protested Brentwood.

"No, old man. Least we can do." Everyone from assistants to the unloading clerk had gathered to welcome the American captain.

"Have you time for tea, Captain?" someone asked.

"Why—er—I've got to be getting up to Oxshott."

The manager was so tickled by the occasion, he couldn't bring himself to correct the American, but he did ring British Rail and ask them what time the next train *down* to Oxshott was. Eleven P.M.

While they were having tea and biscuits, someone brought in a dolly with a carton of at least fifty paperbacks for the officers and crew of Brentwood's ship. The manager saw the captain of the most deadly armed ship in history looking rather nonplussed.

"Not to worry, Captain. We'll have them sent to your ship. You won't have to carry them about." There were a few giggles and polite laughs. "If you'll just give me an address?"

Robert, as security demanded, gave him the U.S. naval P.O. box in Glasgow. Overcome by the warmth, especially after he'd been told so much about the reserved British manner, Brentwood almost forgot to take *Bing* with him.

After tea, there was a pub dinner: pickles, Scotch eggs, and several pints of black Guiness, their brown, creamy heads flowing like velvet down the captain's dry throat. Another pint later, Brentwood asked, "Mr. Harris—can I ask you a straight question?"

"Fire away, old boy."

"This gal—young lady, young British lady—was rather upset with me at a party. Said I was 'worse than Bing Crosby.' You know what she meant?"

"Hmm," said Harris, who was swirling the final ration of Guiness. "Weren't singing, were you?"

"No," answered Robert. "No, I wasn't."

"Romancing then, was it?"

"No—well, I mean, she was kind of annoyed that I wouldn't—"

"Ah—" Harris leaned over to the barman. "Fred—haven't any of the rough red left, have you?"

" 'Fraid not, Mr. 'Arris. I've got a liter of Old Espagnol, though."

"Dry, is it?" Harris inquired about the sherry, Brentwood thinking he'd forgotten completely about his question.

"Mr. 'Arris," said the barman, "if this stuff was any drier, it'd make your 'air fall out—eyebrows, too, most likely."

"How much?" asked Harris, forehead furrowed, ready for a shock. He got it.

"A century."

"Oooh—" said Harris, his head coming back from the bar. "Oh dear—"

"Best I can do, Squire," said the barman. "Rationing and all."

"Oh, quite, quite. Quite all right, Fred."

"You're welcome, Mr. Harris."

Brentwood lifted the last of his Guiness and savored it as it went down. "I like that," he said.

"I think, old man, she was saying you were rather *bourgeois*."

"Straitlaced," cut in the barman, his hand rocking from side to side. "You know—'long the straight and narrow. No 'ankypanky."

"Well," said Harris, "I have to spend a penny. Then I'm off, I'm afraid. I'll take you to the station."

"Going to the loo," explained the barman as Harris made off, a little unsteadily, through the gray-blue haze of cigarette smoke, something you saw much more often these days since the war had begun.

"Straitlaced, eh?" Brentwood said to the barman.

"Yeah. 'Cor, my dad. 'E loved Crosby. Bit of a crooner himself. Always hummin' round the 'ouse. Then I'd be on listenin' to the Who. Drive me mum nuts. Battle royal over that, I can tell you."

"Uh-huh," said Brentwood—it was like listening to a new code.

When Harris returned, they walked out into the chilly night air. They could see the searchlights all around London, in constant crisscrossing, interplaying patterns, reaching thousands of feet and reflecting off the stratus.

"Do no good at all, I'm told," said Harris, looking up at them. "Is that true?"

"More or less," agreed Robert, a cold, bracing breeze coming up from the Embankment. "It's a war of invisible beams," he explained to Harris. "But I guess searchlights give comfort to a lot of folks. Something you can see."

Harris had hailed a cab for Waterloo Station, its headlights two yellow slits. "What you think our chances are? Look here—I don't want to pry—classified stuff or anything like that."

"I don't know," said Brentwood. "Far as I can tell, the experts don't know either."

They got into the back of the taxi.

"How long do *you* think it will last?" pressed Harris.

"Longer than anyone expected."

"That's rather grim."

"Yes."

There was a long silence until they entered Waterloo and Robert Brentwood alighted, turning to pay the cabbie. Harris waved the money away.

"All right," said Brentwood. "You can put off some of the people some of the time, but you can't put off all of the people all of the time." And with that he handed the bookshop manager the liter of Old Espagnol that he'd been hiding under his coat.

Harris was agape.

"Thanks for everything, Mr. Harris. I really appreciate—"

Harris cut in, "I'm—really, this is quite—wonderful."

"Between allies," said Brentwood, smiling.

"Allies *indeed*." Harris put out his hand. He made to say something, hesitated, then dared to go on anyway. "Captain, you might be right. It might last longer than any of us imagined, but if you'll accept a piece of advice—"

"Certainly."

Harris lowered his head. "That gal—any port in a storm, old boy." Then he sat back in the cab, chuckling, shaking his head. "Any port—my God, Captain—don't you tell anyone I told you that. So banal, they'd have me thrown out of the club."

"I won't," said Brentwood. "Good-bye."

"Ta-ta."

When he got to Oxshott, a wind had come up, the oaks and big elms around the station blowing hard, a smell so fresh and clean that despite the distant thudding of antiaircraft guns and the orange scratches against the sky that were the surface-to-air missiles along the coast from East Anglia down, Robert had the sense that he had been to this place before. But not being a superstitious man, and trained in the cold logic of launch mode attack, he decided that it must be the invigorating force of the wind that had cleared the Guiness, heightened his senses, giving him the feeling of déjà vu.

The Spence house, however, looked familiar, too, like the one his parents had in New Jersey—double-storied, semimodern brick. All the lights were out, but flower beds were dimly visible beneath the high silver moon, a dog barking from somewhere behind the house, and a run of big bushes, possibly rhododendrons, giving the whole garden a casually ordered appearance. He rang the bell, realizing that he'd planned this operation badly. But there had been no hotel rooms left in Oxshott, so it was either this or back to the train station to wait until 4:00 A.M. A light came on, then another.

When the front door opened, he saw a woman, her hair in curlers, long, padded dressing gown held tightly by her hand at the throat. He guessed it was the dead boy's mother. He took off his cap. "Mrs. Spence?"

"No, is there something—"

"I'm Captain Brentwood, ma'am. U.S. Navy. Robert Brentwood. My sister is a nurse—she was William's nurse and she wrote me with—"

"Oh—oh." He heard the door chain rattling. "Oh, do come in. Ah—oh, please come in." She switched on the kitchen and living room lights. She switched them off again, explaining quickly, "I haven't drawn the blackout drapes."

"What's—*Rosemary*?" A man in his sixties, tousled head of sparse brown hair, in a tartan nightrobe, was coming down from the upstairs bedroom, peering shortsightedly.

"Oh, Father. This is Captain Brentwood. Nurse Brentwood's brother. He's—"

Richard Spence tightened the belt on his robe and put out his hand. "How kind of you. My goodness, where have you come from at this hour?"

"London, sir. I'm afraid I left it a bit late, and when I reached Oxshott, there were no bed-and-breakfast places, hotels, or anything else. I'm sorry to bother you."

"Bother? No bother. Rose, get Mother quickly." He turned back to Brentwood, tying his robe tighter about his thin frame. "Would you like a cup of tea?"

"Yes, sir. That'd be nice."

Robert Brentwood decided there and then not to tell them

about the damaged tape in his kit. If they asked, he'd say it never arrived. It would be heartbreak for them.

When Mrs. Spence came down slowly, a short, frail lady with soft white hair, she looked dazed.

Richard Spence said softly, "My wife's been on medication, Captain. Ever since—"

"Of course, sir. I understand." Robert Brentwood rose to his feet to greet Mrs. Spence.

Richard Spence left the room hurriedly. The American's manners, his thoughtfulness in coming this far, all the way from Scotland, to bring something of their son's last hours in a foreign place, filled Richard Spence with such gratitude, he had to excuse himself in order to regain his composure. When he reappeared, he was in command of the situation. "I hope you'll be staying."

"If it wouldn't be too much trouble, sir. A bed for the night would be more than—"

"Tonight? When are you due back?"

"Ten days, sir."

"Of course he must stay," put in Anne Spence, the hot, steaming tea Rosemary had made reviving her. "William's room."

There was a quick glance between Rosemary and her father. It was the first time Anne Spence had even considered the idea of anyone entering William's room.

"Perhaps," said Rosemary, who Robert now saw had taken off her scarf and hair rollers, her hair warm and golden, "perhaps the captain has other plans, Mother. I'm sure he has friends."

"No, I don't." He had said it without thinking. Why, he couldn't fathom. First law of defense—never betray your most vulnerable angle of attack. It was Rosemary—her eyes. She was not especially beautiful, but there was a kindness, devoid of any cunning, and in that moment he remembered Lana's injuction about giving love. He had been trained for split-second decisions; his kind of war did not permit anything else. A second lost was a ship lost.

He wanted to stay. The house, astonishingly to him, did not have a different smell from his own home; perhaps it was a spice, something as mundane as a rug cleaner his mother had used with the same odor, or perhaps he'd been at sea so long, he could no longer tell the difference in ambience between one house and another. Whatever the reason, he felt he was in a home he knew and understood. Here there was loyalty and affection. And there was love.

"I'd like to stay," he said.

"Bravo!" said Richard Spence, brightening. "You hungry?"

Brentwood thought about it for a moment. "Why, yes, sir, I believe I am." They all laughed. Even Mrs. Spence showed the trace of a smile.

"Now then, what do you Americans like?" asked Richard. "Wish Georgina was here." He looked over at Brentwood. "She's our younger daughter. Up at LSE—London School of Economics. Political Science—"

"What on earth has that got to do with what Americans eat?" asked the frail-looking Mrs. Spence.

"Haven't the foggiest," replied Richard, rolling up the sleeves of his robe so they wouldn't touch the element. "Well, Georgina thinks she knows everything, I suppose. That's why."

"Americans like hamburgers," said Mrs. Spence.

"Eggs," said Richard. "What's that expression? Easy up?"

"Easy over, Daddy," said Rosemary, chuckling. She shook her head at Robert. "Don't mind us," she said. "I expect we're bombarding you awfully. Perhaps you hate eggs?"

"No, ma'am, I love them." Brentwood also knew that eggs were the least-rationed of foods—much easier to get than meat.

"You see?" cut in Richard happily. "I told you, Rose. How about a Welsh rarebit?"

"Sounds fine," said Brentwood

"Oh," said Rosemary, "how rude we are." She walked over and took Robert's cap. "Call me Rose," she said quietly, and Robert Brentwood did something he normally never did. He looked at her fingers. No rings.

As Anne Spence and her husband busied themselves in the kitchen, Mrs. Spence giving quiet directions, Richard assuring her he knew exactly what to do, Rosemary took Robert Brentwood into the dining room. "Now," she said, "you must tell me all about yourself."

"I'd rather know all about you."

"I'm a schoolteacher."

"Shakespeare," he said.

She brightened, "How—oh," she said, "William, I expect."

"Yes, my sister told me. He talked quite a lot about you— and the family."

"Yes. We miss him very much."

There was an awkward silence.

"Can I ask you about your work?" Rosemary asked. "I mean, they won't put me in prison or anything?"

"No," he laughed. "Ask away."

"This is going to sound awfully silly, but I've never understood why people always say how dreadful it must be on submarines. I mean, I know they're rather crowded, or at least I imagine they are. Even the latest ones, but from the looks of them, I think I should feel much more claustrophobic on the Tube."

"The Tube?"

"The underground," she said, smiling. It was an easy smile, utterly devoid of any pretense. Their banter about the sub and everything else they discussed came as easily to them as if they were old friends—the kind whom one hasn't seen for twenty years or more and yet whose conversation is taken up as if space and time had never existed. He couldn't remember when he had felt more relaxed in the company of anyone outside his family. The house, like that of his parents, was neat but not obsessively so, comfortable but not ostentatiously indulgent. And though he knew nothing much about art, the paintings he saw gave him special pleasure; one in particular, *La Gare du Nord*, had such vibrant colors that at times it seemed to fill the Spences' living room with a sense of life and light. The whole house seemed warm, and Robert felt that ironically it was the death of their youngest that, like the death of a crewman aboard a ship, drew the others closer together. And with Rosemary he felt he had to be honest, even confessing to her that he'd never read much Shakespeare.

"Most people haven't," she said, laughing. "Not really read him. And those who do always try to make him so *dramatic*— and all those flourishes. His language is really very quick. Alive. You know, 'the quick and the dead.' "

Robert shook his head. "Afraid you've lost me there."

She paused. They looked at each other. "I don't think so," she said, and they both knew that it was beginning.

"Will you go away soon?" she asked quietly.

"We'll be casting off in ten days."

"I meant how long will you stay here?"

"As long as I possibly can."

"Good," she said. Her father was coming into the dining room with the tray. "I noticed you have a biography of Bing Crosby with you," he said.

"Yes."

"One of my favorites, too."

Robert Brentwood was about to say that he'd bought it for Richard Spence, but it would be a lie—oh, a harmless one, but there was something about this whole family, something good that made him want to speak only the truth. Ten days might be all that they had. "I'd be happy for you to read it while I'm here—"

"Oh, no, I don't want to—"

"No, sir. Please. I don't think I'll be doing much reading. I'd like to do a bit of walking. Stretch my legs for a change."

"Rose?" Richard Spence said, looking over his cup of steaming Darjeeling. "You're the trail person. Over to the Downs, down to Martin, then over—"

"Yes, yes," said his wife, "but first, where did you put the toast?"

By the time they'd finished the impromptu meal, it was near 3:30 as Richard and Anne retired, Rosemary showing Robert William's room. It was a neat room—in what Robert thought was a very navy way—small writing desk and chair, a bed, a clean, uncluttered Victorian dresser with mirror, and a picture of a young seaman—winter uniform.

"I'll see you in the morning," she said.

Robert Brentwood was tired, but he could not sleep for thinking of her. It was already quite clear to him that they'd be married, but he decided not to rush it. He'd ask her father tomorrow.

In the morning, a Saturday, Robert was surprised to discover, they all enjoyed a late brunch, and afterward, newspapers all round in the sun room. Being the guest, Robert got to take his pick, and while a scantily clad chorus girl under the screaming headline "DOING HER BIT FOR THE WAR EFFORT" caught his eye, he played safe and took the Sunday *Telegraph*. It was a mixed read, for on the one hand, it was clear that the tide had turned in Korea, the NKA in disarray, editorials understanding the American desire to push as far as the Yalu but cautioning against it as part of any long-range solution to the upheavals on the Asian front.

"Who is this awful Freeman man?" asked Rosemary.

"The American general," said Richard. "There's talk of them sending him over to Europe. Jolly good thing, too." He looked over at Robert. "Sorry, Captain Brentwood—"

"Call me Robert, please."

"Yes, certainly. Well, Robert, you must forgive Rosemary's disapproval of this Freeman chap."

"Oh, it's not that I disapprove, Daddy," said Rose. "I've no doubt he's a very good soldier, but he says such awful things about them."

"That's because they're awful people," said Richard. "They blatantly attack South Korea and then expect . . ."

Mrs. Spence excused herself from the table and they tried to steer conversation in other directions, but inevitably it seemed to come back to the war simply because it was worldwide and day by day was affecting more and more people, the *Telegraph* reporting, for example, how so many of the Russian minority groups, from the Georgians to the Estonians to the Mongols, were demanding greater independence from Russian domination and how the Russian tanks had quickly put down any such aspirations, which solved nothing but merely postponed the inevitable bloodshed. And in China the "Martyrs of 1989" were commemorated by students in a silent vigil in Tiananmen Square, watched from a ring of olive-green tanks by steel-helmeted troops of the People's Liberation Army.

"That's why," said Richard, "things have quieted down a bit in Western Europe for the moment. The Bolshies want to make sure their backyard's secure before they move into France."

"You think they will?" asked Robert.

Richard Spence was stirring the tea bag in the pot and squeezing it on the side, something he would never have done were it not for the rationing that was getting more severe all the time. "Attack France? It's inevitable. I'm no strategist, but if you chaps keep doing your job and more of those convoys get through, Ivan's going to have to do something."

Robert nodded. "The French ports."

"Exactly. I'm afraid what we're seeing here, in Europe right now, is a lull before the next storm." It was when Mrs. Spence reentered the room and Richard quickly turned over the war news pages that showed the map of Europe with the three great Russian prongs deep into Germany that he came across the advertisement that had been running for several days and which, like so many, in his opinion, made absolutely no sense. He pounced on it as a diversionary tactic to shift his wife's attention away from all the battlefront news. "Here's this madman again."

Rosemary leaned over to Robert. "This is Daddy's favorite hobbyhorse. Be warned."

Richard Spence was reading it aloud: "It is vital to the national defense that you surrender immediately all your portable hair dryers to the following address. . . ."

"What's it mean?" laughed Robert.

"It means," said Richard Spence, "that some damned old fool called *Dr.* Guy Knowlton is allowed to indulge his eccentricity despite the fact that this country is in a state of national crisis. They're always going on about shortages, paper especially, and here they go allowing . . ."

"It *is* a private ad, Daddy," said Rosemary. "Not the government's."

Mrs. Spence excused herself from the table again.

"Sorry, Mother . . ."

Robert forgot all about the man and the portable hair dryers and everything else about the war as he and Rosemary walked, hand in hand, across the Downs, cycled through the tree-arched byways around Martin, and fell more deeply in love.

Robert had chickened out from asking either Rosemary or her parents about marrying her but gathered his forces and did so on the second to last day of his leave.

Richard Spence was stunned. As he confessed later, it had really been Anne who, to use his potential son-in-law's idiom, "carried the ball." She hadn't seemed surprised at all. But Anne Spence had already lost one child and might lose more if the Russians managed to drive through in the next great offensive and take the French ports. England would be next.

They gave their blessing to Robert and Rosemary, but Richard was still fretting on the last night before Robert would have to leave and go north to Holy Loch. In their bedroom Richard was pacing back and forth, Anne having already taken her pill, trying, despite her jangled nerves, to get some sleep. "Will you stop!" she said finally.

"Too fast," said Richard. "It's all too fast for my liking. Too fast!"

"Perhaps not fast enough," his wife said quietly.

"What do you mean?" he shot back.

"None of us knows whether we'll be here tomorrow. They might as well."

Richard didn't speak for a long time, and not until he was in bed did he concede the point. "Perhaps you're right."

In the blacked-out living room, Robert and Rosemary held each other, not saying much, neither wanting to talk about the cold fact that tomorrow he would be off again to war.

"If you want," she began.

"No," he said, "though I suppose you think I'm nuts."

"I think you're wonderful and I love you and—I'm so afraid for you."

"Maybe we should put it off then until—"

She placed her finger on his lips. "No," she said softly, shaking her head, resolute in her decision. "No. We'll get married as soon as you come back. As we planned."

There was a taping on the front door, and from the living room Robert could see the small red light that came on as whoever was outside also tried the chime bell.

"Chime doesn't work," said Rosemary, getting up and brushing herself down, looking presentable as she walked toward the door.

Outside, a policeman was standing beneath the porch light, rain glistening on his cape.

"Yes, Constable?"

Rosemary came back, her shapely figure outlined in the spillover of the kitchen light, one hand on her chest in relief.

"What's up?" Robert asked.

"The drapes. Apparently there's a slit of light from Mummy and Daddy's room."

As she walked down from her parents' room and he saw her silhouetted again, this time in the faint hallway light, he didn't think he could control it any longer. Nor could she.

He took her by the hand, and in the darkness of the living room they made love.

"Tell me . . ." she said, "promise me, you'll come back."

From the coast came the dull thudding of antiaircraft fire and the high, swishing noise of missile salvos, and he remembered Lana's premonition of danger, her counsel to give love, to have love, while one still could.

"I'll come back."

ABOUT THE AUTHOR

Canadian Ian Slater, a veteran of Australian Naval Intelligence, has a Ph.D. in political science. He teaches at the University of British Columbia and is managing editor of *Pacific Affairs*. He has written several thrillers, most recently *WW III: Rage of Battle, WW III: World in Flames, WW III: Arctic Front,* and *WW III: Warshot.* He lives in Vancouver with his wife and two children.